~THE~ New Eldorado

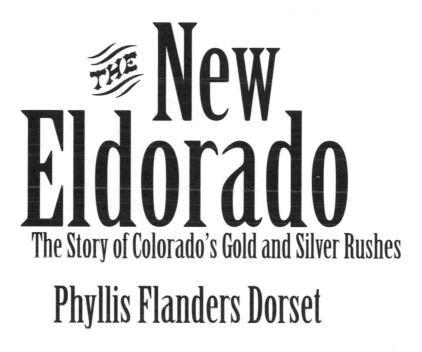

THE New Eldorado

The Story of Colorado's Gold and Silver Rushes

Phyllis Flanders Dorset

FULCRUM
GOLDEN, COLORADO

Library of Congress Cataloging-in-Publication Data
Dorset, Phyllis F. (Phyllis Flanders)
  The new Eldorado : the story of Colorado's gold and silver rushes /
Phyllis Flanders Dorset.
     p. cm.
  Originally published: New York : Macmillan, 1970.
  Includes bibliographical references and index.
  ISBN 978-1-55591-723-4 (pbk. : alk. paper) 1. Colorado--History. 2.
Gold mines and mining--Colorado. 3. Silver mines and mining--Colorado.
I. Title.
  F776.D67 2011
  978.8'02--dc23
                              2011029496

Printed in Canada
0 9 8 7 6 5 4 3 2 1

Design by Jack Lenzo
Cover photograph courtesy of Denver Public Library, Western History Collection, George Beam, GB-8341.

Fulcrum Publishing
4690 Table Mountain Dr., Ste. 100
Golden, CO 80403
800-992-2908 • 303-277-1623
www.fulcrumbooks.com

To my mother, Rhea MacDougall Flanders

Contents

Maps

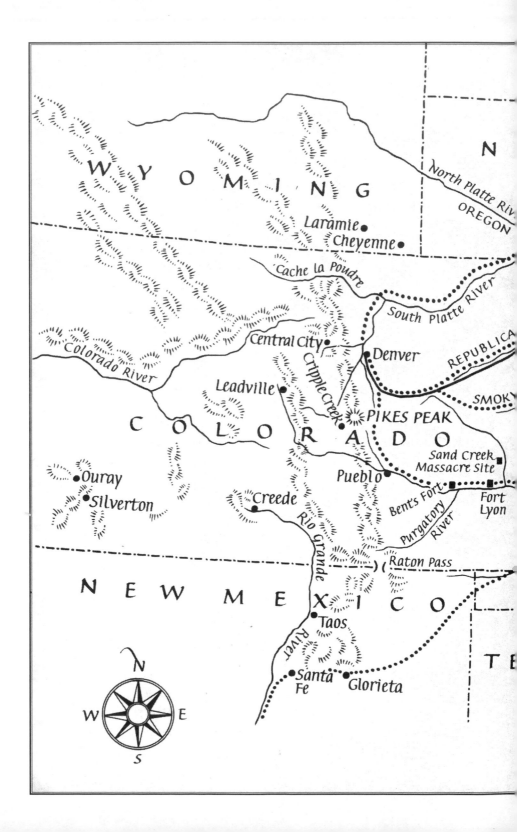

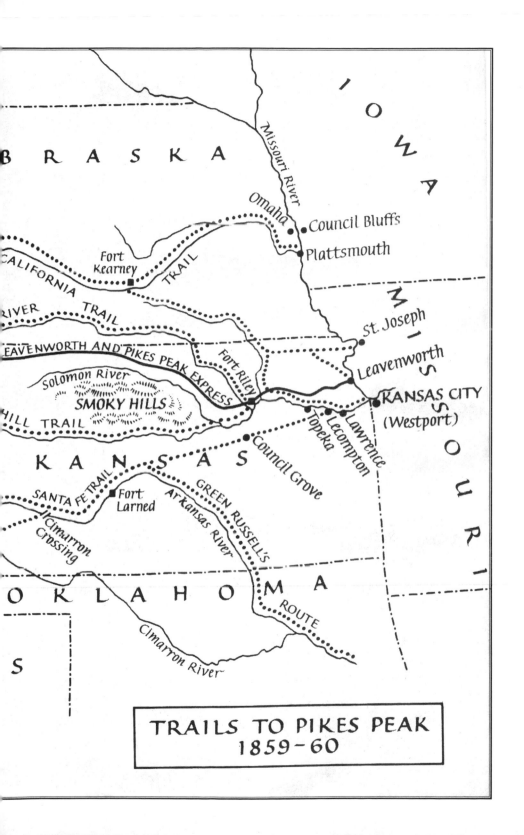

TRAILS TO PIKES PEAK
1859-60

Acknowledgments

THE BENEFICENCE OF MANY IS THE TOUCHSTONE OF THIS ENDEAVOR. I AM INDEBTED TO MY EDITOR, Peter V. Ritner, for offering me the idea and encouragement throughout its execution; I am also indebted to Joan and Wege Anderman for opening to me not only their own convivial doors in Littleton and Silver Plume but countless others in my quest to find remnants of early Colorado; to Virginia Johnson and her son, Eugene, of Cortez, for their unstinting hospitality in the San Juan country, including the key to their cottage in Telluride; to Dana and John Crawford for privileging me inside glimpses of Denver and Larimer Street, then and now; to Nancy Heubach, who saw to it that I met Bobbie Fitzsimmons, who made possible interviews with longtime Leadville-ites Charles and Florence Fitzsimmons and the late peeress of Leadville history, Marion "Poppy" Smith; to Poppy Smith's son and daughter-in-law, Fred and Doris Smith, who precisely at the right moment and quite by chance became my next-door neighbors and who made available to me their wealth of knowledge and extensive library on Colorado; to Madge Barclay and her father, Tim "Bud" Sullivan, whose reminiscences brought alive the latter days of Leadville's boom; to the William M. Wells, whose introduction to the Rae Lairds, publishers of the *Central City Register*, gave flavor to that important camp; to Mrs. Frank E. Rutherford, who graciously opened for my inspection the Hotel de Paris in Georgetown after it had been closed for the season; and to James V. Betts, MD, who put the fine touch of verity on the misfortunes and maladies to which miners are subject.

I also owe thanks to Alys Freeze, head of the Western History Department of the Denver Public Library, to her aides Jim Davis and Opal Harber; to Kay Pierson of the State Historical Society of Colorado; to Cecil Chase, reference librarian of the Bancroft Library at Berkeley, California, and his staff, for their unflagging help in leading me to bonanzas of textual and illustrative material; and to Jacqueline Minta and Sandra Matheson for their typing skills at hand when I needed them most.

And, not least, I am deeply grateful to my husband, Donald, who cheerfully ploughed through file after file of musty newsprint when he might have been jeeping over the slopes of the Rockies or basking in the summer sun of Colorado and who not once during my long preoccupation with this project complained of alienated affection.

—Phyllis Flanders Dorset
Menlo Park, California
September 6, 1969

Author's Note

INEVITABLY, CONTRADICTION CROPPED UP IN THE RESEARCH FOR THIS BOOK, PARTICULARLY IN MINE production statistics and in the stories surrounding the more conspicuous men and women of early Colorado.

As sources of mining statistics, I relied chiefly on William Baillie, "Silver in Colorado," Colorado School of Mines *Mineral Industries Bulletin* (Golden, Colorado, 1962); Frank Fossett, *Colorado: It's Gold and Silver Mines* (New York, 1879); Charles W. Henderson, *Mining in Colorado. A History of Discovery, Development and Production* (US Geological Survey Professional Paper 138, Washington, DC, 1926); and Ovando J. Hollister, *The Mines of Colorado* (Springfield, Massachusetts, 1867).

In selecting stories involving personalities I used two approaches. If the story was told by an eyewitness, I used that version in preference to those based on hearsay. If, however, all versions I found of a particular story were based on hearsay, I chose the one that most nearly fit the character of the individual or the temper of the time, depending on which I wished to emphasize. The era of Colorado's gold and silver rushes was a turbulently creative period in which contradiction and exaggeration took on a truthfulness all their own when placed beside documented fact.

—P. F. D.

Part I

Gold, 1858—1870

1

Raising Color

MEN CAME LOOKING FOR GOLD IN COLORADO. ITS DISCOVERY IN BONANZA QUANTITIES
was no accident as it had been in California in 1848, ten
years before "Pikes Peak or Bust!" became the rallying cry of
a 100,000-man, gold-hungry army that stormed the Rockies.

For three centuries before 1858, the idea that the
Rocky Mountains held stores of gold and silver was continuously nurtured. It first took firm hold in the New
World in 1530 when Nuño de Guzmán, governor of Mexico's northernmost province, sat entranced in his palatial
quarters in Compostela on the Pacific side of the country
listening to the story told by an Indian who had come to
the city from the north. Tejo, the Indian, described in full
and graphic detail the trips he had made with his father
some forty days' travel north of their village to a country
called the Seven Cities where the streets were lined with
shops of silver workers and where his father traded feathers for great quantities of gold and silver. With visions
of treasure crowding out all else in his mind, Don Nuño
immediately mounted a four-hundred-man expedition for
the north. But after months of wandering in the wastes of
the desert, he failed to find the riches or the Seven Cities
described by Tejo, and with his ragged and hungry army
nearly on the verge of mutiny he returned to Compostela
in deep disillusion.

For five years the idea of the Seven Cities lay dormant.
Then in 1536 it was vigorously revived when Cabeza de
Vaca, a survivor of an ill-fated Spanish exploration party
that had landed on the Gulf Coast from Florida, staggered
wearily into the presence of Antonio de Mendoza, the
viceroy of Mexico, to tell what he had seen on his six-year
jornada back to Spanish-occupied territory.

De Vaca told of being held captive by Indians who
lived in very large houses among mountains showing
"many signs of gold, antimony, iron, copper, and other
metals." The Spaniard went on to say that his captors were
great hunters and traders who carried on a lively commerce in much sought-after feathers, exchanging them for

gems and hides among their neighbors to the north. Mendoza was impressed. Here were echoes of the story told by the Indian Tejo. With only this much to go on, the viceroy called for another expedition to be sent into the desolate north to find the Seven Cities. De Vaca was too weak to travel after his ordeal, but his aide, the Moor Estebán was more than eager and able to undertake the task. He went as second-in-command to an adventurous Franciscan, Fray Marcos de Niza, who had been with Pizarro in Peru and who was well grounded in the techniques of running and exploration expedition.

After a long and arduous trek from Culiacán up the west coast of Mexico, Fray Marcos and his party reached the Gila River where they camped for several weeks to rest and to plan the remainder of the journey. The friar's Indian guides assured Marcos that from where they were camped their goal was barely sixty leagues off. Fray Marcos then dispatched Estebán and a small party to prepare the way for the entry of the Spanish into the first of the Seven Cities. Estebán's route took him across the Colorado Plateau and to the northeast where he came to the pueblo of Zuñi, claimed by his Indian guides to be one of the Seven Cities of a country they called Cíbola. Unfortunately, the Moor used the wrong approach in paving the way for the arrival of the main group of Spaniards. His swaggering demands for gems and women insulted the Zuñi and they killed him, sending one of the guides back to Marcos warning him not to come closer. On hearing the news of Estebán's murder and the threats of the Zuñi, Marcos was assailed by indecision. Should he push on to verify the wealth of Cíbola supposedly there for the taking, or should he turn back in the face of the natives' hostility? He compromised. He decided to take two Indian guides and travel to some vantage point from which he could at least gaze on the reported splendor of Cíbola and then withdraw so that he might safely carry the word of what he had seen back to Mendoza. With his two reluctant companions, Marcos trudged doggedly over the hot dusty sands until he came to a ridge from the top of which he could see across a barren plain to the large, dun-colored, multistoried village where Estebán had met his death. The Franciscan peered for a long time through the shimmering haze at the scene in the distance. Search and squint as he might, he saw no streets of gold or silver sparkling in the sunlight, and no glints of light from sun-touched gems shone out from the walls of the buildings. Instead, the irregular pile of stonework appeared uncannily repelling. Remembering the styles of punishment the Indian practiced on interlopers and recalling also that the provisions of his expedition were nearly exhausted, Marcos wondered no longer about what was contained in the mysterious village across the plain but quickly turned tail for camp. When he arrived, he ordered an immediate return to Culiacán under forced march, traveling the six hundred miles "with more fear than food," as he later put it. Once in Culiacán, Marcos disbanded the expedition and with a small retinue hurried on to Mexico City to report his findings.

Marcos's vision of what he had seen was apparently transformed by the trip, for when he was safely in the viceroy's palace he declared to Mendoza that he had indeed found one of the legendary Seven Cities of Cíbola, "larger than the city of Mexico," with its riches fully obvious. All that remained was to conquer it.

Once more the viceroy was impressed and once more an expedition set forth to the north, this one under the leadership of the young and ardent Francisco Vásquez de Coronado, who had replaced Guzmán as governor of the latter's province. Coronado's force was the largest yet gathered for the quest—two hundred helmeted and armored young men of rank on spirited, hand-picked mounts, seventy foot soldiers well armed with arquebuses and crossbows, shouldering brightly painted shields, and a thousand Indians, their skins daubed with red and black, equipped with bows and arrows. Added to the train were extra horses, pack animals carrying artillery pieces and boxes of ammunition, and herds of sheep and cattle for food. By late February 1540, the juggernaut was ready. Mendoza rode the hundred leagues all the way from Mexico City to Compostela to see them off, and it was with high hopes that the cumbersome expedition got underway amid the braying of mules, clanking of armor and harness, and the shouts of officers and drovers.

Following the route of the Marcos expedition, Coronado got to Zuñi in July, but to his dismay the village produced no hoards of gold and silver. All that he could find in the way of gems were a few pieces of turquoise. His disappointment was reflected as rage among the rest of his company, and they rained curses down on the name of Fray Marcos who, fortunately for him, was not at hand. Less fortunate were the hapless Zuñi. Frustrated and angry, the Spaniards took out their fury on the people of the pueblo, subduing and occupying the village, to use it as a base for further explorations. According to the legend they followed, there were seven cities of Cíbola. If Zuñi was one of them, there were six more to find. Perhaps the next one would be the storehouse of gold and silver. With a new set of guides, Coronado dispatched his lieutenant, Cárdenas, to probe northwest of Zuñi. Cárdenas explored as far afield as the Grand Canyon, marveling at its magnificence but reporting no gold- and silver-encrusted villages.

To the east, Coronado sent Pedro de Alvarado with another group of soldiers. Alvarado reached all the way to the pueblos on the Rio Grande, where he found no gold but great stores of food. When Coronado heard of this he decided to move his main force to join Alvarado's for the winter since he had just about depleted the food reserves of the Zuñi.

But he had another reason to travel to the pueblos on the Rio Grande. Alvarado had sent word that one of the Indians in the pueblos, whom the Spanish soldiers called Turk, knew of a rich land to the northeast named Quivira. As soon as Coronado arrived at Alvarado's camp he called Turk before him and listened intently as the beguiling Indian vowed he had seen with his own eyes the lord of Quivira take his afternoon nap beneath a great tree hung with golden ornaments that tinkled in the breeze, lulling the monarch to sleep. Nor was it only the ruler of Quivira who lived in the splendor of wealth, said Turk, but everyone in the village dined off golden plates and drank from golden jugs.

That was enough for Coronado. As soon as spring came he set out eagerly for the fabled Quivira with Turk as his guide. Their path wound northeastward over the high plains and sharp-edged mesas, across the southeastern corner of Colorado into the valley of the Arkansas River to western Kansas where once again the conquistadors were to be disappointed. Quivira turned out to be but another settlement of lackluster Indians, Wichitas this time, whose only suggestion of mineral wealth was a single copper ornament hung around the neck of their chief. For his fantasies Turk paid with his life at the hands of the Spaniards, and Coronado, after wintering on the Rio Grande, brought the bedraggled units of his once proud expedition back to Mexico in gloomy depression.

The great quest for the Seven Cities of Cíbola was over and, from the point of view of finding rich deposits of gold and silver, fruitless. If it seems incredible that a government as shrewd and sophisticated as that of sixteenth-century Spain gave evidence of being in other matters would finance such large and costly exploratory expeditions on the basis of what seemingly was out and out fantasy, it must be remembered that Spain had indeed found a profusion of riches in the New World, notably Cortez in Mexico among the Aztec and Pizarro in Peru among the Inca, on the strength of stories no less flimsy than the ones that captivated Guzmán, Mendoza, and Coronado.

With Coronado's return, the Spanish dream of another Mexico or another Peru lying to the north was dissolved, but with ever an eye on expansion, Spain proceeded to colonize the fertile region of the pueblos of the Rio Grande. In time, the colonizers turned up some placer gold deposits on the river, reawakening interest in the prospect of mineral wealth locked in the "shining mountains" of the north. In 1765, Juan Maria Rivera sought signs of gold in the fastnesses of the Sangre de Cristos and the San Juan Mountains of southern Colorado, and in 1776 the San Juans were once again explored by two Franciscan friars. Neither of these parties turned up any noteworthy traces of gold.

After the Louisiana Purchase in 1803, the US government sent several expeditions west to determine the boundary of the new acquisition, to map the area, and generally to report on its resources. In 1806, on one of these expeditions, Lieutenant Zebulon Pike attempted unsuccessfully to climb the magnificent peak to which he gave his name and that later became the single most important landmark in the rush to the Rockies. When Pike gave up his try to climb the peak and came down the mountainside, he and his party camped on what they thought was the Red River. In reality, Pike was camped on the Rio Grande in Spanish territory. Shortly after he and his men bedded down, the lieutenant was arrested by Spanish soldiers and carted off to jail in Santa Fe. In his account of the experience, Pike tells of hearing from his cellmate, a trapper named James Purcell, that he had found gold in 1805 on the South Platte in Bayou Salade, the alpine park on the west side of the Rampart range of the Rockies. Land, however, not gold was the preoccupation of the period, so there was little official interest taken in Pike's report of Purcell's find.

Pike was followed into Colorado by Major Stephen H. Long in 1820 and by Captain John Charles Frémont, of the US Topographical Corps, who led five sorties into the region between 1842 and 1853.

Included in the party of Frémont's second expedition in 1844 was quixotic young William Gilpin. At thirty, Gilpin could look back on a career that already included enough variety to last the average man a lifetime. A graduate of the University of Pennsylvania, Gilpin entered West Point in 1834 only to resign in good standing the following year. But the Army still held its appeal for adventure and he reentered its ranks, serving with distinction in Florida in the Seminole War of 1836. He next turned to journalism, editing a newspaper in St. Louis until the processes of the law attracted him and he became an attorney. Now, bored with the briefs and torts of the courtroom, he borrowed $100, bought a horse and gun, and joined Frémont's westbound troop. Once in the Rockies, he was fascinated. The great stands of pine, the groves of slender aspen clinging to the silent and sheer-sided mountains cleaved by coursing streams fired his imagination for the future of this ruggedly beautiful country. He went on to Oregon with Frémont but turned back alone to the Rockies, crossing them in the occasional company of trappers and stopping now and then to pan the stream beds. He could not be sure, but mixed in with the sand at the bottom of his pan he thought he saw a flicker of color not unlike that of gold. For William Gilpin, this was a prophetic find. His fortunes would become fatefully tied to the discovery of gold in Colorado.

Gilpin was not the only one at this time to give credence to the probability of masses of gold in the Rocky Mountains. Verification of Gilpin's find came from a group of men who knew the region of the Rockies better than any white man alive. Scarcely distinguishable from Indians with their long flowing hair and fringed buckskin garments, scores of trappers and traders had for some thirty-odd years followed the trail of the beaver, otter, and fox into the remote canyons and peaks of the mountains. Many of these men came from Kentucky and Missouri; many were descendants of the French voyageurs who had explored west of the Mississippi in the eighteenth century. Some

came as employees of large fur companies, some as independent hunters. They moved among the Indians as friends, taking the women as
wives, and living simply but well by the dint of their accurate gun
eye, the acuity of their tracking sense, and their savvy of bartering.
No region of the mountains was closed to them. In their roamings
through ravines and over ridges looking for their prey the mountain
men often ran across traces of gold, but as long as the demand for
pelts was high, the adventure of the hunt, not the search for gold,
drew their energies.

Once a year, in the summer, the mountain men—men like Jim
Beckworth, Uncle Dick Wootton, the Sublettes, Louis Vásquez, Ceran
St. Vrain, and the four Bent brothers—gathered at a rendezvous to
deliver their pelts to company representatives and to restock for
another go at the hunt. The rendezvous was a time of letdown for these
hard adventurers who roamed the mountain wilderness for months
at a time in near isolation. In a few exuberant days they haggled for
their price over bundles of sleek skins, downed vats of whisky, stuffed
themselves with sizzling venison, and caroused with their friends,
white and red, gambling away half their profits, and an occasional
gold nugget, on horse races and wrestling matches, spending the rest
of trinkets for their squaws and on a few supplies, and finally, with
throbbing head and bloodshot eyes, moodily dragging themselves off
to the mountains for another year.

Then fashion changed. Fur hats gave way to silk ones. Commerce
in skins shifted from fur skins to buffalo and deer hides, most of
which were gotten by trading with the Indians. Mountain men turned
from trapping to trading. Trading posts sprang up along the ancient
hunting trails of the Indian. And the doughty hunters whose rovings
among the mountains were at an end found they had an abundance of
time on their hands to recollect where and when in their travels they
had come across trace gold.

Purcell, Pike's cellmate, was only one who remembered. Frémont
reported that "Parson" Bill Williams and a trapper named Du Chet
had supposedly picked up nuggets from streams they crossed in their
wanderings on the headwaters of the South Platte. A man named
Poole recalled he had found traces of gold on the streams that fed
the Arkansas River, and one Norton showed some gold flecks he had
reputedly panned in the Sangre de Cristos. Antoine Pichard, another
trader, told of finding traces of gold at his campsite near what is now
Golden, Colorado.

The clearing house for many of these stories was Bent's Fort, one
of several key trading outposts built by the brothers Bent and their
partner, Ceran St. Vrain, on the Arkansas River, a little over one hundred miles east of the Rocky Mountains. William Bent, in particular,
was aware of the presence of gold in the area, having been shown
samples by the traders and Indians who frequented the fort. Long a

friend of the Arapaho and Cheyenne, Bent would have liked to quash the stories of gold strikes because to him it was all too clear that if a gold rush developed in the Rockies, the Indians' "last and best home and hunting-grounds will be appropriated by the white man and they themselves [will] be finally exterminated."

Fate stepped in to at least postpone what Bent feared would happen when in 1848 John Marshall, excavating the site of Colonel Sutter's mill on the western slope of the Sierras, uncovered a vein of pure gold and the nation's gold-hungry hit the trail for California. As they flocked west, many prospecting parties moved along the edge of the Rocky Mountains. Some, unable to resist checking the possibility that gold lay closer at hand than California, side-tracked from the main routes west to search the streams that fed the South Platte and Arkansas rivers.

One such party, made up of 150 Cherokee Indians and white people, left the Cherokee Nation on April 20, 1850, bound for California. A number of members of the party were originally from Georgia, including Lewis Ralston and his brother. For the Georgia Cherokees, looking for gold was not a new experience. Twenty years before, they had been in the thick of the gold rush that swept their lands in North Georgia but they realized little out of it save experience since the Cherokee, once the word of gold got out, were hustled off to a reservation in Oklahoma. Now their chance had come again. The Ralston party followed the Santa Fe Trail to Cimarron Crossing just west of what is now Dodge City, Kansas. Here, instead of turning south as the trail did, they continued along the Arkansas River past Bent's Fort to the region of Pikes Peak where they wheeled north, panning all the streams they encountered on the way. On the morning of June 20, two months to the day of their departure, the Ralston party arrived on the banks of the south fork of the Platte. The party pitched in to construct a raft and began floating their heavily laden wagons across the river. At 2 PM the next day the transfer was complete, and the Ralstons pushed north, traveling about six miles along the river bank to the mouth of a sandy creek. Here, according to the diary kept by one of the party, they found "good water, grass, and timber. We called this Ralston's creek because a man of that name found gold here." So did the Ralstons' fellow prospector, John Beck—$5 worth of gold dust to a pan of gravel, he said. But for the rest, that was not enough, and on the Ralston party went to California.

For years after he returned to the Cherokee Nation from California, without having found the lode he hoped to find there, Beck still savored the memory of that productive pan of dust he had turned up on the South Platte back in 1850. Finally, in 1857, he heard of a party of Cherokee who were going into the western part of Kansas Territory looking for new buffalo grounds. Beck and his son joined them. To Beck, a quest for buffalo was secondary; he was going to find that gold-rich stream locked in his memory. But he was denied the chance—the hunting party was rudely run out of Kansas Territory by irate Arapaho and Cheyenne who were beginning to resent the deepening tide of westbound interlopers, Indian and white. The Cheyenne in particular had been causing trouble all along the major trails west, marauding and murdering. Meeting violence with violence, the US Army sent several companies of cavalry into the field under Major John Sedgwick to subdue the Indians. With the cavalry rode Fall Leaf, an imposing Delaware Indian, who served as a scout. Fall Leaf knew the country well, having ridden with Frémont on his last expedition. During his sojourn in the mountains and plains with the Pathfinder, he had heard stories of gold discoveries in the streams that spilled down the eastern slope of the Rockies, and when Sedgwick's troops met some Missourians who were prospecting along the front range of Pikes Peak country, Fall Leaf came away with a goose quill filled with sparkling gold dust.

So persistent were the stories of gold found in the Rockies that made their way to the border towns of the Mississippi and beyond, that an editorial published in the *Missouri Democrat* in 1857 urging that the federal government send out a large party of reliable and respected judges to verify the rumors was enthusiastically copied and seconded by the influential New York *Herald*. Nothing however came of the recommendation, and John Beck, for one, wasn't willing to wait for the ponderous wheels of government to do his prospecting for him. He sat down and wrote to his friend William Green Russell, a Georgian who had married a Cherokee woman, telling him of his find on Ralston Creek seven years before.

Green Russell and his brothers, Levi and Oliver, had also succumbed to the lure of the California gold rush, but after desultory luck in the Sierras they too had returned to Georgia in 1852. When Beck's letter found him in the fall of 1857, Green Russell was trying his hand at farming. At the prospect of another chance to find a gold bonanza, he quickly gave up the plow for the gold pan and turned to organizing a party for a trip to Ralston Creek.

By the spring of 1858, with characteristic gentle persistence, Green Russell had persuaded several groups of Cherokee and white men to undertake the journey west. The Russells and sixteen others started out from Georgia. In Missouri, they picked up another twenty-seven willing argonauts, and on June 3 they were joined by John Beck and fifty-seven men in Kansas. It was a formidable train: 14 wagons, 33 yoke of cattle, 24 horses, and 104 men. Formidable enough, Russell assured Beck and some of the others who hadn't forgotten the scare the angry Arapaho and Cheyenne had given them the year before, to remove all fear of being attacked.

The combined party now moved forward, striking the Santa Fe Trail at Pawnee Fork, and clipping off a solid twenty miles a day westward over the short buffalo grass along the banks of the Arkansas. They broke the monotony of the days on the hot and windy trail with a cheerful stopover at Bent's Fort in the Big Timbers where they got a good meal and a bracing, if expensive ($1 a pint), dollop of whiskey. Setting out again on the high bluff country of the fort they traveled over terrain dotted with stubby pine trees, turning north a few miles east of the site of Fort Pueblo, following a creek the French voyageurs had named Fountaine qui Bouille. Near the headwaters of this creek they crossed over the divide and descended to the edge of a northward-flowing stream overhung with chokecherries. On the US topographical map he carried, Russell lettered the words *Cherry Creek*. The gold-seekers then panned the creek all the way down to its confluence with the South Platte, without any promising results. Once at the river, they took advantage of the shade of a grove of cottonwood trees to rest. Some urged that they make camp, but Green Russell was anxious to move on to Ralston Creek. After a brief discussion with the rest of the party, Russell won his point and the caravan lumbered on down the six miles to the site of John Beck's original find.

At first light the next morning, the men were at the gravel beds with their gold pans. They worked likely looking segments of the stream banks all day, and to Beck's chagrin only a few grains of gold showed up in the pans. That night and every night after each fruitless day there was grumbling around the campfire. Where, the men asked each other, was the $5 per pan Beck had raved about all the way across the plains?

At last, on July 3, after nearly a week of unrewarding effort, Russell called the company together to decide the next move. In the flickering light of the fire, he stated the situation as he saw it. The trace gold they had found, little as it was, meant that there was indeed gold in the region, and that what they had found had undoubtedly been washed downriver from large deposits locked in sandbars farther up the Platte. He caught the dubious looks of his followers as he suggested that they start prospecting upriver. Luke Tierney, one of the original sixteen of Russell's Georgia group, who would later write a guidebook to the region, recorded with growing disgust that most of the men complained that the trip had been a wild goose chase and to waste any more time was futile. Especially foolish, they said, was to go up the river toward the mountains which would bring them into the hands of the Ute whom everyone knew to be the most ferocious of the local Indians. Green Russell protested that to have come so far only to give up at the first disappointment was senseless. At this, a man jumped to his feet and cried, "To hell with it, let's go home!" The words he blurted out were what most of the men were thinking and cheers of approval rang out over the camp. After the hubbub died down, Green Russell, standing tall but noticeably shaken, stepped slowly to the center of the group.

"Gentlemen," his voice was composed, "you can all go, but I will stay if two men will stay with me."

One by one the company, including the one-time enthusiast John Beck, voted to give up the search and go back to their homes. Only twelve of the 104 stood up to take their place at the side of Green Russell, and all of these were from his original group of Georgians.

It was with some trepidation that the little band that was left watched the rest pack up and head out. They knew full well that thirteen trespassers in a wilderness belonging to the hostile Ute would not have a chance if the Indians chose to attack them. The Ute danger was not a figment. At Bent's Fort, Green Russell's men heard how the Ute, but three years before, had gone on the rampage, harassing settlements in the Sangre de Cristo Mountains, and finally on Christmas Day dropped into Fort Pueblo where in the midst of the festivities of the season they slaughtered eighteen of the twenty-one inhabitants, taking three as prisoners to an unknown fate.

But as the sun shone on the camp the next morning, casting the mountains behind them in sharp relief and warming muscles stiff from the night's chill, the future looked not at all foreboding and the spirits of the thirteen who remained rose with the sun. They ate heartily of flapjacks and hogback and started up the Platte with fresh hope.

Eight miles upstream, one of the company turned up some promising color. Quickly, the rest dug in with their pans and by the end of the day they had washed out nearly $10 worth of gold dust per man. Not a bad beginning. Russell's men were elated. This was what the long trip across the plains had been for, and since the purpose of their coming was to prospect promising sources of gold for later development, they moved on upstream after marking the site of their strike on the map. Ten days later, about three miles above the first strike, they found another sandbar that produced between $12 and $18 a day per man. A third strike, paying about the same as the second, was made a few days later. Green Russell shared the delight of his friends at the accumulating bags of gold dust, but he was at the same time frustrated that they could not do a first-rate job of prospecting because their company was too small and the danger from Indians too great to allow the men to separate for any period of time. But despite his frustration he took particular pleasure in one thing: his deduction had been correct. The closer they got to the mountains, the more gold they found.

The Russell party had been prospecting on the South Platte for about three weeks when on July 31, John Cantrell, a trader en route from Fort Laramie to Missouri, rode into camp and was invited to share the miners' evening meal. Over broiled chunks of deer meat, Green Russell showed the admiring trader the sacks of gold dust they had taken out of the river. Cantrell, who had been a miner for a time in California and who knew something about gold, bought a bag to show the folks back in Missouri.

But if Cantrell thought he was going to be the first man to prove to the river towns that there was gold in the Rockies, he was wrong. Unknown to either Russell or Cantrell as they sat by their campfire was the fact that someone else had already spread the word.

After the Army had dealt with the Cheyenne and returned to Kansas, Fall Leaf, the Delaware Indian scout, proudly showed his goose quill filled with the precious metal to enough people that thirty of them caught gold fever. The man who had the worst case of it was John Easter, a grocer of Lawrence, Kansas. Shortly, he became the self-appointed leader of the thirty adventurers on the strength of Fall Leaf's promise to guide them to the source of the dust in his quill. The Indian was to get a per diem allowance, payable when the party arrived at their destination. But when the Easter party was all set to leave on May 20, 1858, Fall Leaf was not. Some say he reneged on his end of the bargain because some plains Indians threatened his life if he brought any more white people into their territory. Others tell a more likely story, that the amiable Fall Leaf, on the eve of departure, set out to celebrate the beginning of his newest foray into the West, got roaring drunk, took a fall, and ended up an invalid, unable to travel.

Undaunted by the loss of his guide, John Easter gathered up his company and with eleven wagons set out on the Santa Fe Trail toward the only landmark he knew to look for, Pikes Peak. On the way, they added to their party three more men and a woman. The woman, Julia Holmes, was described by one of the party as "young, handsome, and intelligent...a regular woman's righter, wears the bloomer, and was quite indignant when informed she was not allowed to stand guard." The lady proved her prowess however by doing what the intrepid Zebulon Pike himself could not do. When the Easter party got to Fountaine qui Bouille Creek at the foot of Pikes Peak, they made camp, and while the men got to work panning the stream bed, Mrs. Holmes climbed all the way to the summit of the peak. Her success was not shared by the men in their endeavor. Not one sign of gold lined their pans, and finally the Easter party packed up and headed south. At Jimmy's camp, a much frequented rendezvous of mountain men, the Easter party split up. Some of them headed home to Lawrence; others, including John Easter and the Holmes family, set out for New Mexico. Those who headed east were soon overtaken by John Beck's returning party and heard the disappointing news of his efforts on Ralston Creek. This only strengthened the resolve of the homeward bound that it would be of no use to go farther north than they had gone. A few weeks later, Easter and those who had turned south were overtaken by a trader, perhaps one who had come across Cantrell, who contradicted Beck's story and told of seeing the glittering evidence to prove that the Russell party had found gold. Again the Easter party split. Holmes and some of the others continued toward Taos, while the rest, led by John Easter, wheeled around, revictualed at Fort Garland, and hastened north.

Meanwhile, the Russell company explored the South Platte for nearly a hundred miles in rugged terrain, with very little more color to show for their pains than the first three strikes produced. So they reversed their direction, heading north toward Fort Laramie, prospecting earnestly throughout the rest of August only to be stopped short of the fort on September 5 by a raging blizzard. The portent of an early cold season combined with what Russell decided were signs of Indian danger persuaded him to return his party to Cherry Creek for the winter.

When Green Russell's men approached the grove of cottonwoods that marked the junction of Cherry Creek and the South Platte, they were surprised to see smoke drifting lazily above the treetops. And then a squaw appeared and for an uneasy moment the Russell men feared they had blundered into an encampment of Utes. But they relaxed in relief when behind the Indian woman walked a lean and whiskered white man, whose clothes of buckskin marked him as a mountain man. He came out to meet Russell, smiling broadly, and giving his name as Jack Jones. On the frontier aliases were common. Many a man did not want his friends and family to know what he was up to; others fled their pasts in the anonymity of the West where not a man's name but his actions were what counted. Sometimes, however, despite the subterfuge of an alias, a man's real name caught up with him. So it was with Jack Jones. Reputedly, Jones's name was really William McGaa. Son of a lord mayor of London, McGaa had been scheduled to enter orders but he had entered too many pubs before being ordained, bringing a certain amount of disgrace on his lordship and his family. So William did the gentlemanly thing—he ran away, came to America, to the wilds of the West, where he fell in with Blackfoot John Smith, trapper and sometime honorary chief of the Cheyenne. Both Smith and McGaa had taken Arapaho women as wives, which accounted for the first face the Russell party saw when they entered camp.

The two mountain men made the Russell men warmly welcome, helping them to set up their tents and tether their animals. Later, over steaming bowls of antelope stew prepared by their wives, the two trappers surprised the Russells by telling them that they were not the only prospectors in the neighborhood. About six miles upstream, on the east bank, atop a grass-covered knoll, was the Lawrence party of John Easter. The Easter party, so Smith and McGaa said, were optimistically and busily building some twenty cabins on a town site carefully and ambitiously laid out by a civil engineer in their group. Acknowledging the pervading presence of the majestic Rocky Mountains at his back, Easter had given his town the Spanish name of Montana.

The Russells asked if Easter's party had found any worthwhile deposits of gold. "Just enough to keep them here till spring when they can look some more," was the answer. Green Russell laughed and showed Smith and McGaa the total take for his own party's summer efforts, about $800 in gold dust. "So did we," he said.

Not to be outdone by the zeal of the Kansans upstream, the next day Russell and his friends moved their wagons across the creek to the south bank and got down to the business of organizing. The first thing they did was form three groups. Green Russell, his brother Oliver, and another man took $500 of the $800 and struck out for Georgia to gather more supplies and men. A second group set to work building rough cabins. And a third group led by brother Levi Russell took $200 and headed south to Fort Garland for provisions to last the Cherry Creek camp through the winter. When Levi Russell got to Fort Garland he had to sell his watch to meet the cost of all the supplies he needed. One of his companions who later wrote about the incident felt compelled to comment that it would have looked like humanity in government officers to have given us rations rather than to have taken the last cent of money that we could raise and then strip a man of his watch. But we were not beggars; we simply were working to develop the resources of the great unknown West."

So confident were the Russells of these resources that before Green Russell left for Georgia he gave a name to the town his men were building. Even more optimistic than John Easter, he called it Auraria, after his hometown in Georgia, a name derived from the Latin word for gold.

When John Cantrell reached Westport (Kansas City) some time in late August, he found a number of people more than interested in the sack of gold dust he had brought from Cherry Creek. Here was hard proof of the story spread by George Simpson, an Army teamster, who claimed he had panned gold from Cherry Creek back in May.

Up and down the Missouri River the word was passed: "Gold, at Pikes Peak!" Within the men in the streets of the river towns and on the adjacent farms was born a rising excitement. The news leaped across the land. A telegraphic dispatch in the Sunday morning St. Louis *Republican* stopped coffee cups midway to mouths with a headline that read "The New Eldorado!!! The Pikes Mines. First Arrival of Gold Dust at Kansas City!!!" Beneath the headline were the stories of Cantrell and two other trappers, Bordeau and Bissonet, who had also just come from Cherry Creek with pouches of gold dust. By the next day the story was on the wire to all the major newspapers in the East.

To professional men, tradesmen, and working men, the news was hope sprung as a phoenix from the ashes of the nation's charred economy. Just a year before, in 1857, financial panic had consumed the country, climaxing a period of exorbitant spending and overexpansion. Banks closed, credit was immobilized, and factories shut down, throwing droves of unemployed into the streets, It was also a time of drought and as crops failed the farmer joined in spirits his brother on the cities' pavements parading for "work or bread." Little wonder then that the news of a gold strike, as promising as that in California, electrified all who heard, read, or saw the evidence.

Particularly exuberant were the citizens of the border towns whose prosperity had abruptly declined in the '57 panic. The Omaha *Times* reported with some abandon that three to four dollars a day could be panned from the South Platte riverbed "not over one hundred miles westward from here." Downstream, the Leavenworth papers bally-hooed the torchlight street meeting held to tell the public that Leav-enworth stood ready to become the supply center for any gold-seek-ers bent on reaching the South Platte diggings. Key speaker was the mayor himself, who outlined the advantages of his town as the start-ing point for the emigrants. His words were followed by the "glorifica-tion speeches" of two prominent citizens, Judge Perkins and Colonel William Larimer. The meeting, said the newspaper, lasted "until a late hour and was the very soul of enthusiasm." The Leavenworth *Journal* chronicled cavalierly that the meeting had demonstrated unequivo-cally that the town could furnish outfits and transport for ten thousand men on very short notice, and that in fact Russell, Majors, and Waddell, the renowned western freighting company, was at that moment ready to outfit and deliver two thousand men to Cherry Creek. When Rus-sell, Majors, Waddell and company read the story they shuddered to think what would happen if two thousand people suddenly presented themselves at their offices demanding cartage west.

No less imbued with the prospect of the new Eldorado was the nationally respected journalist, A. D. Richardson. Crowed he in the conservative Boston *Daily Journal* of September 21, 1858, "The excitement which I predicted more than a month ago is now at its height...It stirs men's hearts...Politics are forgotten. Speculation is ignored, and the latest news from Pike's Peak is the universal theme of conversation."

Pikes Peak became the universal theme of more than conversa-tion. In the Missouri river towns, barber shops advertised haircuts in the Pikes Peak style, restaurants featured beef *alamode* Pikes Peak, and pudding topped with Pikes Peak sauce. Pikes Peak prints showed up in the fashions of women who trundled their children in Pikes Peak baby carriages. The addiction caught up hundreds of willing enthusiasts, and before the end of September the overland trails were crammed with adventurers headed west to the magic mountains at whose base they firmly believed lay a fortune for the taking, just as did the Spanish explorers three hundred years before them. Some whose eagerness outran their good sense started out on foot with only the clothes on their backs and a bit of flour and bacon for provi-sions, planning to subsist on the game they could hunt on the trail. Others pulled handcarts heaped helter-skelter with dearly bought picks, shovels, and gold pans. Mixed in with these were an extra pair of boots, a shirt, a pair of trousers, and to live on, a handful of staples.

A few groups pooled their resources to buy wagons, ox teams, and horses with which to make the trip. On the fresh white canvas of their wagon tops they splashed the words "Pikes Peak or Bust!" None of these travelers, the prudent or imprudent, listened to the conflicting stories that now began to appear after the initial excitement of the Cherry Creek gold strike died down.

Army men, traders, and those from the Russell and Lawrence parties who passed through the border towns en route east for reassignment or supplies were eagerly interviewed for the latest word from the gold fields. The military men were not very encouraging, but the traders and the settlement people generally exuded reassurance. One cavalry officer, passing through Leavenworth, said the miners on Cherry Creek were getting only a little float gold for their trouble and blamed the poor showing on the lack of proper tools. Another doubted that even with proper tools could the region ever prove to be a profitable gold-producing area. A few days later, two agents of the Lawrence party arrived in Leavenworth and contradicted the Army. Of course there was gold at Cherry Creek, and lots of it. Why else would they come back for more men, tools, and supplies? The press was perplexed. What was the real story of Pikes Peak? Could it be that the whole business was "an unmitigated humbug," as the Richmond, Missouri, *Mirror* of October 2, 1858, hotly declared? Caution crept into some editorials. One paper suggested that Pikes Peak fever could be abated by "taking a slight dose of reflection." Another indignantly decried the deceit of those who deliberately made false statements of the amount of gold to be found on Cherry Creek and caused normally level-headed men to throw over whatever steady, if dull, work they had, to follow an adventure of uncertain reward. Such warnings went mainly unheeded, so firmly was the idea of ready riches waiting in the Rockies planted in the minds of the argonauts. Even when Green Russell, on his way home to Georgia that fall for new recruits and supplies, publicly cautioned the people of Leavenworth that the Cherry Creek gold deposits had yet to be proven bonanzas, the town turned on him in fury. He was accused of playing down the wealth of the gold fields in an attempt to hoard it all for himself.

For those who had already headed west, no doubts beleaguered their optimism. Left behind was the stifling atmosphere of the depressed towns, ahead lay the great open prairies in which a man could soon feel "free as the air," so wrote gold-hunter Woodward from his campsite some 240 miles west of Kansas City. To Woodward, as it was with other argonauts, walking beside his team with the freshening wind on his face and the sun shining down on him from a cloudless sky, the jangle of the harness and the creak of the wagon were music, the chores of camp—cooking, foraging for game and fuel, washing and mending clothes, and watering the livestock—were but part of the romance of the road at whose end lay the pot of gold.

Writing by the light of a lantern in his tent, Woodward paused over his letter to listen reflectively to the cheerful strains of a fiddle and the rousing songs coming from a nearby camp of Germans, also bound for Cherry Creek. One of that party, jovial "Count" Henri Murat, who claimed to be the nephew of the king of Naples, would have heartily agreed with Woodward's postscript that the trip had been so free of hardship that "merry times" were the order of the day on the trail.

One reason for the easy passage was the fact that the weather on the plains that fall was exceptionally fine. Only two thunderstorms marred what otherwise would have been a perfect trip for the Murats. There was plenty of the short, succulent buffalo grass for forage, adequate water in the streams, and the only Indians they met were friendly. Both the count and his wife thrived on the journey across the country. Katrina gained weight and looked fresh as a maiden to the admiring eyes of her husband and doubtless to the eyes of all the men at Cherry Creek since she was the first woman they had seen since laying eyes on the squaws of McGaa and Blackfoot John Smith.

When the Murats' wagon party rumbled into the tiny settlement on a sunny crisp morning in early November, the world of Pikes Peak looked bright indeed. Both John Easter's Montana City, where they had settled, and Green Russell's Auraria were the picture of industry. There was a trace of snow left on the bottomland of the South Platte from a storm that had struck on October 30, but it was no hindrance to the work of mining and building. The Murats saw about thirty men scattered in groups along the banks of the river busily working the gravel beds for pay dirt, some using gold pans, others using crude wooden boxes fitted with rockers to wash out the heavier gold particles from the lighter weight sand. Murat was told that the prospectors were averaging from $4 to $7 a day at the best locations. No one, however, told him that a sandbar was soon panned out and that a man might have to prospect for weeks before he found another bar that would pay that much in one day. Green Russell had spoken the truth: there was gold on Cherry Creek, but it was only enough to indicate that somewhere nearby were the real lodes. Nevertheless, the work went on, with the ever-optimistic miner certain that tomorrow he would make the big strike. And if not tomorrow, then surely the day after.

Those who were not mining were busy constructing lean-to's and cabins. Logs for the walls were laid up to just over a man's head. The spaces between the logs were chinked with mud and then a flat roof of split logs was laid across the top and covered with six inches of dry grass and sod. Canvas cut from a wagon top covered the door. The more ambitious builders cut windows and tacked oiled cloth over them. Leaky and dark, the cabins nonetheless offered a warmer shelter than did a wagon or tent.

Responding to the tonic of the invigorating dry air of Cherry Creek's six-thousand-foot altitude, the Murats lost no time in becoming a part of the bustling community. Henri, with two men to help him, drove his ox team across the river and climbed the broad plain west of Auraria, trudging through knee-high grass some twelve miles to the foothills that rose abruptly in three tiers against the snow-dusted peaks of the Rockies. Here in six hours they cut and stripped enough pine logs to build a cabin. By the next day, Katrina's rustic chalet was ready for occupancy. It was by no means palatial—bunk beds covered with straw mattresses clung to one wall, a trestle table made of knocked down wagon wood stood in the center of the room. Around it were a couple of rope-seated chairs and several three-legged stools. At the hearth of the sod fireplace sat a cast-iron pot in which Katrina could boil up the week's mess of beans and bacon.

With winter coming on, the Murats decided to defer any prospecting until spring and turned their thoughts instead to an enterprise to fill the intervening months, preferably a profitable enterprise. They looked around them at the straggly-haired, ragged, bearded men in mud-caked clothes smelling of sweat and tobacco, and immediately saw their mission. Countess Katrina got out her washtub and Count Henri stropped his fine German razor blade—garments washed and ironed, 50 cents each; shave and a haircut, $1.

If these were exorbitant prices, at least they were no more exorbitant than other prices on Cherry Creek. With scores of men and wagons arriving each day, the small stock of commodities already in camp dwindled alarmingly as the demand for them grew greater. A typical day brought 110 men and twenty wagons rolling to the new diggings, but in the wagons were no staples to spare. Few had been put aboard on departure and most of those were consumed en route. In camp, a hundred-pound sack of flour went for $20, if one could be found to buy. Sugar, bacon, and coffee cost a man 50 cents a pound, if anyone would part with his precious store. In December, the food situation in Montana City and Auraria became critical. The diggings on the river were abandoned as men shouldered their rifles and went out to hunt for game to keep the camp from starving.

The number of people in camp reached nearly three hundred and the job of killing enough game to feed them kept hunters in the field every day. When the snowstorms drove the hunters in, oxen, the valuable work beasts, had to be slaughtered for food. But the scarcity of food was only one of the basic problems of survival the argonauts faced with the coming winter. Although there were some seventy-five cabins in the two camps, there were not enough bunks for everyone to sleep in at once, so people took turns, keeping the bunks warm around the clock with their sleeping bodies. For those on their feet, night or day, keeping warm was just as difficult. Only a few cabins had fireplaces, and without them the settlers relied on the campfire as the source of warming and cooking heat. With few places to sleep and only a few fires to warm by, many a night saw groups of miners huddled together around campfires, wrapped in their trail blankets, awakening after fitful sleep with their noses, fingers, and toes frostbitten.

As the days of December wore on, accompanied by freezing temperatures and biting winds, the dreams of quick riches on which the Cherry Creek people had fed with such relish in the warmth and brightness of the summer began to fade in the bleakness of a plains winter. It was no consolation that there was not a drop of whiskey in camp to kill the chill in a man's bones and to keep kindled the hope of easy wealth.

On Christmas Eve 1858, at the very moment when the spirits of the argonauts were at their lowest, the settlers heard the crack of a whip shatter the cold air and the crunching of wagon wheels on the frozen ruts of the road. The men of Auraria got up from their desultory card games and went out to see a wagon lumbering down the bank of Cherry Creek toward camp. It looked at first glance to be but one more load of gold-seekers. With everything in such short supply except people, the miners looked with a jaundiced eye on the newcomer. The wagon halted and a massive black-haired man, his face nearly hidden by the slouch hat crammed down on his head, swung down from the driver's seat and walked over to a sullen group by a smoking campfire. The man slapped his gauntleted hands together to shake out the numbness and glanced around him.

"Any Indians camped near here?" he asked.

"Downstream, about ten miles, is a bunch of Arapaho a-wintering," answered the man nearest him, motioning northward where ribbons of smoke from the Indians' campfires reached lazily skyward.

The man touched his glove to his hat and walked back to his wagon and started to climb back up to the driver's seat. With one foot on the step, he turned back to eye the group once more. His searching look traveled carefully over the solemn faces. He could guess the cause of their dejection.

"I come up from Taos to trade with 'em," he drawled, looking from one man to another. "Unless, of course, you all'd be interested in a little swappin'."

Richens Lacey Wootton, trapper trader, mountain man extraordinary, companion of Jim Bridger and Kit Carson, confrere of William Bent, and respectfully known as Cut Hand to a half dozen Indian tribes of the plains, knew exactly what effect his words would have on the dispirited men before him. He knew well the rigors of a winter in these parts and what it was like to be short of food. And besides, the trader in him relished nothing better than to turn a quick profit.

Wootton's eyes grew bright as the settlers crowded around his wagon, fingering the sacks of flour and sugar, the slabs of lean bacon, the baskets of dried apples, and bulging bags of coffee. Small pouches of gold dust were waved in front of his twinkling eyes and here and there a rifle or pistol was offered in trade. Men shouted their offers for a pound of coffee, a plug of tobacco, a peck of apples, jostling each other to get closer to attract the trader's attention. At last Wootton's booming voice quieted the excited crowd.

"Steady, boys," he said. "Give old Uncle Dick a tent and he'll set you up a proper store."

"We'll go you one better than a tent," shouted a man at the edge of the crowd. "You can use my cabin."

In no time, with a dozen or more willing hands to help, Wootton's wagon was unloaded. A shout of approval rang out when the settlers uncovered three large kegs beneath the sacks of flour in the bed of the wagon They rolled the gurgling barrels into the cabin where Wootton set them on end, all in a row. Then he broached the first one, set a tin dipper on the top and motioned for the men to help themselves.

"Have a Christmas drink on me," called Uncle Dick, "but don't come a-gunning for my hide tomorrow if you feel like you got hit by lightning." And here he broke into a rumbling chuckle, "Because you did!"

It was the miners' first taste of Taos Lightning, a regional liquor of doubtful chemistry and fabled kick. Among the mountain men it was legend that no man lived long enough after drinking the lacerating liquid to become addicted to it. But that night a lot of men tried. Word of Wootton's arrival was carried up the Platte to Montana City and soon miners from Easter's camp scurried down to purchase supplies and slake their thirst at the trader's keg. Every time a man raised a dipper of Taos Lightning to his lips, his spirits rose as well. Before dawn the bleakness was dispelled, and the two little settlements, Montana and Auraria, set about to have a merry Christmas.

The next day, in the midst of preparations for a Christmas feast, the miners looked up from their fires in surprise to see a dozen Arapaho braves riding pell-mell toward camp towing a string of galloping ponies behind them. Through Blackfoot John Smith, the Indians, with broad grins showing their eagerness, told the miners they had come to help them celebrate the holiday with a little horse-racing. In tribal society there were three accepted ways of increasing a herd of horses: by gambling, trading, and stealing. To the Indian, the sporting aspect of each method was about the same. The difference was that an Indian preferred to gamble with his friends. So far the men on Cherry Creek were too few to pose a threat to Arapaho life and they had done nothing to anger the red men, so they were friends. Moreover, the Arapaho were very sure they could beat any candidate the white man put up.

The miners, whose very presence in this wilderness was testimony to their gambling natures, were quick to take up the challenge. The Arapaho bet one hundred ponies that one of their horses could run a distance of a quarter of a mile faster than any horse the Aurarians ran against it. The settlers goggled at the recklessness of the wager, and not so confident as the Indians they talked the stake down to eight ponies. The race was run along the sandy bank of the river, with stamping, shouting miners lining the course. To their own and the Indians' astonishment, the white man won handily. Now in the eyes of the Arapaho the settlers' stock rose appreciably. Anyone who was a judge of horses was to be respected. Next to whiskey and tobacco, the Indians valued good horseflesh as the most important thing in life (guns and women, by their own declaration, ranked fourth and fifth).

Christmas day was bright and crisp. In Montana City the aroma of roasting meat wafted out of cabin doors to mix with the pine wood smoke of campfires. Knots of men here and there cleaned freshly caught trout, tended spitted sage hens, and stirred bubbling pots of applesauce. Others kneaded bread and cracked and shelled the sweet-flavored nuts gathered from the piñon trees in the foothills. At the appointed hour, the rapid clanging of a pickhead against the iron rim of a wagon wheel brought the hungry miners clamoring to the table. Between mouthfuls of succulent venison and gulps of Taos Lightning they sang "The Star Spangled Banner" and "The Girl I Left Behind Me" intermixed with lively toasts. One toast neatly summed up the situation on Cherry Creek that December day in 1858.

To the miners and mines—may the latter be as prolific of treasure as the former are pregnant with hope.

After their sumptuous Christmas dinner, the Montana City men lay about the camp dozing and exchanging stories until night-fall when the camp dynamo, A. O. McGrew, who had the singular distinction of having pushed a wheelbarrow all the way across the plains to Cherry Creek, rousted his companions from their torpor and persuaded them to walk with him down to Auraria to see how the village was celebrating the day. When they arrived, they found a great bonfire blazing on the square set aside as the town meeting place, and around it danced scores of shining-eyed miners dressed in getups as wild as they could rig—some in skins of beaver, capes of fox tails, with horns of mountain sheep or antlers stuck on hats. The Germans' fiddles squealed out romp-stomping tunes and the thunder of boots on the hard ground combined with the crackle of the fire sounded to McGrew like a pack of animals fleeing from a forest fire. As the night wore on, groups of curious Arapaho sidled to the edge of the firelight to watch the dancers, their amused faces cast in eerie relief by the warm yellow glow of the fire. It was midnight before the fiddlers put down their instruments and the Indians moved silently away and the last dancer stumbled off to bed down wherever he could find a bunk or bit of floor space, to dream once more of the bonanza awaiting him in the spring, leaving only the occasional cry of a coyote and the sound of snapping embers to break the brittle stillness of the starlit night.

2

Rush to the Rockies

WITH THE FIRST SIGNS OF THAW, THE CHERRY CREEK AND SOUTH PLATTE CAMPS rang with the sounds of shovel and pick as the miners got to work again pursuing their dream. Those who had spent the long winter in camp were soon joined by a new and eager contingent of prospectors who had responded to the glowing letters of promise printed in countless newspapers across the country and in the rapidly appearing "authoritative" guidebooks to the new gold regions. But neither the newcomers nor the men who had come out the previous fall found any gold deposits large enough to warrant commencing large-scale mining operations. Disgruntled and down right indignant, scores of new arrivals who had done little more than poke around the edge of streams expecting to find the glittering ore in piles at their feet threw down their tools and headed back home, claiming it was all "humbug as to the plenitude of gold...Some say that they are making from $5 to $20 per day, but they can never show it." Every morning in camp the people awoke to the street cries of self-styled auctioneers haranguing the public to buy cast-off rifles, pistols, shovels, picks, and spare shirts and boots at one-tenth their original cost. The "go-backs," as the rest of the camp derisively called them, then set out for the East, a few lucky ones on horses and wagons with the once arrogant banner "Pikes Peak or Bust!" now crossed out and replaced with the bitter words "Busted, By God!" Most, however, were on foot with a blanket roll and the few supplies that they could buy with their auction money slung on their backs. So anxious were several groups to get out of the diggings that they built flatboats on the shore of the Platte intending to sail the treacherous, sandbar-studded river down to the Missouri and then to Council Bluffs, Leavenworth, or Westport. Only one of these boats made it. The rest of the adventurers either drowned in the upsetting of their craft or lost all they owned in the river currents, barely managing to save their skins.

The traffic on the cross-country trails during those early spring weeks of 1859 was composed of the disillusioned go-backs, most of whom were destitute, and a westward tide of hopefuls fired up by the stories they had heard all winter of gold at Cherry Creek and still unaware that the diggings had so far proven somewhat less than productive. It was a hard crossing, going and coming. For the experienced transcontinental traveler, crossing the plains was bad enough at best—the distance was long, six hundred miles; the roads were generally poor, the major ones being little more than ruts carved by the wagon wheels of the US Army and the California forty-niners following the age-old trails of Indians and traders. Settlements of any kind and good watering places were few and far between. There was game, but more often than not a man had to pursue it to kill it and that meant leaving the trail, a habit that invited disaster, for it was deceptively easy to lose one's bearings and become hopelessly lost. Then there was the constant and unpredictable danger of Indian attack off the trail and on it. None but the foolhardy set out without an adequate number of companions, wagons, teams, foodstuffs, and arms to ward off the myriad dangers of becoming lost, running out of supplies, and of running into war parties. Above all, none but the foolhardy set out before warm weather came to the plains. Those whose business took them across the vast and lonely stretches of desert and prairie had a deep respect for the explosive blizzards, paralyzing windstorms, and torrential rains that could make a hell on earth for the luckless traveler from December until late May. They knew firsthand what could happen to a man who became lost in a blinding snowstorm, whose oxen froze in the cold or starved to death from the lack of grass, or drowned trying to ford a swollen stream. But the two groups that set out that spring, one from the East, and one from the West, were inexperienced and each was motivated by a single-mindedness that allowed little room for caution. As a result, some died and were peremptorily buried beside the road; the crude crosses and headboards making grim signposts for those who came after them. For some of the westward bound, seeing men die on the trail and hearing the bitter words of the go-backs they met was reason enough to turn homeward without ever having reached Cherry Creek, contemptuous of all who championed the story of Pikes Peak gold. Others, unwilling to be discouraged so easily, pushed onward with a new sobriety to "see the elephant" for themselves.

The man on whom the go-backs vented their outrage was busybody D. C. Oakes. Oakes had come to Auraria in the early fall of 1858 and he spent the winter there cheerfully filling in his time by writing and sending off to the printer portions of a guidebook for goldseekers that praised to the sky the mineral opportunities of Cherry Creek. Between chapters he dispatched similarly glowing accounts of the mining prospects for publication in Kansas newspapers. All the while Oakes was writing about the opportunities to garner mineral wealth that lay in store for others who came to Cherry Creek, he was looking around for a sure enterprise for himself. As he watched the cabins and stores being laboriously built of hand-hewn lumber that winter, he quickly saw in which direction lay his own opportunity on Cherry Creek. In the spring, he made a quick trip to Missouri to buy some sawmill machinery. On his way back, Oakes met a vengeful pack of go-backs who recognized him, roughed him up, and prepared to lynch him. But the persuasive facility Oakes had with words, so effective in his guidebook and letters, saved the day for him and he was left to go on his way, bruised but alive. Not to be done out of their revenge, the go-backs made a mock grave in plain view beside the Overland Trail for all travelers to see. On the wooden headboard they crudely lettered the words:

Here Lies D. C. Oakes
Killed for Starting the Pikes Peak Hoax

When word of the exodus of the go-backs and the story of Oakes's close call reached the Missouri River towns, there was consternation bordering on panic. These towns that all winter and even now were flooded with handbills, hawkers, and newspaper advertisements, all aimed at enticing the starry-eyed gold-seeker to shop at their emporiums for his western outfit, quickly set up a dire protest against the coming throngs with "broken hopes and blasted fortunes." Editorials in Kansas City and St. Louis urged that additional police be sent into the streets to protect the property of the towns' citizens while the surge of returning gold-hunters passed off downriver. In Atchison, Kansas, a reporter buttonholed a cross-country mail carrier who had come through crowds of angry go-backs and promptly wrote his story around the carrier's ominous warning: "...no acts of violence have been committed so far as I can learn but as...this crowd of starving wanderers increases, what assurance will there be against...rapine and plunder?"

When at last the hundreds of go-backs clawed their way through the plains spring and poured into the border towns, the inhabitants were unspeakably relieved to find that the horde was too tired and discouraged to be any real threat to civil order. At the sight of the ragged, hungry mobs, the very newspapers that had so smugly promoted the Pikes Peak enterprise the previous winter did an about-face to call the gold rush "the humbug of humbugs." The vacillating press would change their tune again, however, because the irony of the disillusionment of the go-backs and the indignant righteousness of the border towns was that the new Eldorado was in truth no hoax.

In Auraria, as soon as he recovered from the Christmas celebrations, George A. Jackson, a lanky trader from Missouri, cousin of Kit Carson, and devoted gold-hunter, set out to do a little prospecting with two pals, Jim Saunders and Tom Golden. They headed toward the mountains, in the general direction of Clear Creek canyon that

cleaved the flat-topped promontory just west of Auraria. On the way,
Golden and Saunders spied some elk tracks in the snow and went off
to hunt down their quarry. Jackson, with his two dogs, headed on up
the northern fork of Clear Creek, plowing through deeper and deeper
drifts as the canyon narrowed. It was cold going. For Jackson's dogs,
it was especially tough and they quickly tired as they porpoised their
way through the two-and three-foot drifts in the rock-strewn creek
bed. To rest his dogs, Jackson made camp often, and each time he
stopped, he dug down through the snow to the frozen ground to try
his luck. But nary a flicker of color did he turn up. On a day when he
and the dogs returned to the camp from a prospecting hike to find
a mountain lion had stolen all his dried deer meat, Jackson decided
to head back to Auraria. He had not gotten very far when the lure
of one more try pulled him south, stumbling and slipping up over
a steep ridge and down among the bare-branched aspen and snow-
sodden pines on the other side to the south fork of Clear Creek. Here
he came to hot mineral springs and gratefully made camp near the
welcome warmth of the steaming pools. Then he surveyed the creek
bed but saw no signs of color. Discouraged and hungry, Jackson once
more resolved to break camp and go back to Auraria. Suddenly, out
of the corner of his eye he saw something move on the hillside. It
was a mountain sheep. Jackson took careful aim and fired, dropping
the animal in its tracks. With a new food supply, Jackson decided to
make one more search. Cutting thick chops for himself and his raven-
ous dogs, Jackson ate heartily, drank the last of his coffee, and slept.
The next morning at dawn he set out. All day he cruised the frozen
creek bed in his high leather boots, with no result. Then, near dusk,
he sighted a sandbar that looked promising. He jabbed his knife into
the gravel but it penetrated only a fraction of an inch. The ground was
frozen and impossible to work. With his hatchet, Jackson chopped
enough wood from the fallen trees around his sandbar to build a blazing fire. All night he kept the fire going, and on the morning of January
7 he set to work in the thawed gravel using his belt knife as a pick. By
the end of the day, he had panned eight cups of dirt out of which he got
one nugget and a small vial of gold dust, amounting to about $10 in all.
After the work of the day he sat by his fire feeling the freshening wind
whirring through the trees and sensing that a storm was working its
way toward him. Before he crawled into his buffalo robe for the night,
he took up his pencil in numb fingers and scribbled in his diary, "...
will quit and try to get back in the Spring. Feel good tonight."

Small wonder Jackson felt good. In the spring the world would
know he had unlocked a gold placer bed of bonanza proportions.
Meanwhile, he marked the spot by topping a lodge-pole pine exactly
seventy-six steps west of the hole in the sandbar. Then he filled the
hole with charcoal and slogged his way back to Auraria through a raging blizzard with his lame but faithful hounds. Only to Tom Golden,

whose mouth was "as tight as a No. 4 Beaver trap," did Jackson reveal the secret of his tantalizing discovery.

While he waited calmly for spring, Jackson took a job with Jim Saunders expressing the mail down to Cherry Creek from Fort Laramie. On his first trip or two north to that bustling frontier crossroads, Jackson could well have noticed a carrot-topped, wiry Georgian who idled on the streets of town, but at that time Jackson had no way of knowing that the two of them would share the distinction of having uncovered the mineral treasure of the shining mountains that had eluded men for three hundred years.

A gold-smitten muleskinner, John H. Gregory left his hometown in northern Georgia on August 8, 1858, headed for the mining excitement on the Fraser River in British Columbia. The trip proved to be more difficult than he had counted on, and by late that fall he had gotten only as far west as Laramie. As he thought about the rugged miles ahead over the Rockies and the Cascade Mountains to Canada, his better judgment told him to stay put until spring. While he lounged away the winter in the fort, he picked up the news of Green Russell's gold discoveries on the South Platte, and he thereupon decided that there was no reason to tramp all those miles to the Fraser when his fortune might well lie a little way south on the Platte. Early in February, he left Laramie alone, carrying his clothes, food, and prospecting tools with him. Working his way south along the base of the Rockies, he panned every important stream between the old-time traders' favorite rendezvous, Cache la Poudre Creek, and Fountaine qui Bouille, to no avail. Finally he retraced his steps as far north as Auraria, where he stopped to wait for better weather.

In late April, Gregory and a small group of men from Indiana, whom he had met in camp, followed the north fork of Clear Creek almost to its source. Here, on May 6, 1859, on either side of the stream, among the pine trees lining the sides of the gulch, they found places where the snow had melted, exposing veins of decomposed quartz that looked to Gregory much like the rock in the gold mines of Georgia. The party dug in with picks and shovels, hacking away at the "blossom rock." A few feet down, Gregory's pick sank suddenly into softer material. Quickly he dug out a pan of the crumbled rock and earth and ran to the stream with it. His hands shook as he washed the dirt in the icy water. Gradually, as the light sands slipped away, grain after grain remaining in the bottom of the pan shone gold in the chilly spring sun. When the glitter weighted out at $4, Gregory and his Indiana friends let go with yelps of triumph that echoed all along the gulch sending the ever-present ravens cawing frantically into the air. Immediately the men set about laying out claims.

George Jackson was also on Clear Creek when Gregory made his find, but he was on the south fork stepping off the seventy-six steps from the lodgepole pine to his glory hole of the winter. By early April,

Jackson decided it was time to return to his secret cache. But there was one difficulty. Despite his working all winter, the trader was too broke to buy what he needed to prove the value of his discovery. So he cast about Auraria's saloons until he came across a newly arrived party of twenty-two men from Chicago, well stocked with money, supplies, and zeal. Once he had shown them his carefully harbored nugget and vial of gold dust, they were ready then and there to grub-stake him if he would guide them to the site of his lucky sandbar. Jackson insisted on making some ground rules first, the chief one of which was that he should be allowed a double-sized claim as dis-coverer of the mine. The Chicagoans quickly agreed and on April 17, with wagons loaded with provisions and tools, Jackson led his party toward the beckoning hot springs on the south fork of Clear Creek. It was hard work getting the cumbersome wagons over steep ridges and across ravines cluttered with man-size boulders and fallen trees, but finally on May 1 they made it. They set up camp, relocated the placer deposit, staked out claims, tore down the wagons, using the wagon beds for sluice boxes, and started mining the sandbar. By May 7, Jackson and his party had netted $1,900.

Gregory, on the north fork, was stymied after the first few exhila-rating days of work by a localized ice storm that prevented him from working his claim again until May 16. But from that date until May 23, he and two other men mined the quartz vein for five days wash-ing out $972.

If the combined outputs of the Gregory and Jackson strikes for one week were apportioned among the miners involved, each man would have received about $120. At a time when the average laborer was lucky to make $10 a week, the news of this kind of money waiting to be brought out of the earth ignited the fire of hope all over again in the breasts of the go-backs and of those teetering uncertainly on the brink of taking the plunge westward. There was not yet a tele-graph line connecting Cherry Creek with the East, but the word of the Gregory and Jackson bonanzas needed no telegraphy—it rode on the wind. And within a matter of days, the second stampede to Pikes Peak was on.

A hundred thousand strong, they came from all over the Mid-west, and from every section of the East, from New England, from the mid-Atlantic states, and from the South. Some came on foot, some came on horseback, by stagecoach, and even by train as far west as it could take them. At St. Louis and other Mississippi River ports, they crowded aboard steamboats headed up the Missouri to Westport, Leavenworth, St. Joseph, and Omaha-Council Bluffs, all jumping-off places to the new Eldorado by virtue of their proximity to the main transcontinental trails.

Typical of the gold-fired mob was a nineteen-year-old youth named Ryan who worked his way to St. Louis hoping to make enough money in the river town to finance his trip the rest of the way to the Rockies. But the town was overrun with itinerants by the time he got there and there were no jobs to be had. Still, the excitement of the stampede west kept him there, hanging around the boat landings and outfitting stores. One day his luck changed and he struck up an acquaintance with another westbound young man who loaned him money to buy passage on the steamer *New Monongehela*, destined for Leavenworth, with the ornate artwork on her paddle-wheel box covered with a banner proclaiming "Ho, for the Gold Fields!" In the perilously overloaded steamer, the young men got their first taste of claim-jumping as they tried to find a place to sleep. Cabins and decks crawled with seven hundred other souls bound for Cherry Creek, and the law of squatter's rights prevailed. After a miserable five days on the river, the *New Monongehela* disgorged her human cargo on the landing at Leavenworth.

The border towns were soon seething with manic gold-hunters who darted from one outfitter to another, clutching in their eager hands well-thumbed guidebooks to the new Eldorado. Oakes's guidebook was not the only one available. Some nineteen were on the market by June of 1859. They covered every aspect of a trip to Cherry Creek. In these volumes the prospective miner could find the answers, some right and many more wrong, to all of his questions: what kind of an outfit he would need, where to buy it, how to organize a wagon train, when to start, what trail to take, how to deal with marauding Indians, where to find buffalo and how to track prairie game, what maps to use, how to carry firearms safely, what kind of weather to expect, and not least of all, how to handle displays of temperament on the trail.

Among the authors of these guidebooks, Oakes, Luke Tierney, who was one of Green Russell's men, William Parsons, who was one of John Easter's Lawrence party, and William Gilpin, who lent his name to one such publication, had at least been to the gold fields and knew what it was like to cross the plains. Few others had. From his quickly opened Gold Mine Exploration office in Chicago, W. H. Horner, a real estate agent with a flair for advertising, turned out a few tens of pages crammed with misinformation gleaned from news accounts of the experiences of those who had made the trip to Cherry Creek. He blandly declared that for $60 to $75 a man could outfit himself for the trek west and a stay of four months in the new gold diggings. Tierney and Parsons gave a more likely figure of $200 minimum for just the journey across the country, to say nothing of the cost of staying on Cherry Creek for four months.

Tierney was correct on how much money the trip would cost but he was tragically wrong on another bit of advice. It apparently seemed to him more important that the prospector reach Cherry Creek before the dry season came, when the water in the streams fell too low for mining operations, than not to reach there at all, because he advised emigrants to start west as early as possible in the spring, at least as early as April, even if grass for stock on the road was scarce. To a man half-delirious with gold fever, Tierney's words were what he wanted to read. It would only be some two to three hundred miles west of Leavenworth, when he looked down at the bloated carcasses of his starved oxen and at his own rag-wound and swollen feet and realized that he would have to abandon most of his supplies because he could not carry them all on his back, that the gold-hunter would see what a fool he had been to believe everything he had read and heard.

The most reliable guidebooks recommended that the safest and best way to cross the plains was by wagon, carrying ample provisions for at least a thirty-day trip. But the fifty-niners tried all kinds of conveyances. A. O. McGrew started out pushing a wheelbarrow containing his clothes, food, and tools, A number of others took the advice of the Omaha *Times* and in parties of three or four loaded a hundred pounds of gear for each man in a two-wheeled handcart, snugly laced up their boots, and headed west, pulling their carts behind them, unaffected by the current joke about handcart men:

"If you're going to Pikes Peak, why don't you wait for grass?"

"What do we need grass for, we ain't got no stock."

"Yes, but since you're making asses of yourselves you might need it for food."

If a man could not afford the price of a guidebook, the newspapers offered cheap advice on every late-breaking fad for the trip west. The Council Bluffs *Bugle* blew clamorously in favor of an "entirely new mode of traveling" called the Tebogia, made of sheet iron eight feet long and eighteen inches wide and turned up at the front to be dragged along the prairie by two men in a harness. How far this contraption got is not on record, but another inspired mode of travel at least had more appeal from the point of view of its motive power. This was the "wind wagon" touted by the *Missouri Republican* of April 20, 1859, as the last word in plains transport. Designed by a Mr. Thomas of Westport, the wind wagon resembled a small sloop with lopped-off bow and stern. It had seats for twenty-four passengers. The "craft" rode on four wheels, mounted a twenty-five-foot mast, and carried a mainsail and a jib. Mr. Thomas advertised that his wind wagon would do one hundred miles a day and make the round trip to Pikes Peak in twelve days. Unfortunately, but predictably, on its maiden voyage, when the first squall hit, the wind wagon bit the dust landlubber fashion a few miles west of town, never more to sail the undulating plains.

An honest estimate of sufficient provisions for a well-equipped wagon party was, for each person, one hundred pounds of flour, fifty of sugar, thirty of beans, one hundred of bacon, and twenty of coffee. Salt, rice, dried fruit, soda vinegar, and pickles made up the rest of the food supply. Reliable guidebooks recommended that each wagon be pulled by one or more teams of oxen and have a canvas cover and carry a tent. The emigrant was instructed to take along three pairs of woolen or leather pants, six flannel shirts, three pairs of boots, an overcoat, some cotton drawers, woolen socks, a handkerchief or two, and a broad-brimmed felt hat. William Byers, who had not yet been to the scene of the new Eldorado but who would soon join the stampede, told his readers to leave their razors at home because they would not need them, but "...leave not your character at home, nor your Bible; you will need them both, and even Grace from above to protect you in a community whose God is mammon, who are wild with excitement and free of family restraints." Once on the scene, Byers was to learn too well the sagacity of his own advice.

To Parson, the minimum tools required for mining were a pick, a long-handled shovel, and a gold pan made of tin or copper or sheet iron that could hold three gallons of water. In a praiseworthy but for the most part unheeded effort he also admonished his readers to be prepared to work hard because nuggets of gold did not lie in the bottom of every stream like paving stones. Most emigrants however chose to believe the extravagant assurances such as Tierney's that "... with ordinary tools, and a fair supply of water, from six to eight dollars a day can easily be realized."

On Indians, guidebook advice was sound—be kind but cautious. Curry their favor with gifts and don't provoke them to anger. If a war party threatens, seize a bull whip and thrash the nearest brave. By all accounts this was a proven method of discouraging a plunder-minded band of natives. It presupposed, of course, that the Indians got within reach of the white man's whip to practice their bullying tactics. A war party that swooped down on an emigrant caravan without warning was another story.

The Indian danger on the trail was real. The plains Indians, watching with growing bravado the uprisings of the Navajo and Pueblo tribes during the United States' war with Mexico in 1846–47, began to take out their own grievances against the encroaching white man in the 1850s and 1860s. The Sioux, Comanche, Cheyenne, Kiowa, and Arapaho made commonplace their routs of tiny settlements and of trading, freighting, and emigrant parties, striking suddenly and viciously and disappearing as fast as they came. There were many ways to kill on the plains and the Indian was master of them all. The signature of their handiwork was easily recognized: disembowelment, castration, and scalping of victims dead of bullets, arrows, knife cuts, lance thrusts, or chops of a tomahawk. For captive women there was sometimes a special fate, that of being "passed on the prairie"—raped in relays until death mercifully intervened. It was no comfort to emigrant families to know that the Indian used the same techniques on enemy tribesman as he did on the white man.

Throughout the expansion westward, the Army made an effort to keep the trails safe, frequently sending out punitive expeditions such as that of John Sedgwick and his dragoons to hunt down trouble-making warriors, but the trails were long and there were too few soldiers to police them adequately. Consequently, the wayfarer was left to provide for his own protection. Wagon trains generally followed the axiom of safety in numbers but even then the danger was merely mitigated, not removed.

In early May, Robert Gibson's forty-three-wagon train had just made camp near the Santa Fe crossing on the Arkansas River when a band of Kiowa braves charged into their midst demanding gifts. They were given sugar and tobacco from the train's carefully rationed store. Displeased by the small amounts, the Indians contemptuously threw down the sacks and proceeded to mill around the camp, hounding the edgy white men for any article that struck their fancy. As the emigrants tried to go about their camp duties with feigned nonchalance, the Kiowa boldly climbed in and out of wagons looking for booty, shoving cowering children aside, stopping now and then to finger the hair and clothing of terrified women, and menacing any of the white men who moved to stop them. Finally, after two hours of harassment, one brave leaped upon his pony, gave a bloodcurdling yell, and, followed by the rest, stampeded the emigrants' cattle, cutting loose with a well-placed arrow as a final gesture of disdain to kill one of the panic-stricken oxen.

Another emigrant party, this one numbering some two hundred men, managed to take the fight out of an eight-hundred-man Cheyenne war party by showing the chiefs their Colt repeaters and demonstrating that two hundred men with guns were an equal match for eight hundred with only bows and arrows.

It was not so easy to dissuade another brave, intent not on war but on wiving. He took a fancy to the blond wife of one colonist and offered to buy her. It took a large tribute parcel of blankets, sugar, coffee, tea, flour, tobacco, and all of the woman's jewels to convince him she was not for sale. If her husband had not gritted his teeth a while earlier and smoked a well-mouthed peace pipe with the would-be Indian suitor, the brave might well have dispensed with the formality of the barter system and simply made off with her.

Tribal chiefs were not above the loathsome acts practiced by rancorous braves. Satanta, chief of the Kiowa, frequently awed his young men with special displays of savagery. After leading a raid on a white settlement in which he took captive a woman with a babe in arms, Satanta grew tired of the child's cries on the long ride back to camp, so he pulled up under a cottonwood tree, seized the baby by the feet, and brained it against the trunk of the tree. To still the mother's shrieks, he beat her bloody with a length of knotted rope.

An equally terrifying but invisible menace on the trail was cholera. The disease came west with the white men, reaching epidemic proportions during the mass emigration of gold-seekers who crowded along the trails to California in '49, cropping up sporadically in the years that followed. It struck its victims suddenly and generally fatally. Seemingly fit as a fiddle in the morning, a man could be dead of racking cramps and convulsions by nightfall. Sometimes it wiped out entire parties, sometimes it cut down but one or two in a group. George Shaw, his wife, their two young children, and his mother, set out from Independence, Missouri, in a wagon made comfortable as possible with quilts and rugs thrown over sacks of flour, sugar, slabs of bacon, and boxes, trunks, and the prized rocking chair. They joined a train of emigrants and were making their way happily, if slowly, along the Santa Fe Trail when the children came down with the virulent disease that claimed both of their lives in two days. The grief of the parents and the grandmother scarcely had time to reach its full dimension when the train was attacked by hordes of Comanche. In the running battle with the savages, George Shaw was struck down by an arrow through the neck and died in a few days. When the remnants of the wagon train reached the shelter of the next settlement, the young mother, deep in shock, was tenderly carried into the house of a settler where she lingered comatose for a day or so until she too died. Uncle Dick Wootton, who happened to be passing through the settlement when Mrs. Shaw died, never forgot the poignant sight of the bewildered, despairing grandmother, the only survivor of what three weeks before had been a vibrant, hope-filled family.

But in the rush to the Rockies, few Pikes Peakers were discouraged by stories of hardship and the likelihood of dying on the trail. They continued along throughout the spring and summer of 1859 to pour into the towns along the Missouri River by the thousands, eyes glazed with dreams of quick riches, minds closed to peril.

Once in the border towns young Ryan and hundreds of others found economic chaos. Hucksters met the boats, clawing at the travelers with offers of cut-rate lodging, supplies, and transport. Handbills advertising the excellence of hundreds of items to be found in countless stores were thrust by the dozen into the hands of the dazed new arrivals. When the gold-seeker finally got his bearings after the barrage of advertising, he found that the demand had outstripped the supply. With commodities dwindling, prices soared. A night's lodging at a "one-horse" hotel leaped to an exorbitant $2.50. "Little rats of mules" that ten days before had sold for $20 now went speedily at $150. Sharpers unloaded spavined horses and dropsical oxen on scores of guileless argonauts who were generally too delirious with gold fever to realize they were being cheated. If a man were broke, he cadged a little dried meat, a few crackers, and maybe a handcart, and set out on foot blithely unconcerned that it was six hundred miles to Cherry Creek. Many of those with money to buy their way west in a train of prairie schooners found all bookings taken for weeks ahead. Others found they could buy in, but the price was prohibitive—one year's provisions for the entire party. Nevertheless, they plunked down the cash for passage.

From January to June 1859, an ever-increasing number of wagons rumbled west from the border towns. In Omaha in May in one week alone 584 wagons left for Colorado. For the first twenty-five miles or so out of town, before the distances widened between groups of emigrants with different conveyances, the prairie was alight at night with hundreds of campfires for as far as one could see.

Four trails led to Cherry Creek, two well traveled and two nearly unknown. The most frequently used, and the longest and safest, were the Oregon–California Trail to the north and the Santa Fe–Arkansas River Trail to the south. Between these two lay the slightly shorter but forbidding Smokey Hill and Republican River trails. Guidebook recommendations on trails often depended on which border towns bought the most space for advertising in the publication. The jumping-off places for the Oregon–California Trail were St. Joseph, Plattsmouth, Council Bluffs, and Omaha, and each of these towns vied fiercely for the emigrants' business. One way to attract the attention of the gold-seekers was to place a big boldface notice in a guidebook proclaiming "Plattsmouth, Portal the Gold Fields!" Enough of these expensive, full-page advertisements and the author of a guidebook was more than glad to recommend the Oregon–California Trail to his readers.

The Oregon–California Trail led over undulating prairie for some two hundred miles along the River Platte to Fort Kearney. From there, it was another two hundred miles or so to O'Fallon's Bluffs with the trail following the river, and flanked on each side by a broad tableland scarred with deep gulches. Forty miles beyond O'Fallon's Bluffs, at the junction of the north and south forks of the Platte, the trail to Cherry Creek left the main road to follow the south fork to Cache la Poudre Creek, then south again to Cherry Creek. From O'Fallon's Bluffs to Cherry Creek was roughly another two hundred miles.

From Leavenworth, Lawrence, and Westport (Kansas City), the gold-seekers took the Santa Fe Trail through Council Grove westward to the Great Bend of the Arkansas River, past the ruins of old Fort Atkinson, then over desolate bluff country to Big Timbers. Here, weary, parched travelers, their clothes, wagons, and teams covered with the powdery dust of the trail, gratefully made camp among the cottonwood trees near Bent's Fort, the first shade they had seen in 175 miles. From Bent's Fort the trail wound west through sagebrush and piñon for a few miles, then north across Fountaine qui Bouille Creek at the foot of Pikes Peak, and over the divide, down Cherry Creek to Auraria.

To many who were anxious to get to the gold fields before all the choice claims were snapped up, it was ridiculous not to take the most direct route possible to Cherry Creek. A few miles shorter than the Oregon–California Trail and the Santa Fe Trail, the Smoky Hill and

Republican River trails drew many a hapless emigrant. None of these travelers knew the hardships that were in store for them. The trails lay among broken, barren ridges cluttered with rocks interspersed with long stretches of deep sand that for man and beast made torture out of walking. The grass was thin and reedy, and what little there was of it spiteful Indians often burned off. Not far out of Fort Riley, the trees thinned and finally died out altogether, being replaced by a dwarf cactus and occasional Spanish nettle and prickly pear. Without wood, the traveler had to scour the countryside for *bois de vache*, the chips of dried buffalo dung used by plainsmen for fuel for their fires when wood was unobtainable. Too often the waterholes were alkaline, and the water unfit to drink. No grass and poor water meant no game, and for the emigrant who had counted on living off the land, this was a death warrant. Scores died of starvation, their bodies lying where they fell beside the trail, carrion for the hawk and buzzard. Out of one party of seventeen men who set out on foot with their meager supplies slung on their backs, only three managed to reach Cherry Creek alive.

Grimmer still was the fate of the Blue brothers. Caught up with gold fever, Daniel Blue and his two brothers, Alexander and Charles, left Kansas City on March 6, 1859, with eleven other men and one packhorse, bound for the new Eldorado over the Smoky Hill Trail. On March 22, five men decided to leave the party and go hunting for buffalo. On March 26, as the rest of the party was climbing a brush-covered ridge, their packhorse suddenly reared, pulling free of the man who led him, and streaked off into the wasteland. The men gave chase but the animal escaped, carrying most of their provisions with it. The Blue brothers and the three other men pushed on with what supplies they had left. After eight days on the barest of rations, their strength gave out and they made camp, some seventy-five miles east of Auraria, determined to stay put until they could replenish their food supply with game. The only firearm they had was a shotgun. The rest of the guns they had thrown away on the trail when, in their growing weakness, the argonauts found them too heavy to carry. For ten days the party lay up, attempting to track game. But with their waning strength they could not venture far and the only creatures they managed to shoot were a few rabbits. The desperation of their plight now fully dawned on them and after a council they decided five men should take the little food that was left and set out for the nearest settlement for help. Left in camp were the Blue brothers and a man named Soleg who was far gone in starvation. In a day or so, Soleg died, but before he slipped into unconsciousness he urged the Blue brothers to eat his flesh after his death to sustain their own lives. Hideous as it was, they had no choice if they were to survive. Soon Alexander died of exhaustion, and the two remaining brothers packed some of his remains for food and tried to continue their journey westward, but the going was too much for Charles and they made camp again after struggling forward for nearly ten miles. For ten days they lived by eating Alexander's remains, growing weaker by the day. One morning Daniel awoke to find his brother dead. Left alone, Daniel, now nearly mad, fed on Charles' body until an Arapaho hunter found him, gibbering and unable to stand, and took him to a way station of the newly established Leavenworth and Pikes Peak Express where he was cared for until the stage arrived. In a day or so after Blue was brought in, the second coach out from Leavenworth, bound for Cherry Creek, rattled up to the station and the skeletal figure of the gold-seeker, with his eyesight nearly gone and his body so weak that his limbs flopped grotesquely each time he attempted to move, was placed aboard and taken the rest of the way into Auraria. Under the watchful care of good Samaritans in the little mining settlement, Daniel Blue slowly regained his physical health, but his mind never fully recovered from his ordeal.

The only other passenger in the coach that carried Daniel Blue into Auraria was journalist Henry Villard, who was cheerfully following up an assignment to report on the new gold fields as a correspondent for the Cincinnati *Daily Commercial*. Villard was a German émigré who had banged around the United States for several years learning the American language and taking any job he could find, but preferably one that called for him to use his literary talent and his newly acquired tongue. His account of the Pikes Peak rush launched him on a career in journalism that carried him through the Civil War. After the war he became the agent for a group of Germans who held great blocks of Oregon railroad bonds. Before long the intrepid Villard became an officer in the companies whose bondsmen he represented, climaxing a meteoric career in business as president of the Northern Pacific Railroad in 1881. But on this May day in 1859, the twenty-four-year-old newsman's mind was occupied not with the subtle manipulations of finance but with the adventure of his western trip.

Of all the modes of travel to Cherry Creek, riding in one of the express company's brand new, bright red Concord coaches was the fastest and least comfortable. W. H. Russell, the line's founder, was a recognized authority on how to get anything anywhere by horse, mule, or oxen. He was a partner in the famous freighting outfit of Majors and Waddell, supplying the far-flung trading and Army posts of the West, and would later create the Pony Express. But in the meantime, Russell was quick to capitalize on the money-making potential of a direct line to the new Eldorado, bringing the first coach into Auraria from Leavenworth on May 7, 1859. Normally, nine people were jammed into the stagecoach, seated shoulder to shoulder, three abreast, on hard, horse-hair-covered seats. Swinging hour after hour with a deadly rhythm to and fro on its leather thoroughbraces as it careened over rocks and ruts, through slashing rainstorms and stifling clouds of dust, the coach temporarily turned stalwart men into palsied wrecks and made swooning hags out of spritely women. Every one hundred to two hundred miles the coach mercifully made a brief stop to change teams. Three times in every twenty-four hours it stopped again to dump its passengers at a way station where they bolted down a standard meal of fried salt pork, soda biscuits foundering in greasy gravy, dried apple pie, and muddy coffee. For all this luxury, a passenger paid $125 one way and $1.50 for each indigestible meal.

For Villard, however, as the coach's only passenger until the appearance of the piteous Blue, the trip had been rather agreeable. He could loll by day and stretch out by night on one of the long seats, no stranger's knees interlocked with his, and no alien head drooped in utter exhaustion on his shoulder, and even the constant swaying under these circumstances was not unpleasant. The four Kentucky-bred mules pulling the coach clipped off a respectable thirty to sixty miles a day depending on the weather and the condition of the road, depositing the journalist safely in Auraria on the twelfth day out of Leavenworth.

The $700 in shot gold brought into Leavenworth by the first return coach from Auraria sent Kansans into paroxysms of pride over *their* gold mountains. Their proprietary interest stemmed from the fact that the boundaries established by Congress for Kansas Territory five years before included the area of the new diggings. The arrival of the coach was promptly heralded by a banner emblazoned with the rhapsodic prose of the day, "Leavenworth hears the echo from her mineral mountains and sends it on wings of lightning to a listening world."

One of that listening world was the indefatigable editor of the prestigious New York *Tribune*, Horace Greeley. Greeley had been keeping his ear tuned to the news coming out of Pikes Peak and now, with the arrival of the first gold shipment from Chicago Bar, as Jackson named his diggings, he too joined the stampede west "to see the elephant." His companion was the genial young reporter Albert D. Richardson of the Boston *Journal*. On the morning of their departure from Leavenworth, Greeley caught the festive spirit of the hundreds of cheering well-wishers who crowded around their coach in front of Planters' House, leaning far out of the window, waving his broad-brimmed straw hat at the crowd. Suddenly there was the crack of a whip and the mules, trained to get underway at a dead run, jerked the Concord coach forward, throwing the slightly top-heavy Greeley off balance. As his rotund form arced gracefully toward the street, the quick-acting Richardson snatched a handful of the elder newsman's white duster and hauled him unceremoniously back into the coach as it jolted along its impulsive way.

After that auspicious start the two passengers settled down to what they hoped would be an uneventful trip. But they had yet to learn that the Kentucky mules could not bear the sight of Indians or buffalo. A few days out of Leavenworth, on a windless afternoon, as the stage swayed hypnotically along, a band of Indians in a playful mood suddenly swooped down on the coach, totally demoralizing the mules, who ran terror-stricken off the road and down a steep-sided gully, upsetting the vehicle as they dragged it hell-bent behind them. The portly Greeley rode the coach to its downfall, suffering a nasty crack on his white-shocked head and a gouge in his left foreleg, while the more agile Richardson leaped clear just before the coach rolled over, coming up without a scratch. Richardson and the driver patched Greeley's wounds, calmed the mules, harnessed one team to the side of the overturned coach and righted it, rehitched the teams to the front of the vehicle, and then the party set out on the trail again.

But it was not long until the second of the mules' two anathemas presented itself and they found themselves rapidly approaching a herd of grazing buffalo. At the sight of the coach, some of the animals, led by a great shaggy-maned bull, doubled their legs under them and charged the intruding vehicle. The mules brayed frantically as the driver laid on the whip for more speed. Richardson quickly caught up the driver's rifle and leaning far out of the window began firing excitedly and ineffectually at the oncoming buffalo. Gradually, the coach pulled away from the pursuers and as abruptly as it came the danger was gone. Richardson, his face flushed and his eyes bright with excitement, reluctantly withdrew the smoking rifle barrel from the window to hear the wry voice of his companion coming to him over the wild creaking of the swaying coach.

"Pray continue to fire, Mr. Richardson, since it gives you so much pleasure and so little discomfort to the buffalo!"

That put an end to Richardson's sharpshooting.

On the way west, the journalists met hundreds of people. Some were parties of emigrants on their way to the diggings, others were the disgruntled go-backs who filled the ears of the inquisitive newsmen with such discouraging stories that Greeley decided that Pikes Peak was an "exploded bubble." His spirits rose for a brief time, however, during their short stop at Station 15, about halfway across the plains. The *Tribune* editor was gratified to find that seated across from him at the dinner table was a good-looking, well-spoken young clerk who had spent all winter on Cherry Creek. The youth was shy and reticent at first, but under Greeley's barrage of enthusiastic questions he affirmed that the situation in the new Eldorado was not so black as the go-backs painted it. The clerk assured Greeley that he had seen some gold being taken out of the sandbars of Cherry Creek and that the big strikes of Jackson and Gregory were drawing hundreds into the mountains every day. Asked why he was not following them, the youth answered that he had had no more luck than to freeze his toes in the icy water of the mountain streams and had decided to give up his devil-may-care life to go back to Indiana and finish school. Greeley lauded him warmly for his moral courage, and his good sense.

But the virtuous editor came in for a rude discovery when, a short time later, he was told that his dinner table friend was in reality a girl indelicately masquerading as a boy to keep her identity secret. "I had not dreamed of such a thing," wrote the mortified journalist. "We heard much more of her...quite enough more, but this may as well be left untold." The younger Richardson, whose practiced eye knew well the shape of things, did not have to be told but immediately saw through the girl's disguise. If Richardson got back at his older companion for his remark on sharpshooting buffalo by ribbing Greeley about his poor powers of observation, it is not recorded. However, the *Journal's* crack reporter was as gallant as Greeley and revealed nothing more about the girl's mysterious history on Cherry Creek.

With the first encouraging words about the new Eldorado now highly suspect, the two journalists fell back on the dark descriptions given them by the go-backs. Imagining Cherry Creek to house a collection of a few ragged tents and thrown-together log cabins peopled with starving, destitute, and dispirited souls, Greeley and Richardson were agape at what they found. Businesses of all kinds boomed in hastily but imposingly built buildings with false fronts, real glass windows, and covered porches. Saloons, stables, bakeries, blacksmiths, boarding houses, and mercantile stores lined the neatly platted streets. There was even a barn of a structure that called itself a hotel where Greeley was pained to find that a man brought his own blankets for his bed and the prices were as high as those of the Astor. After two nights, the incessant clamor of gamblers and the erratic revolver shots of drunks drove the journalists out of Denver House and into an abandoned cabin where they could live in some relative peace and quiet during their inspection of the gold fields. The windowless cabin Greeley found unspeakably stuffy and the food at the boarding house where they took their meals was "decidedly limited." But for all their complaints the newsmen could sense an air of progress and permanence in the sleepless little settlements on Cherry Creek. It was clear that the spirited inhabitants were determined that if their destiny was not to manifest itself as discoverers of great deposits of mineral wealth, they could do as well by building a major supply center for the gold regions in the mountains to the west.

3

Gateway to Gold

NOT ALL THE FORTUNE-SEEKERS WHO HOTFOOTED IT TO CHERRY CREEK WERE AFTER gold buried in the earth. Major General (of the militia) William H. Larimer could not have been less intrigued with his own prospects of chancing onto a rich vein of ore. What did excite him was the fact that other people, hundreds of them, would be drawn to the site of real or rumored mineral wealth. To Larimer, a congregation of people was as good as a pot of gold, for his forte was land speculation and town promotion.

The word of Green Russell's discovery found the general in Leavenworth on business, and immediately he got busy organizing a party to make the trip to Pikes Peak. With his son, Will Jr., and four others, Richard Whitsett, M. M. Jewett, C. A. Lawrence, and Folsom Dorsett, all of a like mind to his, Larimer filled a wagon with enough provisions to last for six months and left via the Santa Fe Trail for the new gold fields on September 25, 1858. The general knew the taste of success. Throughout his forty-nine years, the canniness threaded through his Scottish ancestry had helped him turn idea after idea into pure profit. From his home state of Pennsylvania all the way to Kansas he had made money in a variety of enterprises, including freighting, railroading, and coal mining, but his specialty was town-making. He had just founded the little town of La Platte City (every respectable town in the West had "city" tacked on to its name), Nebraska, when the panic of 1854 wiped him out financially, and for a time he sat on his reputation in the Nebraska state legislature, keenly alert to new opportunities to recoup his fortune. The prospect of boom times in Pikes Peak country was just the impetus he needed.

Town-making at this period of western expansion amounted to what could be called a national pastime. Larimer was only one among many who had their eye on the town-booming possibilities presented by a gold rush to the Rockies. And he was fully aware of the importance to a town-maker of being among the first to reach the site

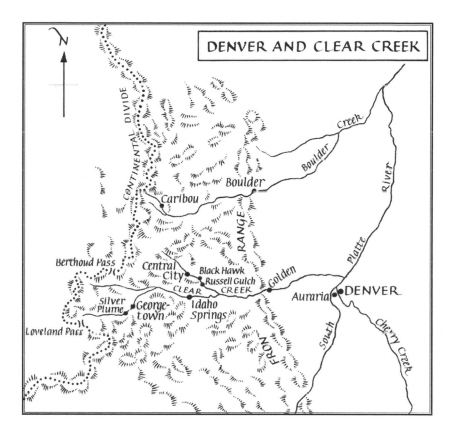

of a prospective settlement. Consequently, he set an unrelenting pace for his little band of boomers, ticking off twenty to thirty miles a day along the bank of the Arkansas River. On the way, Larimer hailed every eastbound traveler he saw and asked for reports on the gold situation at Cherry Creek, but he heard nothing to hang his hat on until about fifty miles east of Bent's Fort when he met Green Russell, who was returning to Georgia for reinforcements. Russell's words of caution about the gold prospects at Cherry Creek Larimer facilely translated as unqualified assurance and his party hit the trail west again "in high glee."

At Bent's Fort, Larimer's party stopped for a welcomed but short respite from the rough going of the trail. Here, the abstemious and frugal general carefully counted out a dollar for a dozen ripe apples but emphatically turned down the offer, at the same price, of a pint of wallipete, a raw, powerful distillation brewed by the Bents' partner, Ceran St. Vrain. Not so fastidious were some of Larimer's companions. As soon as their leader's attention was directed toward the comparative merits of a stack of buffalo robes, they readily partook of St. Vrain's revivifying spirits. It had been a long, dry trip.

Headed west again, Larimer found Fountaine qui Bouille, just as William Bent had predicted he would, and turned north until he came to the ruins of Fort Pueblo. In the distance, among a few adobe huts, Larimer could see some figures moving about. Taking no chances on blundering into a camp of hostile Indians, he sent Dick Whitsett ahead to reconnoiter. When Whitsett approached the ruins, he was relieved to find no Indians but only a handful of Spanish–Americans and a small group of Kansans. But when Whitsett learned that the Kansans were from Lecompton, carrying authority from Governor Denver of Kansas Territory to set up a functioning county at the new gold diggings, his pleasure at finding his fellow countrymen in the wilderness turned sour. Making a lame excuse, Whitsett scurried back to tell the general that their cause appeared to be lost. Larimer was stunned. It was untenable to him that he should have come so far only to be dealt out of a chance to ply his organizational specialty. Falling back on the military axiom that the best defense is a good offense, Larimer wasted no time in trotting over to the camp of his erstwhile rivals to impress on them his long experience in civic and county affairs. To the Lecompton men he explained that he was on his way to found a town at Cherry Creek, that he had already made arrangements with W. H. Russell to start an express service to the region, and smoothly suggested that the two parties join forces since their missions were so similar.

Spokesman for the Lecompton men was handsome Ned Wynkoop who, although not yet twenty-one, had landed the commission as sheriff of the new county-to-be. Wynkoop quickly and astutely decided that he would rather have Larimer as a friend than as an enemy and agreed that it would be to the mutual advantage of each group if they merged. Over a handshake, the two men sat down to sketch out a joint plan of action. As they elaborated their plans, neither man was at all disturbed that the land they proposed to prepare for settlement by white men was, by the 1851 treaty of Fort Laramie, the absolute property of the Arapaho, Cheyenne, and Sioux.

With the details of implementing the new government and establishing a county seat settled, Larimer was impatient to get to their destination and urged that they push on to Cherry Creek. At first, Wynkoop balked—it was now November and already there had been a couple of snowstorms. The trail north, patched here and there with snow, led over the high, exposed watersheds of Fountaine qui Bouille and Cherry Creek and was no road to travel in winter. Wynkoop counseled waiting in the comparative safety and comfort of old Fort Pueblo's rotting adobes until spring came. But Larimer, driven by the fear of others reaching Cherry Creek ahead of him, insisted that they go on. It was a harrowing trip. Just as they got to the top of the wind-swept divide, a vicious blizzard struck. Visibility dropped to zero and the little caravan stopped dead in its tracks. All night the screaming wind tore at the men huddled under buffalo robes beneath the whipping canvas of the wagon tops. Sometime during the night, in their frenzy at the noise and force of the storm, the oxen broke from their traces and disappeared into maelstrom. With the coming of light, the wind abated and the storm petered out, leaving deep drifts of fresh snow banked against the battered wagons. Numb with cold, the exhausted civic planners crawled out of their buffalo robe shelters and staggered through the drifts looking with little hope for their stray teams. But within an hour they found the animals, alive and sensibly clustered together in the lee of a sheltering outcrop of rock. Before noon, the Larimer and Lecompton parties were once more on the road to Cherry Creek, fortified by a breakfast of cold beans and hardtack. The going was slow, but the air was clear and calm and the landmarks called out on their maps, including the majestic white-clad Pikes Peak, were plainly visible.

On November 16, fifty-three days out of Leavenworth, Larimer rode confidently into Auraria, sitting ramrod straight in the snugly fitting, long blue military overcoat that was his trademark. Some of his cocksureness disappeared, however, when he discovered that there were already two town companies organized on Cherry Creek. It seems that the Lawrence party and the Russell brothers also knew a thing or two about town-making.

Charlie Nichols, one of John Easter's party from Lawrence, brooded about the location of Montana City when he found that Auraria, by virtue of its being at the mouth of Cherry Creek, was smack on the main trail that led from Fort Laramie to New Mexico and might well have a headstart in becoming a busy seat of commerce. So he gathered up a group of Lawrence boys and proceeded to cut himself in on whatever future lay in store for their rival downstream by leaving Montana City for a new site on the bank of Cherry Creek across from Auraria. Nichols, as the idea man, got to name the town. Modestly, he christened it "St. Charles Town." And to guarantee against trouble from the nearby encampment of Arapaho, Nichols gave William McGaa and John Smith shares in the new town. With the papers signed and stored, the St. Charles people headed home to Kansas for the winter on October 11 without so much as raising a lean-to on the site of their new town.

Taking their cue from the St. Charles settlers, Green Russell and his brothers organized Auraria into a town company on October 30 with a hundred shareholders and a board of directors to supervise the division of the town into blocks and lots. Each shareholder had until the following July to build a 16 by 16 foot cabin on his lot. Added to the cabins already erected, these would make an impressive display of stability for the next year's emigrants.

Such a challenge to his own organizational abilities as a promoter rekindled the doughty Larimer's resolve to create his own town. The only trouble was the two best sites were already taken. This, however, remained a problem for Larimer for only a very short time. Sizing up the two town sites strategically placed one on each side of the creek, one with structures already in place and one without any at all, the general chose as his own the one that was most vulnerable to annexation. In the dead of a black, frost-ridden night, he folded up his blue coat, took off his boots, waded across icy Cherry Creek, and laid claim to structureless St. Charles Town. The next morning Will Jr. arrived with pine planks, nails, a hammer, and a saw. Within the day, they had their cabin built. For the frontier, it was an elegant residence, boasting a sod fireplace, a window with genuine glass in it, and a door with a latchstring.

Larimer, father and son, were comfortably settled into their cabin a few days later when Charlie Nichols, atop a lathered horse whose sides heaved in breathlessness, reined up short at their front door. Nichols and his friends had gotten only a day or so out of town when they ran into wagon after wagon of Pikes Peakers wending their way toward Cherry Creek. Remembering that there was no one left to guard St. Charles Town from incoming squatters, Nichols raced back to his town site. Now as he gazed down on the benign and smiling face of the general he exploded in outrage and ordered Larimer off his property. But Larimer refused to move. Instead, to strengthen his position, he too formed a town company. Rounding up Lawrence, Whitsett, and the rest of his Wynkoop's group, Larimer platted the town, sold shares, and named his creation "Denver." The name he chose was aimed to gratify the governor of Kansas Territory from whom the general expected a few favors, but unfortunately his gesture was wasted. During the time Larimer was inching his way west, the convulsions of free-soil versus slavery factions in Kansas had forced the resignation of Governor Denver. The formation of the Denver Town Company did not by any means vanquish the St. Charles men. Calling a public meeting, Nichols openly accused Larimer of town-jumping. Larimer righteously claimed that William McGaa had sworn to him that the St. Charles site had been abandoned. McGaa, not to be caught in the middle, countered that if he had said such a thing it was only because two of Larimer's henchmen got him drunk. With matters at an impasse on Cherry Creek, Nichols took his claim to the Kansas legislature. Despite the fact that it had no jurisdiction over lands belonging to the Indians, the Kansas law body was sufficiently carried away by the stream of oratory to pass the St. Charles bill, giving Nichols and his crew title to the disputed land. Far from surrendering when he heard this news, Larimer dispatched a man to persuade the new Kansas governor to veto the bill, but the St. Charles men got to the good governor first, and in exchange for one hundred lots for his son the chief executive signed the bill. The St. Charles promoters exuberantly returned to their disputed townsite, ready to throw out the interlopers, But the fist-shaking, threat-spouting Nichols had soberer thoughts about ousting Larimer when one night he was followed to his cabin, pinned against the rough-hewn timbers by a burly member of the general's clique, and notified that if he pursued his plan to take over Denver City a "rope and noose would be used on him." While Nichols vacillated, Larimer adroitly proposed that they negotiate their differences and merge to found a larger, stronger Denver. The scheme worked, and it was a greatly relieved Larimer who at the end of a long series of discussions handed over to the St. Charles Company $250 in heartbalm money and gave each man a share in the new, greater Denver. If the spring brought hordes of gold-hungry settlers to Cherry Creek, as Larimer gambled it would, the general's fortune would be made, holding as he did four shares in the company, twenty choice lots, and two large tracts of land on the outskirts of the neatly platted town.

Construction now boomed in Denver. On both sides of the major streets—Larimer, Lawrence, McGaa, Blake, and Wynkoop—twenty-five cabins and pretentious, false-fronted buildings were occupied as fast as carpenters could build them. Two of Larimer's cronies were among the earliest to open their doors to trade: C. A. Lawrence and his nephew Folsom Dorsett opened Denver's first commission house. Close behind them were Andrew Williams and his partner Charles Blake, whom A. D. Richardson had last seen in New York dressed in spotless broadcloth and sparkling white linen and was now scarcely recognizable in the "half-Mexican, half-Indian" costume of the frontier. Blake and Williams opened a general store, stocking it with the contents of four wagons of mixed merchandise hauled out from

Omaha. Flour sold for $25 per hundred pounds, bacon for 40 cents a pound, and whiskey, guaranteed by its countless consumers to "kill at forty-five rods" went for $8 a gallon. The enterprising Blake and Williams also threw together the first hotel, the long, narrow, pine-boarded Denver House that Horace Greeley the following year would derisively describe as "the Astor House of the gold region." A large saloon graced the front of the building, and it was never empty. Sun- and wind-weathered, full-bearded prospectors lounged on benches already scarred by hobnailed boots. Trappers in fragrant buckskin, shod in moccasins reaching to the knees, picked their tobacco-stained teeth in deep concentration over a hand of Three-card Monte, while a gambler chanted loudly of the wondrous luck to be had at his idle roulette wheel. An occasional Mexican slipped up to the bar to order a shot of Taos Lightning, his Spanish dialect mingling unnoticed in the constant undertone created by the gutteral speech of German freighters, the frontier French accents of trappers and traders, and the twang and drawl of north-, mid-, and southeasterners. Behind the saloon, the rest of Denver House was partitioned by cotton sheathing into six bedrooms, each with a wooden bunk, an overturned barrel for a table, and a tin basin to serve as a sink. Patrons of this last word in frontier hostelry were instructed to dump their dirty wash water on the earthen floor to keep the dust down.

On the surface, both Denver and Auraria late in 1858 appeared to be thriving in the light of the new construction and the new businesses that opened every week, but it was a prosperity based almost wholly on speculation. The situation was not as William McGaa pictured it in a letter that was printed on the front page of the *Missouri Republican*. No one was washing out gold to the tune of $6 to $10 a day. At best the take was 25 cents a day on the sandbars of Cherry Creek and the South Platte. Most of those who had ranged farther, following for a few miles every stream on either side of the Platte during the first stampede to the region, now retreated to the twin towns on Cherry Creek in the face of freezing weather, with their grub gone, and with only a few grains of gold to show for their pains. They swarmed into Denver and Auraria, ballooning the population and straining the available supply of food to its limit. The four-wagon stock of Blake and Williams was soon gone, and with no reserves, prices went sky high. What was worse, with winter closing down plains-travel, there were no prospects for the arrival of freight wagons with more supplies. Some of the settlers took to hunting to stock the two camps with fresh meat, and just when the food situation was at its bleakest, in rode Uncle Dick Wootton to save the day. The miners could now settle down to wait out the winter with a minimum of hardship. In the saloon of Denver House and in the barroom Uncle Dick opened in conjunction with his store, the pessimistic carried on endless arguments with the optimistic over the prospects of the new Eldorado. What little gold they had was passed around among barkeeps—"twenty-five cents for a drink of dubious whiskey, colored and nicknamed to suit the taste of the customers"—and gamblers. Already Denver and Auraria had a large population of professional gamblers who had come from nearly every frontier town of the West, following the scent of suckers to Cherry Creek as unerringly as a bird dog flushes quail. From the vantage point of his cozy cabin, Larimer found only two aspects to criticize in his otherwise pleasing creation. "You have no idea of the gambling carried on here..." he wrote to his wife, "They go at it day and night...and Oh how they drink! You cannot conceive of anything as bad as they carry on here." If his judgment was prejudiced, his powers of observation were not—the inhabitants of Denver and Auraria were deeply committed to drinking and gambling. Even when the Reverend George Fisher arrived to conduct the first religious services on Cherry Creek, he discovered that he must preach at one end of a saloon and compete for the ears of his congregation with a gambler at the other end who badgered them all the while with his high-pitched spiel of "Ace of hearts wins—turn it up and win twenty dollars. Who'll go me twenty?" But after all, it was too cold to mine and too cold to build any more structures, so there was little else for a man to do that winter on Cherry Creek.

There was one other pastime that rivaled drinking and gambling in popularity. In Auraria, down on Indian Row where the double cabin of the squaw men McGaa and John Smith stood was Cherry Creek's first brothel. It was operated by beautiful, brown-eyed Ada Lamont who had arrived in one of the wagons in the first stampede. She was just nineteen when she and her new husband, a young minister imbued with the mission of saving the souls of the gold-greedy, joined a wagon train headed west in October 1858. Halfway across the plains, the minister disappeared. Also missing was a female of questionable virtue. After an unsuccessful search for the missing man, the wagon train continued on its way. Ada slipped into a morose silence for the rest of the trip. Her faith in mankind turned to cynicism, and as soon as she got to Denver she declared herself in business as a prostitute. In the womanless society of Cherry Creek, she wanted not for customers. Ada soon rented more space, increased her "staff" as each wave of emigrants arrived, and for ten years ran a clean and honest house, noted for serving the best drinks in town. Then one day a friend brought her a badly weathered little Bible bearing a faint inscription in her handwriting. She was just able to make out the name of her missing husband on the flyleaf. The Bible had been found, she was told, near a skeleton on the road to Kansas—in the back of the skull was lead bullet. The revelation than her benedict had met foul play and not lived the life of a libertine with a fancy woman was too much for Ada. She drank herself into bankruptcy and ultimately died of starvation in a mountain camp west of Denver in the 1870s.

But in the winter of 1858, if drink, dice, and women could not fill the chill days until spring, there was always politics. As Denver grew, the Georgians in Auraria awoke to the fact that the enterprise they had laughed off as pure speculation and doomed to fail was suddenly a very healthy rival to their own blossoming little town. What was really galling, however, to the independent-spirited Georgians was the imposition of county officers by the governor of Kansas from six hundred miles away! The firebrands of the Georgia group called a meeting to decide what to do about Sheriff Ned Wynkoop, Judge H. P. A. Smith, and Hickory Rogers, the county supervisor, whose alliance with Larimer and the Denver Town Company made them all the more distasteful to the indignant Aurarians. It was a stormy conference. Some called for tar and feathers, others for lynching if the trio attempted to exercise their functions. Cooler heads finally prevailed, and after passing a resolution repudiating the governor's men, they set a day for the election of their own local officials.

The atmosphere between the two towns was charged with ill will and although the new bridge connecting Denver and Auraria at Larimer Street improved communication, it did nothing to improve relations between the two towns. The feud reached its climax and crystallized over the falling out between two longtime friends. John Scudder and Peleg Bassett were Missouri riverboat pilots who had together joined the rush to Pikes Peak. When they got there, both men quickly became involved in the civic affairs of the two little communities: Scudder as recorder for the Auraria Town Company, and Bassett as recorder for the Denver Town Company. From January to March, the pair worked side by side in apparent friendship on various committees gotten up to promote a railroad to Cherry Creek, to solicit aid against possible Indian attack, and to encourage merchants from the East to move their businesses to Cherry Creek. Whether it started out as a friendly game of baiting, no one knew, but early in April the two men got into a violent argument over the merits of the two towns. Their friends broke it up but not before the two rivermen squared off to fight. A few days later Scudder heard that Bassett was busy spreading stories about his untrustworthiness, particularly in the handling of other people's mail. Enraged, Scudder stalked over the bridge to Bassett's cabin and called him out.

"Did you call me a thief?" Scudder shouted at Bassett.

"No," was Bassett's prompt answer.

"Did you accuse me of opening your mail?"

Bassett's "yes" was equally prompt.

The slight-statured Scudder made a move forward, and as he did Bassett picked up a pick handle and swung it at his former friend. Scudder nimbly jumped back and the pick handle missed him. But before Bassett could swing again Scudder pulled out his pistol and shot the Denver town recorder point-blank. Bassett spun and fell backward from the impact of the bullet, his shirtfront suddenly crimson from a gaping wound in his right lung. The sound of the shot brought Larimer and Wynkoop on the scene. When they lifted the injured Bassett onto his bunk, they could see that he was mortally wounded. After lingering for several hours in agonizing pain, Bassett died. Scudder surrendered immediately to the young sheriff who advised him that the best way for him to live long enough to stand trial was to leave town. With the enmity between the two towns running at fever pitch, Wynkoop was afraid "Judge Lynch" would preside if Scudder stayed to face the settlers. It was good advice and Scudder took it. When the word of the murder got around, men of both Auraria and Denver were ready to carry on the battle, but they simmered down when the fulcrum of their rage was nowhere to be found. For a year, Scudder hid out in Salt Lake where he helped W. H. Russell organize the forthcoming Pony Express. When he came back to Denver, the fires of the feud were banked and he was tried for murder but acquitted on a plea of self-defense.

In the constant bickering between Auraria and Denver, between advocates of squatter sovereignty and those favoring the jurisdiction of the Kansas-sent officials, Larimer and his cohorts suddenly saw that they might be able to end the factionalism and at the same time elevate themselves to political realms well above the level of mere town-making. If they could unite the gold-diggers behind a movement for regional self-government, the cream of the political milk bucket would be theirs. Astir with anticipation, the general got up a meeting of Aurarians and Denverites and proposed the idea of sending a man to petition Congress for the organization of the Pikes Peak region into a separate territory. His audience thought that was a great idea and they eagerly set a day for the election of a delegate. To the utter mortification of teetotaling General Larimer, the Denverites nominated Will Clancy, one of the original town company shareholders of Denver and a man noted for his deep devotion to the bottle. Aurarians nominated H. J. Graham, a friend of the now-vindicated D. C. Oakes, as their candidate. Election Day saw the region of Cherry Creek swept by a howling blizzard and few residents were willing to leave their warm cabins to cast their ballots. Noting the small turnout, Clancy and young Will Larimer decided to nudge the results in their favor. Taking up a Kansas directory, Will read off a list of names while Clancy voted yea or nay until he felt he had a safe majority. But the Aurarians had done a little poll-watching also. After supper, some of them got together and rigged up a ballot box from an old cigar box, and, bundling themselves up in their warmest clothes, they went from cabin to cabin extracting votes from those too lazy to brave the storm. By 9 PM, it was clear that the Aurarians had won the day despite Clancy's chicanery. There was considerable whooping and hollering among the men of Auraria when the results were announced. What tickled them most, as some said, with a jab aimed at Larimer, was that they had stopped the Denverites from sending a drinking man to Congress.

But it might have been in their favor if they had elected a more convivial fellow who knew how to win friends in cloakroom get-togethers. As it was, the petition Graham presented in Washington failed to pass the House and Senate committees on territories. Congress was not yet ready to be convinced that there were enough people or prospects to separate the new Eldorado from Kansas Territory.

For a month or so, the issue flagged, except in the mind of the persevering Larimer. Not long after this setback, he thought of a new ploy and drummed up another convention of settlers committed to popular sovereignty. On April 15, above Uncle Dick's saloon, the convention met to draft both a state constitution and a message to Congress asking again for territorial status. One way or another, Larimer and company were determined to win. The two propositions came up for a vote on September 5. Between April and September the miners had time to reflect about what the referendums meant to them, and their verdict showed up clearly in the election results. Territorial status, with its implication of a lesser financial burden than statehood, won by 358 votes. Now Larimer called a meeting to elect delegates to a convention to draft a territorial constitution. He also called for another election to send a new delegate to Congress. With this "mandate" from the voters, the designers of the new territory included in its bounds not only what is now Colorado but for good measure added the Nebraska panhandle, a chunk of southern Wyoming, and a generous slice of eastern Utah. Jefferson was the name chosen for the territory. Enough members of the convention agreed that one territory carved out of Thomas Jefferson's splendid Louisiana Purchase ought to bear that illustrious name for the motion to easily pass. It readily won out over the suggestion of one buckskin-clad delegate who would have named the territory "Bill Williams" in honor of the wildest mountain man of the Rockies.

By now it was obvious to anyone who lived on Cherry Creek that as a popular pastime voting ran a close race with drinking, gambling, and whoring. In 1859, the people of Denver and Auraria went to the polls no less than nine times. The ninth time was on October 24 when the Territory of Jefferson, extralegal as it was, was voted in with the lawyer R. E. Steele as governor, L. W. Bliss as secretary, Uncle Dick Wootton as treasurer, and Hickory Rogers as marshal. The constitution of this "provisional" body was almost identical to that of the state constitution voted down by the election-happy residents on September 5. Matthew Dale, an argonaut from Ohio who watched the political proceedings with a hawk's eye, was moved to anger at this blatant turn of events. "The political demogogues in Denver and the other valley cities, who got up the State government, not satisfied by its defeat, and the unceremonious shaking off of the public teat it gave them which loomed before them with such inviting fat yielding the prospect now clubed in caucus to euchre the people out of what they could not get honestly and openly."

The first act of the provisional legislature of the new "territory" seemed to bear out Dale's harsh analysis. The representatives of the various mining districts voted themselves hefty salaries to be paid out of a $1 poll tax to be levied against all voters in Jefferson Territory.

Up to this point a number of people had watched the machinations of the politicos with tolerant amusement. But this was too much. The miners rebelled. When the word of the poll tax reached the diggings on Clear Creek, which were crawling with prospectors as a result of the Jackson-Gregory strikes, the free-thinking miners gave notice that any man who came up the gulch bent on tax collection would be summarily shot. The uproar from all of the camps, reinforced by a petition against the government of Jefferson Territory signed by six hundred miners, quickly persuaded the newly elected delegates of the territory to repeal the tax. From then on Jefferson was a paper political body blocked from any effective action by the contentious miners who had come to work the earth for minerals and who were defiantly certain they could govern themselves without the assistance of town-makers or territory-makers whether near at hand or far away.

For three years, until Congress passed an enabling act paving the way for formal, legal establishment of Colorado Territory, the miners proceeded to prove they could handle their own affairs. Following the pattern set earlier by the California gold rush, as soon as a large body of ore was discovered the prospectors who flocked to the area met en masse, in the open, to organize the region into a mining district. In each district were elected officers: a president, a secretary, a treasurer, and a recorder. The size of the district and the size of claims were defined, operating rules were agreed upon, and a simple penal code was adopted. Murder, claim-jumping, and theft, considered to be the major crimes, were generally punishable by hanging, flogging, or exile, depending on the circumstances of the offense. The accused was given a hearing before a jury of his fellow miners, rounded up for the purpose, and his sentence was carried out straightaway. If a man spoke fast enough, he might appeal his sentence before the frontier court of last resort, a meeting of all the people in the district.

Similar alliances sprang up among the settlers in Colorado towns before it was accorded territorial status in 1861. Claim clubs, approximating those popular in the 1830s and 1840s in the South and Midwest were organized to provide for orderly division, holding, and transfer of land for farms and homesites. The claim club was an attempt to bring order to a region where populations built up before formal government machinery could be set up. The claim clubs appeared to work well, and their officers took their jobs seriously, some such as Bassett and Scudder perhaps too seriously. Farm and homesite claim-jumping were generally settled tête-à-tête, but in extreme cases they were settled with the help of that handy arbiter, the Colt repeater. So it was with the first claim club dispute on Cherry Creek.

Many of the emigrants who followed the gold trail west in the spring of 1859 had been homesteaders elsewhere, and they brought with them their institutions, including the claim club concept. Before long, Cherry Creek had a vigorous association of landowners who set rules for the acquisition and improvement of property. Included in the claim club membership were the original shareholders of the Denver Town Company. This shortly became a source of some dissatisfaction for those who were not among the select body of town-makers. The claim club majority grumbled that the Town Company held unto itself an unfair number of lots. After considerable jockeying, the majority prevailed and the Denver Town Company was persuaded to give up about one-third of its original plat. With the town boundaries reestablished, the Larimers decided to reclaim some of the land they lost in the deal and quietly bought up several of the best riverfront lots that the Town Company had relinquished to the open market. The general and his son, apparently forgetting the lesson they taught Charlie Nichols, did no more than start an excavation for a water well on their new lots before packing up and heading back to Leavenworth to spend the winter.

In the spring, when the second stampede of adventurers arrived on Cherry Creek, they found most of the choice land already taken. Lots that were left, lots that had been claimed the previous year at no more cost than a recorder's fee, were now exorbitantly expensive. The town-makers gloated, and the newcomers boiled. Self-appointed leader of the outraged newcomers was Bill Parkinson, a peppery-tempered river pilot of the same ilk as Peleg Bassett. He seized upon the greediness of the claim clubbers as a cause célèbre and promptly led five of the newcomers, with several loads of lumber from D. C. Oakes's sawmill, onto the Larimers' unimproved and apparently abandoned riverfront lots and immediately set to work building cabins. As soon as the first logs were in place, the claim clubbers held a mass meeting and sent an order to Parkinson and his men to move on or they would be run out of town. For an answer, the squatters trained rifle barrels on the deliverer of the message. This development called for another wordy and inconclusive mass meeting of claim clubbers. Meanwhile, Parkinson and his bunch made the mistake of withdrawing from their cabin fortresses for the night. The next morning when they found their handiwork scattered from one end of the lot to the other, it was Parkinson's turn to deliver an ultimatum. He announced that if any claim clubber tried to stop them from rebuilding their cabins, he and his fellow squatters would shoot them on sight. To emphasize his threat, the riverman zinged three bullets past the head of the claim clubbers' dignified spokesman, Robert Bradford. To meet Parkinson on his own ground, the claim clubbers then broke out rifles and pistols and took up strategic positions around the besieged lots. While everyone waited for someone to start the war, an uneasy quiet settled over the town, the streets emptied of people, doors and windows were bolted shut, business fell off until even the saloons were uncrowded, and armed men patrolled the town at night to protect it against a surprise attack by Parkinson's gang. As the tension mounted, it was Ned Wynkoop who stepped into the breach. The young sheriff was a large shareholder in the Denver Town Company but he also believed in fair play, and after carefully scanning the bylaws of the claim club he called a meeting of the opposing parties and pointed out that in the name of justice the Larimers should have to forfeit their claims because they had failed to comply with the regulation that called for a cabin to be built on the land within a year of its acquisition. He then proposed that the lots in question be put on the open market at an uninflated price and that the squatters be reimbursed by the claim club through the agency of the Town Company for the cabins they built on the land. Begrudging agreement by both sides came the next day, with the Larimers in for a surprise when they returned.

Watching the raising of cabins and buildings in Denver and Auraria from their tepee village downstream, it was becoming clear to the Arapaho that the white man had come in numbers and to stay. It was no source of pride to the red men that the settlers had exhibited superior efficiency in organizing Indian treaty lands into a mining district and townsites. Giving the white squatters a dose of their own medicine, some of the more audacious headmen of the tribe sent word to the whites to get out by spring or face the consequences. The threat had little effect on the miners since they calculated there were now enough men in the area to withstand any attack the Indians in the vicinity of Cherry Creek might mount against them. On the other hand, it behooved the settlers to stay on the good side of their red neighbors, so as a gesture of goodwill, the people of Denver and Auraria prepared a huge feast and invited all the Arapaho to attend. Some five hundred warriors and three hundred squaws and children delightedly responded, pouncing on the three roasted oxen, scores of spitted antelope, and piles of bread and dried fruit heaped on blankets spread beneath the cottonwoods down by the river. The white men watched in wonder as a brave named Many Whips systematically devoured nearly an entire antelope. The occasion was a great success and led some days later to a peace parley at Denver House. In solemn attendance was Chief Little Raven and some lesser chieftains of the Arapaho and the officials of the two towns, headed naturally by the assiduous General Larimer. The upshot of the meeting was that the Indians offered to let the settlers live in peace on their lands

in exchange for goods and instruction in agriculture. It was a fair
exchange and the settlers were quick to agree to it. The powwow
ended with much nodding of heads, smiles, and touching of hands
in the sign language of friendship. All other exchanges, when the
meeting finally got started, had to be handled through the interpreter,
Chat D'Aubrey, a white trader, who arrived at the meeting glassy-eyed
and loose-limbed to sit down heavily beside the dignified Little Raven
only to pass out ceremoniously at the Indian's feet a few moments
later, causing the minutes to read that the meeting "was immediately
adjourned for a day until the interpreter should become sober." Once
more it was Larimer's cross to bear a chronic drunk on his team.

Although the Arapaho agreed to let the white man live in peace,
he had no such intentions with respect to his arch foes, the Ute Indians, whose territory stretched from the front range of the Rockies
only a few miles west of Denver-Auraria to the western slope of the
San Juan Mountains and into Utah. The Arapaho put to use their
new alliance with the white men every now and then by leaving their
squaws and children in the safekeeping of the settlers while they sent
war parties to storm their enemy's stronghold. After a victorious sortie, they returned triumphantly to Cherry Creek, parading and dancing up Larimer and down McGaa streets, waving freshly taken scalps
and spattering fascinated buckskinned and calicoed bystanders with
scarlet drops of Ute blood.

If by some chance a resident missed the Indians' scalp dance, he
could read all about it in the fledgling but vigorous *Rocky Mountain
News*. Its publisher and editor was William Newton Byers, a surveyor
by training who had caught the spirit of territorial expansion some
years before and had plied his trade westward from Ohio to Iowa and
along the Oregon Trail to the Northwest. At the end of his perambula-
tions, Byers settled in Omaha and dabbled in politics. As a member of
the Nebraska House of Representatives in 1855, it is not unlikely that
he became acquainted with his fellow legislator, William Larimer.
Perhaps if their paths did cross, some of their discussions revolved
around the increasing number of rumors circulating about gold discoveries near Pikes Peak. Byers could add to the stories from firsthand knowledge; on his surveying junket west he had talked with
some trappers in Fort Laramie in 1852 who had picked up some float
gold in the headwaters of the South Platte.

To restless men such as Byers who saw the border states torn by
the factionalism of slavery and abolition and wallowing in a morass
of depression, the opportunity to make a fresh start in virgin territory
held an undeniable fascination. When the rumors were borne out by
Cantrell's display of some of Green Russell's gold dust in the towns
along the Missouri River, Byers put together his guidebook for plains
travelers. His writing experience may have been limited, but his
trail experience was extensive, and of the many guides that flooded

the scene at the time, Byers' was one of the most reliable. With the proceeds of his first publishing venture, the handsome twenty-eight-year-old surveyor hired a printer, bought a Washington handpress, had it shipped to Cherry Creek, and went galloping after it "with his shirttail full of type."

On April 17, 1859, at the end of six blustery weeks on the Overland Trail, Byers and Tom Gibson, his printer, rode into Denver from the north, stopping on a bluff at the edge of town to take in the scene before them. Denver, its cabins and clapboarded buildings in neat rows, reached from the river bank nearly all the way to the bluff they stood on. Auraria, across Cherry Creek, lay mostly on river bottom-land. After a few minutes' pause the men spurred their animals forward. As his horse delicately sidestepped potholes and avoided the wagons, handcarts, and ox teams clogging the dusty streets, Byers' eyes missed nothing. There were people everywhere, and they all seemed to be in a hurry. Full-bearded miners, the tools of their trade hung about their clothes military fashion, jostled mountain men and trappers in greasy buckskin who unloaded bales of fur and buffalo robes under the portal of the trading post. An occasional Arapaho touched Byers' knee and pointed at his saddlebag, begging for tobacco. Knots of agitated men clustered outside the open door of the cabin housing the town recorder. A drunk weaved out of a saloon and fell into a horsetrough beside the boardwalk. A passerby dragged him out and dumped him soddenly onto the dirt street. Bullwhackers filled the air with their shouts and the cracking of their whips as they moved lumbering freight wagons through town. Byers crossed over the bridge to Auraria, ambling up and down the streets until he came to Uncle Dick Wootton's saloon-store on the bank of the creek. Over a glass of Taos Lightning, Uncle Dick agreed to rent the loft of his building to the fast-talking neophyte publisher.

Now all Byers needed was his press, and he could produce the first newspaper on Cherry Creek. Day after day he hounded the express office looking for his shipment. His impatience flowered into full-bloom anxiety when he found out that over in Denver "Jolly Jack" Merrick was feverishly readying the first edition of the *Cherry Creek Pioneer*. Merrick had arrived four days earlier than Byers and swiftly set up shop in a rented cabin. His printing establishment consisted of two green assistants and one little hand-lever press that already had a long and colorful history in journalism. First used by the Mormons in the 1830s to print their underground newspaper in Independence, Missouri, its career was interrupted for a time when one dark night an anti-Mormon mob set a torch to the sect's hideout and dumped the offending press into the Missouri River. Not long after it was fished out of the river, cleaned and oiled, and put to work printing the St. Joseph *Gazette*. The *Gazette* surplused it just in time for Merrick to pick it up and trundle it west to the gold fields of Pikes Peak.

At this point, the historical record of Merrick's press was of absolutely no interest to Byers. Pacing the floor of Uncle Dick's drafty attic he waited tensely for the arrival of his own press, periodically sending Gibson over to Denver to see how close Merrick was to publication. Downstairs in the saloon the inveterate gamblers chalked up odds and took bets. In no time half of Denver and Auraria had a stake of gold dust on one or the other of the budding editors. Finally, on the evening of April 21, a wagon rolled up to Wootton's saloon and Byers and Gibson wasted no time lugging its welcome cargo up the narrow stairs to the little loft. They worked all night assembling the press. The next day and night Gibson set type while Byers wrote. By the evening of April 23, in the midst of a late spring snowstorm, and a bare twenty minutes before Merrick's *Pioneer* hit the slushy streets of Denver, Byers pulled the first edition of the *Rocky Mountain News* off his press.

That first edition set the pattern for the *News*'s long and energetic lifetime (after a dozen or more owners and a hundred years of continuous publication, the *News* was a powerful spokesman of the mountain West until it closed in 2009.) Printed on good white rag paper, it carried a wide variety of stories covering the local, national, and international scene: within its six pages was a feature article on the opening of Japan five years before by Commodore Perry, one on the convention held to organize Jefferson Territory, an interview with Thomas Carlyle by an American minister, an editorial advising farmers that they could grow as rich by staying home from the mines and growing produce to sell to hungry miners in exchange for gold, and an unadorned advertisement offering the services of A. F. Peck, MD, from Cache la Poudre Creek, "Where he may be at all times found when not professionally engaged or digging gold."

Byers' weekly was not cheap: it sold for $25 a year or 25 cents a copy. The costs of production were high—news of the outside world came to the *News* office from Fort Laramie, two hundred miles away to the north via pony relay; supplies of paper and ink cost $20 per hundred pounds to transport across the plains and were on a low priority basis among freighters compared with their cargoes of urgently needed food, mining, and household supplies. Another reason for the dear price tacked onto the weekly was that few merchants of Denver–Auraria, before the strikes of Gregory and Jackson, had much capital to spend on advertising space, or for that matter very many goods to advertise. Nevertheless, since it was the only paper (Merrick had retired from the field), it was immediately popular among the news-hungry settlers.

The birth of the *Rocky Mountain News* and the peak of the go-back exodus from Cherry Creek coincided. As the streams feeding the South Platte were found to yield but a few ounces of scale gold, more and more men disgustedly abandoned their claims; some headed for the mountains to try their luck once more, others headed back to their homes in the East. One by one the cabins that had been so hopefully built in the fall of '58 were abandoned and the twin towns took on the forlorn aspect of depression. Among the people who stayed on there was a kind of inertia born of not knowing where to turn next, and of harboring a last hope that the new Eldorado would be discovered in the mountains to the west. Into this troubled atmosphere Byers brought a new confidence. From the first he predicted that the region of Pikes Peak, with its fine climate, and abundant water, timber, and minerals would prosper, with or without discovery of a great money-metal bonanza. It did not bother him that the embittered go-backs paused at the office of the *News* on the way out of town to spit invective laced with threats on his life for helping to sustain what they firmly believed was a hoax. With a developing fearlessness that became the editor of a frontier newspaper, he continued to run stories of any new gold strikes, however inconsequential, that came across his desk.

Byers had only two weeks to wait after the appearance of his first edition before his faith in the future of the area and that of Larimer and a few other diehards was handsomely rewarded.

On Sunday, May 8, three men sat around the office of the Leavenworth and Pikes Peak Express discussing the unpromising results so far in the spring surge of prospecting. One of these men was Henry Villard, who had arrived the day before in one of the company's spanking new Concord coaches. Since the express office was a clearinghouse for local news, he had come there to begin his report of the situation on Cherry Creek. A short, heavily bearded miner, just down from the mountains, who came into the office inquiring of mail, interrupted the discussion. Villard asked the miner if he was discouraged over the lack of finding gold in paying quantities. The man hesitated a moment, his eyes darting quickly over the faces of the three men before him, and then he pulled out a bottle of glimmering gold dust.

"Here's forty dollars of dust I got me out of forty pans of dirt up on Clear Creek. Me and John Gregory and some other boys from Indiana is all up there," the man blurted out. "And that ain't all." He dove his hand into his pocket and produced several hunks of quartz shot through with shining gold streaks.

When the quartz was assayed, the result electrified the people of Denver and Auraria. A few days later the future of the area was sealed when another man brought in $80 he had washed out of thirty-nine pans of gravel a few yards upstream from Gregory's claim. About this same time, two of George Jackson's Chicago friends came down to Cherry Creek for more supplies. In Doyle's mercantile store they heard about the Gregory strike.

"That's nothing," said one, tossing a full pouch of gold dust on the counter, "here's a sample of our stuff. We're taking out nearly two thousand dollars a week up on the south fork of Clear Creek."

The twin towns burst into a new life at the news of the Gregory-Jackson bonanzas. Gone was the depression. Townsmen greeted each other with jovial slaps on the shoulders, and gleeful shouts of "We're all right now!" "Praise God, the country's saved!" The clear, bright air never felt so invigorating, the mountains took on new majesty, and everyone set about their work with new enthusiasm. Real-estate values skyrocketed, as much as 100 percent at one leap. Merchants contracted for new buildings, and the sounds of construction filled the air day and night. Workmen in Denver and Auraria pocketed $5 to $8 a day, as good or better than a man could make in the gulches working in somebody else's mine. Every new group of arrivals brought those with needed skills who shortly found themselves with more business than they could handle. Two more brickyards opened at the outskirts of town, and six more sawmills were started within a radius of twenty miles of Cherry Creek. Lumber sold for $5 per hundred feet. Cheaper were bricks at $6 a thousand, but the brick kilns could not produce enough to satisfy the sudden demand. Sod roofs gave way to shingled ones, and plank floors and ceilings became common features of the estimated three hundred buildings that were put up by the winter of 1859. The few isolated ranches that had been settled in late 1858 now found company in a rash of small farms that sprang up around the booming towns of Denver and Auraria. An overwhelming hunger for fresh vegetables and fruit fostered part of this sudden agrarian interest, and even the busy editor of the *News* staked himself out 160 acres of rich bottomland two miles from Denver. Watering mouths waited impatiently as seeds of watermelon, tomato, cucumber, cabbage, green beans, and beets imported from the East came to fruition.

While they waited, the populace could read the exciting stories of the gold rush in the two-week-old New York *Tribune*, the St. Louis *Republican*, and the Leavenworth *Times* that the express brought into town on every trip west. Freight wagons hauled in other articles of interest—French worsteds for "gents" and French calico for the several women who by midsummer made up the female population of Denver-Auraria. If a woman could cook, she never wanted for company. No one refused the opportunity to forego his own monotonous mess of beans and bacon for a bowl of squatters' soup, a savory potage of venison and corn, accompanied by the great delicacy of the plains, smoked and roasted buffalo tongue, or perhaps by the gelatinous but tasty white meat of beaver tail, its strong flavor of willow tempered by a piquant vinegar sauce. To quench the dry desert thirst of a Pikes Peak summer, yeast, malt, and hops were delivered to the newly opened Rocky Mountain Brewery for quick conversion to lager beer that sold for "ten cents a glass—froth included."

It was in June and July that the great mass of emigrants who had heeded the admonition to start west no earlier than May began to arrive in Denver-Auraria. They came in droves, some alone, some in parties of two or more, and some with their entire families. Most paused in the twin towns only long enough to buy fresh supplies and then whip their teams to a trot headed for the diggings on Clear Creek. For a second time within a year, business boomed on Cherry Creek. But this time a genuine prosperity overlaid the area. Gold, the money metal, was pouring out of the mountains.

Since there was no mint close at hand and it took a long time for the gold to be returned in coin form to Denver-Auraria, everyone carried his freshly mined gold dust in a small buckskin pouch as elsewhere coin and currency were carried. To buy an article in a store, the purchaser selected what he wanted and passed over his pouch to the clerk who carefully weighed out the correct amount of dust on a set of scales mounted on the counter. A "pinch"—the amount of dust taken up between the thumb and forefinger—was the smallest "denomination," equated at 25 cents. Clearly, hardly anything in Denver-Auraria, including the *Rocky Mountain News*, in those halcyon days of the first genuine boom cost less than 25 cents (except lager beer), and many a merchant made a tidy profit from the grains of gold swept up in the dust of his floor.

$$\sim$$

If the merchants of more reputable goods and services were making money, the merchants of 40-rod whiskey and deuces-wild poker got more than their fair share of the contents of the buckskin pouches. Saloons and gambling halls on Cherry Creek opened their doors with the sequential frequency of a stack of falling dominoes. Once open, they never closed. All day and all night the raucous cries of dealers drifted over the towns, and gambling fever was nearly as epidemic as gold fever. Miners came down from the gulches and streams with more money in hand than they had ever had, determined, it appeared, to get rid of it as fast as possible. Some set a record doing it. Their weekly earnings blown in one night's spree, they shrugged off their losses, turned a bucket of creek water over their aching heads, and trudged back to their claims to make enough for next Saturday's fling. It could be worse for the newcomer—he could lose his poke to a lightning-handed dealer in a smoky, dimly lit saloon before he ever reached the diggings. Nor was the contagion restricted to the unsophisticated. The town-makers, with the exception of the abstemious Larimer, of course, for all of their preoccupation with high-toned problems of the law, government, and commerce, were not above carrying on a fast shuffle in deeds over a few hands of faro. With the price of some town lots easily within the reach of a miner whose claim was paying well, any man with an inclination toward property ownership in Denver-Auraria could afford to buy into a game where deeds were the stakes. He might even win, because the town-makers were known to lose. On a Sunday afternoon, his day off, Judge Smith dropped into Wootton's saloon and lost thirty lots of Denver real estate in less than ten minutes.

Men bet on anything—on Merrick and Byers in their race to issue the first newspaper, on horse races with the Indians, on footraces, prizefights, billiards, and even on chess games. Stakes were high: $150 a side for billiards. Uncle Dick ran a lively business on the side in pawned articles to feed the need for ready cash for the tables. Even Sheriff Wynkoop pawned his gun for some betting money. Fortunately, no reason arose for him to use his weapon before he was able to redeem it.

In October, Denver-Auraria was treated to the gambling opportunity of the year—a formal duel, complete with all the old-world trappings of besmirched honor, outraged principals, solemn-faced seconds, and a pastoral setting. Richard Whitsett, the quiet, unassuming Irish ally of General Larimer, had been selected to succeed the murdered Peleg Bassett as Denver's town recorder. Whitsett was filled with forward-looking ideas to advance his town's fortunes and he went about the business of putting his schemes into practice with a certain facility that began to rankle his onetime friend, William Park McClure. McClure, known around town as "The Thunderer," for his brash assertion of authority, tried with compulsive frequency to wangle his way into the directorship of every important enterprise undertaken by the officials of Denver. As the criticism of his behavior grew more and more vocal among the townspeople, McClure focused his animosity on Whitsett, the most popular man in town. Accusing Whitsett of character defamation, McClure drafted a coldly formal challenge and sent it to his former friend. The terms were Colt Navy revolvers at ten paces. Whitsett was aghast. Revolted by guns, he would not even own one. However, honor had to be served, and reluctantly but dutifully Whitsett agreed to McClure's terms. The hour set for the event was 5:30 PM, at a remote location about a mile from town on a stretch of bluff land. The principals and their seconds arrived a few minutes early, followed by Dr. Peck, who had foregone his gold digging long enough to attend to the victim, or victims as the case might well be at ten paces. Around the field of honor stood nearly five hundred men, their eyes alight with anticipation and their wagers quickly made as the gamblers moved among them quoting odds and noting names and the numbers of dollars. McClure and Whitsett took their positions, the referee gave the count, and two shots reverberated over the heads of the spectators. Whitsett, unscathed, opened his eyes to see dimly through the smoke. The Thunderer lurched sideways and fell heavily at the feet of Dr. Peck. The crowd cheered and jeered, depending on their fancy, while money rapidly changed hands, and McClure, with a nonfatal hole in his groin, was carted off to repair his health and his tattered image. Whitsett, having gambled and won with the highest stake of all, his life, smiled feebly and quickly walked away. Later, in writing about the affair to his friend Larimer at home in Leavenworth for a visit, Whitsett allowed himself a small expression of pride. "General," he wrote, "I plumped him."

In the profusion of gambling halls and bawdy houses and in the confusion of bodies claiming to have legal jurisdiction over Denver and Auraria—Arapaho county officials, Jefferson Territory electees, and the people's courts—crime found a ripe culture in which to grow. Drifters, looking for an easy buck, slipped into town and moved unnoticed among the roughly dressed miners, watching their chance to mug and roll their liquored-up and loose-talking victims. Pickpockets, their fingers sensitized to buckskin pouches and smooth-sided gold watches, were as numerous as flies. Mayhem and murder stalked the burgeoning twin towns. The sensation of the season was the Stoefel case. On April 7, about seven miles up Clear Creek from Denver, some settlers stumbled upon the body of a young German named Antoine Beingraff. He had been shot once through the head. An investigation turned up the fact that the young man's father was also missing. The prime suspect was Antoine's brother-in-law, John Stoefel. The posse caught Stoefel the next day, and Judge Smith, with a hurriedly gathered roster of seven jurors, heard without much enthusiasm General Larimer's impromptu defense of the suspect. Stoefel was accused of murder and consigned to jail to await trial at the first term of the criminal court. But the crowd saw no virtue in waiting and the prisoner was promptly bound and taken in a wagon to the nearest cottonwood that had branches large enough to hold a rope. With his neck firmly in the noose, Stoefel raved and swore as they pulled the wagon out from under him. As the rope tightened, his gyrations stopped and he died, according to onlookers, with scarcely a struggle. Uncle Dick Wootton, who was a spectator, later commented that "It was as neat and orderly an execution as ever took place anywhere."

The citizenry of Denver and Auraria who flocked with such eagerness to the unscheduled but real-life dramas of duels and hangings were by October of 1859 able to attend scheduled performances that were as dramatic if not as authentic. Colonel C. R. Thorne, the self-advertised renowned theatrical impresario, lately of Australia by way of California (and actually of Kansas), brought his star company of actors into Denver at the end of the summer in a trail-tired but jauntily daubed red-and-yellow wagon. His coming to Denver was no accident. His friend General Larimer had written to Mrs. Larimer in Leavenworth telling her to urge the Colonel to come because the populace on Cherry Creek was ready for a touch of the arts. Forthwith, the Colonel came. As soon as the company arrived Byers quickly printed up handbills announcing, appropriately enough, that the troupe's first production would be:

THE CROSS OF GOLD, or THE MAID OF CROISSEY,
Apollo Hall, 7:30 PM Front seats reserved for the ladies.

On the night of the opening, 350 miners, merchants, farmers, drovers, and gamblers paid out $2.50 a ticket and sat themselves down on the pine benches in the upper story of Libeus Barney's two-story combination saloon-theater, ready to be moved to tears and laughter. With the illustrious Colonel himself in the lead as a dashing French soldier and the beauteous Mlle. Haydee as his ingénue, the play got off to a fine start. But the evocative emoting on stage was soon interrupted by the after-dinner bedlam of noise coming up from the saloon below and the loud and uncomplimentary contributions of a curious inebriate who tottered up the stairs to see what all the applause was about. But the rough and ready patrons of the drama in Denver-Auraria didn't mind a few interuptions and they took to theater-going with almost the same fervor that marked their interest in gold-mining and gambling. The Colonel however found the inartistic atmosphere too much for his sensibilities. He pulled out after two weeks in the raw frontier town and headed for the amenities of the East, leaving the company in the capable hands of Mlle. Haydee.

To keep the company together, Mlle. Haydee needed money. With good intentions, but little talent, a group of stage-struck settlers decided to produce a benefit for her, in the common practice of the time. At least once during a season a very popular actor or actress appeared in one of his most famous roles and at the end of the performance all the proceeds were turned over to him as a kind of annual bonus. The benefit produced by the citizens on Cherry Creek for the popular Mlle. Haydee, however, was a play written and acted by themselves. Its author was robust Captain A. B. Steinberger who, in addition to being interested in mining engineering, fancied himself, quite inaccurately, a poet. When the opus was at last finished, it turned out to be a deadly dull romance entitled *Skatara, the Mountain Chieftain*, written in heinously stilted, pompous doggerel, with a plot that hinged laboriously on the settlement of the gold country by doughty white men in the face of mortal opposition of the noble red savage. Steinberger himself played Skatara, the unyielding Indian chief. None other than Sheriff Ned Wynkoop, his handlebar moustache waxed to perfection for the occasion, played Hardicamp, the white mountaineer in search of gold. Later, in the Indian wars of the sixties, Wynkoop would play a real-life role in that drama of the plains settlement. But meanwhile, no matter how seriously Wynkoop and the others took their stage roles, the play flopped gloriously. However, it came to floor-stamping, hand-clapping, cat-calling life when the irrepressible Mike Dougherty, a professional comedian-turned-miner, rewrote it as a farce, calling it *Skatterer, the Mountain Thief*, and boldly presented on stage the thinly disguised characterizations of the most notorious citizens of Denver-Auraria. The audiences loved it, cheerfully pouring quantities of gold dust into the ticket office hopper to return Mlle Haydee's company to solvency. What the townspeople did not know was that they were at the same time funding a dowry for their vivacious darling so she could elope with Tom Evans, one of the diggings' most infamous gamblers.

Shortly after this memorable theatrical season, Steinberger left Denver, taking various minor posts in the government. He next turned up, to the fascination of Denverites, in the pages of the *Rocky Mountain News* in 1875. The *News* reported that Captain A. B. Steinberger, late of Denver, while on a survey mission to a Pacific island in 1873 had quickly ingratiated himself among the natives and to his gratification was soon elevated by the islanders to be their king. There, in the sensuous atmosphere of the tropics, the romantic surveyor held forth in much the same regal fashion as his protagonist Skatara, with his devoted subjects showering tribute at his feet. Included in the tribute were eligible daughters, all of whom Steinberger, gallant as he was, accepted. Inevitably, jealousy crept into the royal household and in a moment of spiteful passion one of his inamoratas betrayed him and his kingdom to the captain of a passing warship. Dethroned and dismissed, Steinberger, the island chieftan, disappeared forever.

❧

For all of its appreciation of the drama, Denver-Auraria was not quite prepared for the one-man show that thundered into town in the form of Professor Owen J. Goldrick. His appearance popped eyes on every street corner as people stopped dead to stare at him, standing clean-shaven, frock-coated, and top-hatted at the front of his wagon with one of his striped trouser legs braced against the brake and his yellow doeskin gloved hands masterfully manipulating the reins of his straining ox team. He brought his frontier chariot to a dust-billowing halt in front of Uncle Dick Wootton's saloon, where someone brought himself to ask the imperious young man standing in the wagon what brought him to Cherry Creek.

With a flourishing wave of his black silk topper and in a voice heavy with an Irish brogue, Goldrick answered, "*Majura verum iritia!*"

"Aw, hell, what kind of Indian lingo is that?" asked his questioner.

"That, my dear man, is Latin," Goldrick bowed slightly, "and it means that what brings me here is 'The beginning of greater things.'"

With that, Goldrick's place in the history of the new Eldorado was assured. Much earlier, the twenty-six-year-old Sligo-born Irishman had figured out that a man with his flare for languages, philosophy, and mathematics ought to do well in the wide-open West, and he was not at all flustered when some of the people on Cherry Creek interpreted from some offhand remark of his that he was a graduate of both Trinity College, Dublin, and Columbia University in New York. He was actor enough to sound as if he had been to those illustrious seats of learning, and as a matter of record he had had enough book learning to be a teacher. It was by taking various teaching jobs that he hop-skipped westward, finally ending up as a private tutor to the children of prosperous rancher J. B. Doyle, down on the Huerfano River in northern New Mexico. It was in the home of this good Irishman that young Goldrick heard his employer's friend Uncle Dick Wootton talk about the gold rush to Cherry Creek. After persuading Doyle to let him drive a team and wagon of goods to Doyle's mercantile store in Denver, Goldrick packed his books and his wardrobe and headed north. When he arrived in Denver, he had $500, all of it tied up in fancy clothes, 50 cents in cash, and a very simple formula for sucesss: settlers have children; children must be taught; I can teach; therefore, I will open a school.

And so he did—the first one in the entire Pikes Peak region. In a drafty, leaky-roofed log cabin, he admitted thirteen pupils at a tuition of $300 a year for each child. The profit was easily enough to provide the "Adonis of the Rocky Mountains," as he was quickly nicknamed, with his fondest desires: good cigars and fancy clothes.

As the little school prospered, Goldrick gradually gave over the teaching duties to the wives, sisters, daughters, and sweethearts of gold-hungry men who welcomed the wages their women made to supplement the family grubstake. The professor then turned to a latent love: journalism. In reporting for the *News*, he adopted a dozen or so pen names and scoured the countryside for stories with a touch of the sensationalism to them. His subjects and style drew scores of readers, and soon Byers promoted the flamboyant and much admired reporter to assistant editor. Goldrick was a prodigious writer, and as his journalistic career developed he became a prodigious drinker. His associates could almost measure his output by his intake. On a day after a particularly liquid night, it would be well after noon that Beau Goldrick made his appearance in the newsroom, and in two or three hours of uninterrupted penning he could fill his page in time for the 5 PM deadline. Reformation set in when the estimable Goldrick, now the editor of his own weekly, the *Rocky Mountain Herald*, met a middle-aged widow, married her, and touched nary a drop more while she lived.

❧

Within a year and a half after Green Russell's party first made camp on Cherry Creek, the cottonwood grove at the junction of the creek and the South Platte had grown from a few tepees and tents to a busy center of commerce, with a population of a thousand or more, and with all the necessities and some of the amenities a civilized community requires. Gold was coming out of the mountains in ever-increasing amounts, and the twin towns on Cherry Creek got their share before it passed on east to the money markets. The future of Denver and Auraria with their "bold shrewd, go-a-head style about everything" looked bright, indeed.

4

Bonanza on Clear Creek

ON A JUNE MORNING, FOUR WEEKS AFTER THE STRIKES OF JACKSON AND GREGORY, Henry Allen, Auraria's postmaster, sat in the doorway of his combination cabin and post office contemplating the coming day. As he watched the cloud-streaked horizon ignite with tongues of vermilion that quickly faded into bright blue sky as the rising sun edged upward, he was put in mind of the irony of recent events. A bare month ago, gangs of go-backs stomped into his post office to call him a liar for saying there was lode gold somewhere in the mountains, and even his friends ridiculed his exuberant confidence that a great bonanza would be found. Now, like the dissipation of the angry-hued early morning sky with the coming of the golden sun, gone were the foul curses and the arch disbelief. In their place was the same bright optimism Allen had held all along. Vindication, he decided, was too sweet to enjoy alone. Fishing out pen and paper from the battered trunk that contained his few belongings, he sat down and ripped off a letter to his hometown newspaper, the *Council Bluffs Bugle*.

> Foster, Slaughter, and Shanley are in the mountains and are making money—not by the dollar but by the hundreds of dollars!...It is quartz diggings and there is no knowing how extensive they are. One thing is certain, I have traced some leads nine miles and there is plenty of them. I have found them from 25 yards to 30 rods apart and running parallel for miles...Now you can tell whether there is any gold in the country or not and how near the truth we have been writing...

By the time Allen wrote his letter, the narrow defile cleft by Clear Creek swarmed with three thousand gold-hungry souls. Denver-Auraria was all but deserted as grocers, doctors, bullwhackers, blacksmiths, bakers, timber-cutters, and bartenders headed for the mountains armed with picks, shovels, and gold pans. On the plains, hundreds of go-backs wheeled around and raced back to get in on the bonanza that only a few weeks before they had roundly damned as the figment of unscrupulous promoters. Prospectors spilled over into both the north and south forks of Clear Creek, testing every sandbar on the streams and poking into every niche and crevice of rock in the narrow gulches. In the gulches they looked for "blossom rock"—the decomposed quartz outcrop of a vein of ore that varied in color from bright orange to brown. The test was to crush a panful of the quartz, wash it in a stream, and watch for bits of flour-like gold particles to appear in the bottom of the pan. If a pan showed good color, the miner staked out a claim and scurried on to look for another promising lead. The veins, as Allen wrote, varied in width: some were only a few inches wide; others were hundreds of yards wide. They extended in all directions up the sides of the gulches and over the ridges in a network of riches. The fever of discovery swept along Clear Creek like wildfire. In Iowa, Nevada, and Illinois gulches, all named for the home states of the men who first prospected them, the sound of picks and shovels crunching against rock and the rush ing of water in the creeks and the shouts of those who struck a high-grade vein combined in a cacophony of incongruity against the mountain wind and silent sky. Millionaires were made in a matter of moments, as here and there the born lucky swung their picks and felt the rock give way to a softer interior. This was pay dirt—a pocket of mixed sandstone, clay, and quartz, impregnated with flour gold. Out of pay dirt a man could commonly wash $100 a day. Villard reported that one of Gregory's original claims produced $1,500 in one day.

Most miners soon found that panning out the gold took too long. To speed up the extraction process, they turned to the methods used in the early days of the California gold rush. They hollowed out logs or tore down their wagons and handcarts for lumber to build rockers and "Long Toms." These were three-sided boxes mounted on barrel staves or lumber cut to form a curve. At intervals across the width of the boxes they hammered strips of wood to catch the heavy gold particles that were washed out of the sand by the constant rocking action as water was poured through the box. If there were a stream near his claim, the more ambitious miner built a slightly inclined long wooden box, also fitted with riffle boards nailed across the bottom. The stream was diverted to run through the sluice box and the miner shoveled the blossom rock into the box, letting the flow of the stream wash out the gold into small pools of mercury placed in front of the riffle boards. No matter how the gold was washed out, mercury was always mixed with the loose particles to form an amalgam. After the day's work of mining was done and the miner had eaten his supper of beans and hardtack, he put the amalgam in a cast-iron retort over his campfire. Under the heat of the fire the mercury vaporized, leaving a lump of native gold. This could be the best or worst moment of the day, since the day's take in dollars and cents was measured by how much the lump of gold weighed. But whether the day's take was exceptionally high or abysmally low, the miner harbored his precious supply of mercury, letting it condense in the coil of the retort and collecting it for use in the next day's ore separation.

Into this hive of industry rode three weary and saddle-sore journalists, Greeley, Richardson, and Villard, and their guide. They had been on the trail to Clear Creek for two days, having left Denver in a wagon drawn by four mules on June 6. At the bank of the South Platte they took their place in the long line of emigrants waiting to cross the river on William McGaa's rope ferry. Once across the river, they moved northwest for fifteen miles to Clear Creek where they left their wagon and climbed aboard their mules for the rest of the ascent to the diggings. Greeley was exhilarated at the prospect. "When we reached Clear Creek on our way up...though the current rushing from the mountains looked somewhat formidable, I charged it like a Zouave, and was greeted with three ringing shouts from the assembled Pikes Peakers, as I came up gay and drippping, on the North shore."

As the party made their way laboriously over precipitous, windswept foothills and down steepsided ravines into lightless canyons, Greeley began to regret he had been so frolicsome—his head and leg wounds from the coach accident on the plains ached fearfully under the constant jarring of the mule's heavy step. Alongside the foursome, however, few were better off. Caravans of prospectors struggled toward Clear Creek, forcing their teams to drag wagons bulging with tools and provisions up the giddily steep grades. When the oxen balked, four or five men grabbed the spokes of the wheels to inch the wagons upward, their lungs nearly bursting from the lack of oxygen in the thin air. Some seven miles short of the Gregory diggings, on a small piece of level land, Greeley and his friends made camp for the night. The *Tribune*'s dauntless editor was so exhausted he had to be lifted from his mule and placed on a bed of pine boughs to recover his strength. By six the next morning, however, he was again in the saddle, his white duster drawn closely about him to keep out the morning chill of the mountains. By noon, they reached the diggings.

Fording Clear Creek once more, but this time at low water and by way of the dry stones in the stream bed, the journalists found themselves in a narrow valley surrounded by high rising hills from five hundred to fifteen hundred feet high, covered with a dense growth of tall yellow pine. They paused to inspect the site of Gregory's original lead, and then moved upstream a half mile or so to where the greatest number of miners were at work. As they rode slowly up the gulch, men stopped their work to stare at the curious trio of obvious greenhorns.

"Why, I'll be damned if it ain't old Horace Greeley!" whispered one man loudly.

It was no wonder that the young men on the stream bank, most of whom were in their twenties and early thirties, thought him old. In his white duster, with his white slouch hat pulled down firmly to his ears, his spectacles well down on his well-formed nose, and his ruff of whiskers sprayed out like a cock's tail above his string-tied shirt collar, Greeley looked far older than his forty-nine years. Nor was it surprising that he was recognized. For ten years cartoons of this intrepid champion of the cause of western expansion had regularly appeared in the pages of every major newspaper in the country.

Greeley's arrival marked a great day in Gregory Gulch. Despite his odd get-up, the miners were flattered that a man of his stature would come all the way out to a remote gulch of the Rockies to see how they were getting along. And since most of them were well acquainted with the senior editor's admonition to "Go West, young man," they set out to prove to him that his advice had been sound.

On the very afternoon of his arrival, they led the dog-tired, limping, aching editor over boulders and logs, upstream and down, showing him every facet of the diggings. It pleased Greeley to see that some miners had already seen the prudence of building shelters. Although there were few mud-chinked log cabins strung along the narrow gulch, they appeared to be "commodious and comfortable." Sprinkled among the cabins were tents, and wagon beds tilted on their sides with canvas stretched from the top to the ground to form a kind of tent, and lean-to's of logs covered with pine boughs. Not a chair or a table was to be seen. A few white women and some squaws, seated on stumps or fallen logs, tended cooking pots over open fires. Here and there were tents with crudely lettered signs reading "Groceries" or "Saloon" hung on them. The newsmen poked their heads into the store tents to find that what flour there was on hand was dreadfully coarse and sold for $44 per hundred pounds, while the beef on sale turned out to be freshly slaughtered oxen, their meat tough and stringy after the long hard pull across the prairie and over the mountains to the gold camp. To make absolutely sure he was impressed with the prospects of Clear Creek, one group of miners secretly filled a shotgun with chunks of high-grade ore and shot it into a crevice of quartz and then nonchalantly invited Greeley and his two friends to wash out a pan of this "typical" ore. From what was probably the first but certainly not the last mine to be salted in the Rocky Mountain area, Greeley took an impressive $4 out of one pan.

"Gentlemen," exulted Greeley to the delighted miners, "there can be no doubt about the immensity of your discovery. I've seen it with my own eyes and you may be sure the news will reach as far as my newspaper can carry it."

With this, Greeley, Richardson, and Villard sat down to write out a joint statement verifying the proportions of the Clear Creek bonanza.

That evening, amid the flickering campfires of the three thousand people in the gulch, Greeley was persuaded, without much difficulty in spite of his exhaustion, to speak to the assembled miners. Speak he did, in his high-pitched, shrill voice and with great eloquence on the virtue of temperance, and the evil of gambling, and the sacredness of the family and home ties, and the importance of statehood for the new mining region. For weeks afterward, his impassioned words were the subject of conversation over uncounted hands of faro and nips of Taos Lightning in all the saloons and brothels from Clear Creek to Cherry Creek.

A curiousity though he may have been, Greeley saw clearly the pattern that would be set for the region: "...within ten years, the tourist of the Continent will be whirled up to these diggings over a longer but far easier road winding around the mountain tops rather than passing over them...in utter unconsciousness that this region was wrested from the elk and the mountain sheep so recently as 1859."

On the way back to Denver, Villard fell off his mule and temporarily immobilized his left arm, their guide took a fall that left him badly shaken, and when they finally got to the ford on Clear Creek, Greeley had lost all the Zouavelike spunk he had on the way up. He got off his mule and walked to a place downstream where it was shallow enough for a wagon to cross and pick him up. Of the four, only the urbane Richardson seems to have escaped accident on the trip to and from Clear Creek. But once in Denver aching bodies were forgotten as the enthusiasm over what they had seen carried the three men as fast as they could walk to the office of the *Rocky Mountain News*. Byers took one look at the dispatch signed by three of the most reputable journalists in the country and set the little Washington hand press humming with the paper's first extra. The fact that he was temporarily out of newsprint did not faze the *News*'s publisher for one minute. To meet the emergency, he sent his printer's devil down the streets of Auraria from store to store to comandeer all the brown wrapping paper that could be spared.

By June 20, the story reached Kansas, and that day the Leavenworth *Times* carried a banner headline:

UNPARALLELED RICHNESS. GOLD! GOLD!! GOLD!!!
A few hours since we published a large extra *Times* containing a most satisfactory statement from Horace Greeley, A. D. Richardson, and Henry Villard in reference to the immense gold deposits at the Gregory diggings and the general richness of the whole gold district.

The next day the story hit the New York *Tribune* and other eastern papers, and if the nation did not already believe there was a genuine gold rush to the Rocky Mountains, it did then. A new wave of treasure-hunters bought guidebooks, scraped together funds for an outfit, and headed for Clear Creek.

One of those who came in this new wave was Horace Austin Warner Tabor, whose name would be forever linked with the mineral fortunes of Colorado. Tabor was born in Vermont, but he left home while he was still in his teens after quarrelling with his stepmother, and went to live with a brother in Massachusetts. From his brother, young Horace learned the stonecutting trade. In 1853, when he was twenty, Tabor set out for Portland, Maine, to find a job. On the train, the slow-spoken but gregarious young man met plain-featured but friendly Augusta Pierce. As luck would have it, Augusta's father was a stone contractor and was looking for stonemasons to hire. Tabor promptly went to work for Pierce and did well, so well in fact that he felt privileged enough to propose to the boss's daughter some two years and innumerably rehearsed speeches later. Augusta made up some of the time lost by her swain's inarticulateness by accepting quickly.

Their marriage, however, did not take place immediately. The effects of the depression that struck the nation in 1854 filtered down to the stonecutting business, and, despite his engagement to the quarry owner's daughter, Tabor suddenly found himself out of work. The young stonecutter knew from his avid newspaper reading that the opening West offered a man scores of opportunities to better himself. Now he decided to try his luck in that direction. Promising the tearful Augusta that he would come back to claim her as his bride as soon as he could make enough money, Tabor joined a Kansas-bound party of homesteaders sponsored by a colonizing company called the New England Emigrant Aid Society. The party left in April 1856, carrying with them clothes, food, farm tools, and packets of seed. Many also carried "Beecher Bibles," as New Englanders had come to call the Sharps rifle after Henry Ward Beecher's impassioned speech at Hartford championing the cause of free soil. "Give each man a Bible in one hand," exhorted the fiery preacher, "and a Sharps rifle in the other and send him to Kansas!" Tabor was not drawn to the West solely as a place to better himself and his bride-to-be. Kansas, where the issue of free soil versus slavery was moving toward a climax, was where the action was, and, like many another young man, the Vermonter harkened to the call of adventure.

Near Manhattan, Kansas, Tabor appropriated an abandoned cabin and the 160 acres that went with it and proceeded to put the land under cultivation. Between plowings, he worked with other New Englanders who had founded the anti-slavery colony of Lawrence, Kansas, not far from Manhattan. When Lawrence was sacked and burned by a pro-slavery mob, Tabor was among its besieged defenders. Soon afterward the personable young man was elected to the extralegal Free Soil legislature. When that body gathered in Topeka for its first meeting, the US Army under orders from President Pierce ran the lot of them out of town.

With his political career abruptly ended, Tabor returned to his fields and gazed with satisfaction on the bumper crop that met his eyes. But his success as a farmer was also rudely cut short. Tabor's acres of ripened corn went begging as the country slipped into another slough of depression. Discouraged, Tabor gave up both politicking and farming and was about to return east empty handed when a windfall came his way. The Army began a building program at nearby Fort Riley and advertised for stonecutters. Tabor's skill landed him a job and by the end of 1856 he had enough money to go to Vermont, marry Augusta, and bring her to the homestead.

The newlyweds packed their meager possessions, bid goodbye to Augusta's parents, and started west. They traveled as far as St. Louis by train. From there they took the steamer to Westport on the Missouri River where Tabor bought a yoke of oxen, a wagon, some more seed, and set out for Manhattan to try his hand at farming once more. Unlike the first year, a disastrous drought claimed all of their crop. To help make ends meet, Augusta boarded two neighboring hired hands, but her table was skimpy. To add to her trials, Augusta was pregnant. By the time Nathaniel Maxcy Tabor was born late in 1857, his parents were in dire need of money. Once again, Tabor took up stonecutting at Fort Riley. All day Augusta and the baby were left alone in the 12 by 16 foot cabin. Through the long winter, Augusta's amusement consisted of reading the old New York *Tribunes* they had used to paper the cabin walls to keep out the slicing Kansas winds.

In the spring, Tabor bought a new batch of seeds and a wagonload of provisions with the money he got for his winter's work. They sowed another crop on the homestead and waited hopefully for the harvest. The drought of the previous year broke, and their fields returned a bountiful crop, but the Tabors' joy was short-lived. When they got their produce to market, they found it glutted, and they made barely enough to break even. Bleak days followed one after another as their small store of root vegetables and stringy dried beef grew smaller and smaller, and the incessant wind blew mournfully over their depleted land, whining around the drafty cabin, plunging the young couple into deep despondency. With little hope for the future, Tabor turned again to stonecutting at Fort Riley.

In the midst of their desolation, Tabor happened to meet one of Green Russell's men as he passed through Fort Riley in February 1859, en route to Georgia for more supplies and miners. Tabor rushed back to the farm and burst in on Augusta with the news of Russell's gold strikes.

"Augusta, I'm going to Pikes Peak!" His fist came down hard enough on the rough-hewn table to rattle the tin plates Augusta had laid for supper. "They've got gold there, and I'm going. I heard it from a man who just come from there. You can take Nat and go home to Maine and I'll send for you as soon as I get enough gold to do it."

Without waiting for his wife's reaction, Tabor turned and strode to the corner of the room to take hold of their trunk, pulling it bumping and rattling over the uneven floor to the center of the room. Augusta watched him impassively, and when the room was quiet again she said softly, "Tabor, Nat and I won't leave you."

At his wife's words, Tabor paused, frowned, and eyed her solemn face for some minutes. Then his eyes suddenly grew bright and he smiled as an idea abruptly burst upon him.

"All right, my dear, come you shall," Tabor said. "With you to cook and do the wash and mending, I'm sure I can get our boarders to come along with us. And with what they pay us we can buy food for all of us for the entire trip."

Elated, Augusta took over the packing of utensils, staples, bedding, and clothes while Tabor beat it back to Fort Riley to make a few more dollars before they left. While he was there he was cautioned not to make the trip before early May. Tabor grudgingly heeded the warning, but as soon as May 5 rolled around, the Tabors climbed into their wagon, tied their milk cow onto the back, and struck out with their infant son and the two hired men for the Republican River and the dreaded Smoky Hill Trail.

The trip was excruciating. They found no habitation until they were within eighty miles of Denver. For a road they followed creek beds, more often dry than not, guided by an old Army topological map Tabor picked up at Fort Riley. Often, they had to travel twenty-five to thirty miles a day to find enough buffalo chips with which to build a fire. On Sundays the men foraged for game from sunup to dark while Augusta, left in camp alone, washed their clothes and cooked up a mess of beans and bread for the next week's meals. Between chores, she tended baby Nat, who squalled constantly in the misery of teething, and trembling with fear and ague, she fended off curious Cheyenne who came repeatedly to beg for sugar and tobacco.

They arrived in Denver in the middle of June, the strain of the journey plainly showing in their gaunt, weathered faces. Augusta was down to ninety pounds and was so weak she could scarcely lift herself from the straw mattress that served them as a bed. It was with a great sigh of relief that she heard Tabor say that they would camp in Denver until the oxen had regained their strength.

In a few days, Tabor went off to Clear Creek to prospect. When he returned, they once more loaded the wagon and moved out. Just as Greeley had found it, the trail was narrow and steep and several times the Tabors had to unload the wagon and push it up the mountainside. On the downgrades, their only brake was a pine tree lashed to the rear axle, and each time they started down, Augusta had visions of their hard-bought belongings lying crushed and broken under a runaway wagon at the foot of the slope. Sometimes darkness came while they were still laboring up or down a grade so steep that in order to keep from rolling downhill when they stopped to sleep they had to brace their feet against logs laid horizontally against stakes driven into the earth.

It took nearly three weeks for the Tabors to reach Idaho Springs, the site of Jackson's diggings. As it was, however, theirs was the first wagon into the area and Augusta was rather proud to find that she was the first white woman there as well. Tabor and the hired men cut logs and laid them up for a cabin, roofing it with their canvas tent. Then they went prospecting. Tabor staked out a placer claim in a few days and worked it diligently but with very little return. Augusta, on the other hand, in her usual role got to work baking pies and bread to sell to the miners. It was the money from her enterprise that kept them going all summer.

Soon after their arrival on Clear Creek, Tabor was invited to attend a meeting got up by the miners for the purpose of organizing the area into a mining district and to lay down some rough rules as guides for themselves and any newcomers to the gulches. To prevent one man from securing all the best claims for himself, they decided that no one except the man who first discovered a particular gold field could hold by right of discovery more than one creek (placer), one gulch (patch), and one mountain (lode) claim. Any number of claims could be held through purchase. But discovered or purchased, each claim had to be worked within ten days to establish title. The dimension of a placer claim was set at one-hundred-feet-square. A lode claim was restricted to one-hundred-feet along the vein and fifty-feet in width. A gulch claim followed the meander of the creek for one-hundred-feet and extended from bank to bank.

As more and more men poured into the gulches looking for pay dirt, it wasn't long until nearly every foot of likely looking real estate was claimed. It was then that the hordes of newcomers began to agitate for a reduction in the size of claims. "Old-timers," those who had been on Clear Creek for all of three weeks and who were already realizing a tidy profit from their claims, naturally enough wanted the rules left alone. But they soon were outnumbered by the influx of miners eager to have their chance at striking it rich, and in the middle of June the issue came to a head. The newcomers knew what they wanted, but they were not sure how to get it. A mass meeting of the district to discuss the question seemed like a proper beginning. At three o'clock on the afternoon of the appointed day of the meeting, the miners, all dressed in variations of flannel shirts, wool trousers, and leather boots, began collecting in a clearing on a hillside above the diggings. Among them were the old-timers, not in a group but carefully and unobtrusively scattered among the newcomers. So many men had flocked to the diggings in such a short time that, except for a few prominent figures such as John Gregory and his partner Wilk Defrees, a man was hard-put to recognize the man next to him as an old-timer or a newcomer. But the majority knew two things—they wanted the size of the claims changed and they knew there were enough of them to override the old-timers. The leaders of both sides discussed the question of how to select a chairman amid shouts of "Let's get on with it!" and cheers and whistles from the restless gathering. At last the leaders decided the chairman would be drawn from a handful of names scribbled on cartridge paper and tossed into a hat. A newcomer was to do the drawing. He reached into the hat and, as luck would have it, pulled out the name of Wilk Defrees. Defrees was roundly booed by the newcomers, but they quieted down when they were told that the chairman had to remain impartial. He was not given the power of casting a deciding vote in case the issue became deadlocked. Mollified, the crowd let Defrees proceed. The chairman opened the meeting, calling for statements of the grievance. At this there was bedlam. Defrees' shouts for order were drowned out and it was only after he fired off a pistol that the miners let him be heard. In somewhat more orderly fashion, several men then proposed that the chairman select at random a twelve-man committee to decide on a solution to the claim-size question. The crowd voiced its approval and Defrees promptly carried out the will of the majority, looking over the assembled group and pointing to this man and that until there were twelve. The fact that he had carefully plucked out twelve old-timers mingled throughout the company of newcomers went unnoticed. The committee removed itself to hold council at the edge of the woods, well away from the smug crowd who were certain their cause was won. After an interval of apparently much heated discussion and show of vehemence, the committee returned to recommend that claim sizes remain unchanged. Before the stunned newcomers could recover, Defrees called for a voice vote and the prompt, loud "ayes" of the old-timers carried the day. Defrees announced the verdict and quickly adjourned the meeting, leaving hundreds of bewildered miners with a dawning realization of the power of political strategy, even at the grass-rootiest of levels.

Later, when surface gold was played out and it was necessary to dig deeper for the ore, the miners of Colorado defined a tunnel claim, a unique contribution to general mining law. Under this rule, if a prospector found an outcrop that dipped below the surface, he and his partners could establish a series of claims running as long as two hundred feet along the surface of the ground if the boundaries of the claim were staked out *before* tunneling began. It was up to the miner to gamble on the direction the vein took. If his guess was correct, he might make a fortune. If he was wrong, he might get enough to pay the $2 recorder's fee.

With claim rights established and copious amounts of gold coming out of the gulches, the question of the price of the precious metal became a vital issue within a very few months. Shipped east to the mint in Philadelphia, retorted gold brought about $17 an ounce. In the towns on Cherry Creek, gulch gold was pegged at $14 to $17 an ounce, depending on whether it was in an amalgam or in pure form.

But soon dissatisfaction over these prices spread among the miners and they sought buyers who would pay them more for their hard-earned product. They found these buyers among the gulch merchants and nearby farmers who were more than glad to up the price to $16 for amalgam and $18 for retorted gold in order to get the miners' business. Around this nucleus of trade, small towns developed all along the gulches of Clear Creek. More and more permanent buildings were built as the winter of 1859 approached. Soon these clusters of cabins and stores clinging shakily to steep hillsides, some of them hung on stilts, took fancy names—Nevadaville, Central City, Mountain City, Black Hawk, Missouri City—and like their older sisters down on Cherry Creek, the towns developed fierce rivalries. When some of the citizens of Nevadaville had the nerve to declare themselves free of the mother town and to elect their own officers, the authorities in Nevadaville promptly arrested the upstart rebels on a charge of secession. When the people of Black Hawk heard that Central City was going to incorporate, they rushed to do the same before their neighbor up the gulch could claim the honor of being first.

All of the camps looked alike, with tents and cabins and store fronts raggedly ranged along terraced streets housing groceries, bakeries, restaurants, and lodge halls but, commented one resident in a letter home, "nary a church." This shortcoming was removed, however, with the arrival of mission priest Joseph Machebeuf in the fall of 1860. Sent by Bishop Lamy of Santa Fe to bring Catholicism to the Pikes Peak region, Father Machebeuf established one of his first beachheads on Clear Creek. Holding mass every Sunday in any shelter he could find, the gentle priest drew together an attentive and devoted congregation. For all their devotion, however, the miners quickly scattered whenever the soft-spoken prelate mentioned the subject of money to build a church. For weeks the issue stagnated. Finally, one Sunday the good priest said mass as usual, this time in one of the camp's meeting halls, and as he gave the benediction Machebeuf walked slowly down the aisle toward the door at the back of the hall. At the last "amen" he suddenly slammed shut the door and bolted it.

"And now, my children," thundered the priest in a tone his parishioners had never before heard him use, "we shall all stay here until we can see our way clear to finance a house of God!" Stunned miners looked at one another and at the forbidding countenance of the custodian of their souls and then one by one they sheepishly produced pouches of nuggets and gold dust. Construction on the new church began the following morning.

Until the campaigns of Father Machebeuf and other dedicated men of the cloth produced churches in the gulches, saloons and theaters provided all the pageantry and ritualism to be had on Clear Creek. Each camp had its favorite saloon. In Mountain City it was

Jack Keeler's. He had no glasses for his patrons, only a bottle on the bar and a tin cup. But the miners weren't fussy about sanitation. They did mind a little having to wait for a man to finish a drink before another could have his, but Jack let a man buy a drink on credit and that was worth waiting for. In Nevadaville, thirsty miners crowded into "Mountain Charley" Forest's place. The smoking, swearing, drinking Forest ran an uproarious establishment, catering mostly to strapping muleskinners who favored razor-edged knives over revolvers in settling their frequent disputes and who found a complement to their virility in the fact that "Mountain Charley" was a woman.

Some years before, back in the Midwest, Eliza Jane Forest had been a happily married housewife. Then one day her conventional life collapsed around her when a sneak thief murdered her husband and disappeared. Not one to bury her grief in weeping, Eliza Jane vowed vengeance. She then climbed into a pair of her husband's trousers, clapped his broad-brimmed hat on her head, strapped on his pistol, and started out in search of his killer. The trail led from the Mississippi to California and eastward again, with Eliza taking odd jobs on riverboats, railroads, and wagon trains. As the years passed, the trail grew colder and colder and Eliza Jane, contentedly settled into the role of a roustabout, gradually lost sight of her mission. By the time she got to Colorado, the desire for revenge gave way to a desire for revenue, and with the few dollars she had left from her last bull-whacking job, she gave up nomadic life to open her popular grog shop in teeming Gregory Gulch.

If a man tired of seeing himself in the person of his drinking companions, he could drop around to the makeshift theater to see himself caricatured on stage. In Central City, miners flocked to the National, a converted stable, run by self-styled impresario George Harrison. Next door was Charley Swirtz's hell-for-leather saloon and variety hall. Before long, competition raged between the two barons of entertainment and it was by no means friendly. Harrison complained that Charley's loudmouth audience disrupted performances on the stage of the next-door National. Swirtz countered by proselytizing Harrison's actors. The feud erupted in a knockdown, drag-out fight that bystanders felt compelled to jump into and stop when Charley's pugilistic prowess reduced Harrison to a bleeding, tottering hulk. But the scrap did not end there. Embittered and unrelenting, each vowed to draw on the other the next time they met. Harrison dragged himself off to his rooms to poultice his bruises, and when the various swellings on his face and neck receded he took the stagecoach east to hire more actors. On the day he was to return, Swirtz got out his revolver, cleaned and oiled it, loaded it, stuck it in his holster, and dropped around to Barnes and Jones's saloon for a loin-girding gulp of whiskey. As the stage carrying Harrison rattled and swayed up the gulch toward Central City, the impresario had the foresight to drop off at Black Hawk, just below

Central, and make his way on foot over a ridge to the east and into Central City. On the way, he picked up a double-barreled shotgun and the information that Swirtz was at Barnes and Jones's saloon. From a balcony overlooking the street, Harrison began his vigil, and as soon as Swirtz came through the swinging doors of the saloon, Harrison let him have both barrels. By the time the variety hall operator slumped to the boardwalk he was dead. His body lay in state in his establishment for several days while business went on as usual, with an occasional patron, lachrymose from drink, pausing briefly over the body to drip boozy tears over its waxen face. Harrison was arrested and tried but a laggardly jury refused to convict him. He finished out the season, but there were too many reminders of his crime to let him live in peace in Central City, so he sold his theater to Jack Langrishe, a roving thespian, and left the gulch.

Langrishe opened the newly redecorated and renamed theater while stagehands were still applying some of the finishing touches. His first offering at the Montana was a romantic tear-jerker called *Alice, or the Mysteries*. On opening night the pit was filled to capacity and the patrons hung on every word spoken onstage. Near the close of the second act the distraught hero stood in center stage with his head thrown back and his arms outstretched, and with tear-filled eyes turned heavenward he beseeched in a tremulous voice, "Alice, why don't you speak to me?" In the emotion-charged pause that followed, the irate voice of a workman bounced down from the rafters:

"Why, you damn fool, Alice ain't up *here!*"

From April until November, the Langrishe company played everything from *Macbeth* to *The Pikes Peak Gold Fiend,* a melodrama that never failed to draw deafening applause at curtain call time. Everyone agreed that for comedy, Langrishe himself was tops. The tall, lanky, double-jointed man could convulse his audience without so much as opening his mouth. And his greatest asset in this regard according to some was the "size, shape, and mobility" of his very prominent nose.

So popular was the theater that Langrishe often counted more than $300 for one night's receipts. Much of the popularity of the group stemmed from the broadly comic performances of Mike Dougherty. Mike had given up the muse to follow Lady Luck west to the mines of Clear Creek. He struck a sizable vein of lode gold near Central City, but his taste for alcohol ate up most of his profits and before long he was penniless. Undaunted, Dougherty fell back on his old talent and teamed up with Mlle. Haydee and her dramatic troupe down in Denver. His acting ability and original songs filled with rib-tickling local allusions were an immediate success. When Denver was rocked by the elopement of their favorite danseuse, the lovely Mlle. Haydee, with that no-good Tom Evans, the star company of players disbanded and Mike went off to prospect the Black Hills of South Dakota. He came back to Denver without having found his bonanza and just in time to read of the appearance of Langrishe's traveling dramatic company for "six nights only" at the Apollo Theatre. The genial and able Langrishe found in the puckish Dougherty a man after his own heart. They formed an inspired partnership and the six-nights-only engagement stretched into a tour of four triumphant years as the team of actors presented the latest hit shows from abroad and the East, liberally intermixed with burlesques written by themselves.

Of all their collaborations, the one that became a favorite of the gold camps, and in particular Central City, was *Pat Casey's Night Hands*. It was a rollicking musical that told the rags-to-riches story of Pat Casey, a young Irishman who could neither read nor write but who stumbled onto pay dirt and became a millionaire. Casey took to high-spending and boasting, so the libretto revealed, and rarely visited his mine except to show off its workings to visitors to whom the unlettered scamp would explain, "Why, gentlemen, you wouldn't believe it, but I use up ten lead pencils in a day figuring my business!" Casey was pictured as an inveterate drinker who kept a gang of men at work in his mine during the day and another gang drinking with him in the saloons at night so he would not have to drink alone. Spoofing Casey's nonacquaintance with simple arithmetic, the tongue-in-cheek playwrights showed him going from mine to mine looking for drinking partners. Stopping at one mine, he would call down the shaft, "How many men you got down there?"

"Five," would come the sepulchral answer, and after a studied pause, the stage Casey would loudly call downstage the line that never failed to bring a roar of laughter.

"Well, half of yez come up here and have a drink!"

One reason for the operetta's popularity was that the real Pat Casey lived in Central City, and the stories of his illiteracy and uncanny luck were a combination of truth and fiction. Like mercury, Pat seemed to have an affinity for gold. At a funeral one day, as the minister conducted the graveside service, he saw Pat stoop and pick up a hunk of quartz, examine it carefully, and then immediately begin to step off a claim. Thereupon the minister, without betraying the solemnity of the occasion, clearly intoned, "For our dear departed brother a place in Heaven—*and stake me out a claim too, Pat*—this we ask for Christ's sake, Amen."

Casey took his religion seriously and when the parish priest of Central City called for funds for a new church, Pat was one of the first to subscribe. So generous had he been that the touch was put on him again, this time to finance the purchase of a chandelier for the church nave. Casey cheerfully turned over the money. "But who," he wanted to know, "knows how to play one of them things around here?"

When Casey heard that Langrishe and Dougherty would open their next season in Central with the farce that made him out a laughingstock, his Irish dander flared. Casey considered himself to be a

paragon of astuteness. He liked being the object of awe and he could
see neither humor nor truth in the Pat Casey of *Night Hands*. Except
for that little affair of $2,000 in damages in New York when he had
blazed a trail of bullet holes along the walls of the hotel hall so he
could find his room after a spree on the town, Casey could recall no
reason why he should be the butt of a public performance. He threatened violence. But the show was billed as Mike's benefit and since he
would get all the proceeds from the popular play the actor was determined to put it on. Casey was equally determined to stop it. On opening night Casey augmented his natural feistiness with a magnum of
champagne, and rounding up some one hundred of his brawny day
and night hands, he started for the theater. Meanwhile, careful not to
underestimate the capability of his opposition, Dougherty scratched
off a note to Frank Hall, captain of the Central City militia, asking for
his help. Hall responded promptly. "Yes," he penned back, "the Guard
will be there. Reserve seats for fifty men." It wasn't often Hall could
treat his men to a free evening at the theater.

It was approaching curtain time when Casey and his gang
rounded the curve of muddy Lawrence Street and marched up the
center of the road to the theater. There they found themselves facing fifty Belgian rifles with fixed bayonets. Casey stopped and his
men stopped behind him. Leaning unsteadily uphill into the grade
of the winding street, Casey pulled his revolver and shot it into the
air. The bullet ripped into the cornice of a building above the heads
of the militia. Hammers clicked ominously along the line of Belgian
rifles. At that moment, Sheriff Billy Cozens trotted out to Casey and
caught the Irishman in his arms just as his champagne-loosened legs
collapsed under him. As the sheriff trundled Casey off to jail, the day
and night hands quietly dispersed. The militia took their seats in the
theater, the curtain rose, and Dougherty regaled a wildly appreciative
audience with every hilarious word of *Pat Casey's Night Hands* while
its hero blissfully slept off his opposition in jail.

When Pat awoke, he had done with the rawness of the frontier. He
yearned for the esteem of his fellow men and the finer things in life
that he had sampled on his frequent business trips east. So he liquidated all of his Gregory Gulch holdings for $150,000, which inciden-
tally started a wave of eastern speculation in the mines of Clear Creek,
and moved to New York where he became a respected dealer in gold
stocks and drove one of the most admired carriages on Fifth Avenue.

❧❦❧

John Gregory, whose lucky pick swing started all the excitement on
the north fork of Clear Creek, also left about the same time. For weeks
after his strike, he could barely comprehend that he was a rich man.

He had never known luxury. Now he couldn't sleep and he scarcely touched food. To anyone who would listen he repeated over and over that now his wife could become a lady and his children could be educated. If there was no one to listen, he talked to himself, sitting with his back to a tree stump and periodically lifting a tin frying pan under which he kept chunks of retorted gold, as if to reassure himself that his good fortune was not a dream. He was a pitiful sight around camp, and early in July, 1859, his friends persuaded him to go see Doc Berkley down in Denver. Before he left the gulch, Gregory sold his claim for a paltry $21,000, to be paid at the rate of $500 a week. In the first day of its new ownership, the Gregory claim yielded $607, but its original finder was mentally too unhinged to notice he had been fleeced. After several weeks in the careful care of Doc Berkley, John Gregory recovered sufficiently to head back to Georgia. But he wasn't there long before the magic of Clear Creek brought him back, this time as the operator of a quartz mill. For a while, he was content to run his mill, taking in $200 a day in clear profit. Then one day he sold it for $42,000, six times its original cost, and once more left the gulch, this time never to return. Whether or not his wife took on the airs of a lady and his children went to school is not recorded.

The coming and going of Gregory was hardly noticed among most of the frenzied miners on Clear Creek. They were too busy making money, or trying to, at least. From May to July that first year of the Clear Creek bonanza, the gulches were alive with activity. A man might prospect for a month and not find any leads. Or, he might find a rich outcrop in an hour. In the matter of discovery, it mattered not whether a man was famous or infamous, rich or poor. Luck was all that mattered. The luckiest hit pay dirt on a lode that outcropped along the hillside, paying from $30 to $100 a day per man. Not so lucky, but still ahead of the game, were those who uncovered placer deposits in the streambed. Running the easily separated gravels through their sluices and Long Toms, placer miners took out neat profits of $3 to $15 a day per man. Gold production from these claims ran as high as $80,000 in the first few weeks of operation. Then some of the rich placers began to play out. In addition, there were often too many owners to one claim, and in their eagerness to get rich, the men stripped the surface deposits and abandoned the claim without developing it further. Not only that, but streams began to dry up and the shortage of water to work the claims became acute. By August, only about twenty-five sluices were at work. Gloom began to settle over the little communities on Clear Creek. To make matters worse, the violent rains that soaked the gulches in July caught many without shelter and the camps were hit by an epidemic of a typhoid-like disease, spread by the bite of the wood tick. Scores of miners died of this "mountain fever" despite the appearance of the gulches' first doctor. When the epidemic was over, the doctor turned to mining but soon gave up in disgust. He made more money doctoring at $10 a treatment than he could at digging out gold.

After the brief deluge of rains, the summer drought returned, and the pinewoods turned as dry as straw. One night on the hillside above the Gregory diggings a tongue of flame flickered and ignited the underbrush. In a few minutes it had shot up nearby trees and leaped to others from bough to bough until the entire slope was afire. Miners dashed from their tents and cabins and grabbing up their shovels dug a firebreak at the edge of the camp. The wind did the rest, carrying the flames up and over the gulch away from the diggings. For a week the fire raged over the area, sweeping from one ridge to another with a roar that deafened men to all other sounds. When it was over, seventeen prospectors who had been gophering the hills and ravines beyond the north fork of Clear Creek were dead, and a score more were badly burned, having jumped on the backs of their mules and run for their lives in front of the fire with the cinders and fiery brands singeing mule and man alike.

Played-out claims, lack of water, mountain fever, and the forest fire sent hundreds of miners down the canyon of the creek to try their luck elsewhere.

Some, however, hung on, determined to make the gulch give up its promised hordes of gold. They turned to the lode claims, following the outcrops of rock and hacking out hard chunks of quartz crisscrossed with veins of yellow gold. To free the premium metal, miners who had been in on the California gold rush built crude mills of the kind used by the forty-niners. These mills, called *arrastras*, had long been used by Mexican miners. The *arrastra* was a circular basin chipped out of stone or built with wooden sides around a flat-topped rock. A heavy millstone was fitted in the center of the basin and turned slowly by a horse, an ox, or a burro, grinding the quartz chunks to the consistency of fine gravel. Crude as it was, the *arrastra* could reduce from one to one-and-a-half tons of ore a day. But even by washing this amount of gold-bearing ore over mercury, tests showed that the miners were able to extract only a small percentage of the gold locked in the quartz. It was soon obvious that larger, more efficient methods of crushing were necessary if gulch-mining was to continue to be profitable.

In September, the component parts of a steam stamp mill, the first of its kind in the region, were hauled on a sled by long-suffering oxen up the trail to Gregory Gulch and assembled there. It was a small mill with only four 400-pound iron hammers (stamps). The stamps were hoisted aloft by steampower and allowed to fall on the chunks of ore, pulverizing them. A flow of water carried the crushed ore over a table covered with copper, coated with mercury, and the free gold was recovered from the amalgam. To increase the amount of gold picked up by the mercury coating, the miners of Gregory Gulch designed a "bumping table" which was fitted with spikes and constantly moved up and down while the ore-water solution passed over it, causing the gold to settle and the lighter materials to wash away. By December two more mills were in operation, and soon the dull thud of falling stamps reverberated up and down the gulch and the once crystal-clear stream of the creek ran milky-white with mine tailings. Despite the improvement in extraction of gold afforded by the stamp mills, over 40 percent of the precious metal was still locked in the quartz residue dumped into the creek.

Winter with its freezing temperatures and sudden violent snowstorms now arrived to hamper mining operations further. The rain and snow temporarily alleviated the water shortage, but they brought other problems to vex the miners. Before the ice-forming water could be used in the sluices and stamp mills, it had to be heated. And as if that development were not discouraging enough, the ground froze, making it impossible to dig out enough quartz to work. Many men grew worried that the mining season in these gulches might be restricted to an unprofitable four to six months a year. To many of the inhabitants who had not even made their expenses after a summer of back-breaking work, the cold drafty cabins and tents took on an ugliness that not even a night of revelry at Jack Keeler's or Charley Forest's could dispel. The "unparalleled richness" headlined in the Leavenworth *Times* was not found by all comers, and it took very little to persuade the unlucky that Clear Creek was no place to spend the winter.

H. A. W. Tabor was one of these. After the first snowstorm swept down the canyon, an old mountaineer came to the Tabors' comfy cabin, where he was treated to a piece of Augusta's succulent dried apple pie, and casually warned Tabor that his cabin was right smack in the path of a snowslide that each year buried the floor of the canyon. Anyone caught in the unpredictable avalanche, cautioned the old man, would be covered by thirty feet of snow. Tabor believed every word the old man said and he wasted no time in packing up Augusta and the baby and moving to the safety of Denver. When they got to Cherry Creek, Tabor rented a room over a store with the money Augusta had made in her summer bakery and waited, like hundreds of others who fled the mountain winter, for spring to come. When it did, the Tabors once more traveled the now familiar steep bumpy road to Idaho Springs. To their dismay they found their claim and cabin had been jumped by the friendly old mountaineer who had warned them of the snowslide.

Others also returned optimistically to their diggings on Clear Creek to again face the problems left unsolved through the winter— the water shortage and the hard-to-work ores.

It was Green Russell who solved the water problem. He had returned from Georgia in the early summer of 1859 to join the rush of prospectors to Clear Creek, locating a rich placer claim in a streambed about three miles below Central City and Nevadaville, up a gulch that soon took the name of Russellville. By 1860, six thousand people were mining in the narrow declivity that was Russell's gulch, and the lack of water to operate the rockers and sluices soon became the source of violent disputes as men fought over water rights. It was the same situation in Gregory Gulch and in all the other gulches that fanned out from Clear Creek. As the shortage increased, more and more miners contracted with muleskinners to have their ore hauled in wagons and sleds to streams that had not been depleted. This costly and time-consuming practice shortly became prohibitive. Meanwhile, Russell was pondering the problem, tramping the tops of gulches seeking a solution. When he located the main stem of Clear Creek and found an abundant supply of water cascading down from the higher reaches he proposed a gravity ditch, some twelve miles long, to bring this water all the way to the mines in Gregory Gulch and southward. The plan was immediately accepted, and miners clamored to buy shares in the joint stock company formed to build the Consolidated Ditch. Not only their money went into it, so did their brawn. Miners by the hundreds turned out to help in its construction, and when the project was completed in July, 1860, Clear Creek got all the water it needed for mining, with some left over for drinking, and even for bathing.

An adequate supply of water, however, could not change the character of the Clear Creek ores. With the alluvial beds of gold-bearing gravel and the soft surface "blossom rock" stripped away, the miners went deeper into the earth in search of gold. At depths less than one hundred feet they suddenly came across ores that not only contained less gold but what gold there was in them was held in a chemical bond with sulphides to form highly refractory, pyritic ores that resisted amalgamation with mercury.

"Rebellious" was the term baffled mine-owners gave to these ores, as if by reducing the problem to one of personality it would be easier to solve. But few miners in the gulches understood why it was that the ores they took from below the surface were not returning the profits they gave promise of doing in their assay. Even those who did understand that sulphur inhibits the action of mercury were at a loss to offer a way to treat the sulphide ores. With profits diminishing, miners borrowed money from the bankers who handled their gold shipments in order to keep their mines going. In each man was the fervent hope that the next swing of the pick would bring them to a layer of soft, high-grade "blossom rock." It was a fruitless and illogical hope. Only quartz near the surface, where it could be acted upon over the centuries by the elements, decomposed to form the much sought-after "blossom rock." Now the meager profits men extracted from their mines could not keep up with their growing indebtedness. Some mines were foreclosed. Others were sold at a loss to eastern business interests who were reacting to a quickening of speculation in gold as the Civil War approached. Once on the speculation market, prices of these mines skyrocketed. Pat Casey's claim on the rich Burroughs lode, along with the more productive mines on the Gregory, Bobtail, and Gunnel lodes, sold for $1,000 a foot. Starched, cuffed, and collared agents representing corporations with such grandiosely symbolic names as The Ophir Mining Company of New York made the hard journey out to Clear Creek to pick their elegantly shod, ankle-spatted way fastidiously over the gravel dumps and rusted mining tools to inspect their new properties. The new owners of the mines in the gulches were more than willing to gamble their dollars on the development of their new possessions. They listened hungrily to all of the self-styled scientific experts who offered ridiculous suggestions, for equally ridiculous compensation, for the solution of the problem of refractory ores. These suggestions, combined with a firm belief held by capitalists in Philadelphia, New York, and Boston that it was only money that was required to get the mines of Clear Creek producing again, led to the costly transport of complicated and ponderous machinery to the gulches, including massive stamp mills, in a futile attempt to liberate the gold held fast in the sulphide ores.

To the flannel-shirted miner, high finance and intricate machinery were conundrums. All he knew and cared about was that his claim no longer paid enough to make him a profit. Many, like George Jackson, whose rich sandbar gave way to the rebellious sulphides, threw down their hand tools and left the gulches for greener, or golder, fields. With them left the symbol of good times—the gambler. In all the ragged settlements up and down Clear Creek, games of chance suddenly were nowhere to be found. "The gamblers," wrote one miner, "could not live...on hope...and so have cleared out to a man."

In the early sixties, hundreds of men in the few square miles of the Clear Creek diggings abandoned their played-out or foreclosed claims that in the three scant years since Jackson and Gregory made their discoveries had produced $2,500,000 in gold. To the residents of the new Eldorado, it looked as if the bonanza on Clear Creek was finished.

5

Fanning Out

THE FIFTY-NINERS KNEW LITTLE ABOUT GEOLOGY AND LESS ABOUT METALLURGY. What they did know was that the place to look first for gold was along the beds of streams. That was how Gregory and Jackson had done it, and with the countless streams and gulches threading the Rockies, who knew what other bonanzas waited to be tapped? As soon as Clear Creek real estate was staked out with all the claims it could provide, prospectors fanned out in all directions along the eastern slope of the mountains and over their crests to the west. Even some of the owners of high-paying claims on Clear Creek left their mines in the hands of day-labor miners and scampered west looking for the tell-tale signs of float gold in the sands of creeks. "Most in the least time" was their motivation, observed newsman Henry Villard. For others, it was the compulsion of discovery that propelled them, and once they located a claim they sold it and pushed on looking for another.

About twenty miles north of Denver-Auraria and fifteen miles north of Gregory diggings over a lodge pole trail well worn by Indians and trappers, avid gold-hunters panned their way until they turned up substantial patches of blossom rock on Boulder Creek, a branch of St. Vrain's Creek. By 1860, Boulder City had some fifty log houses, several stores, and two hundred inhabitants.

Small as it was, Boulder looked around at other prosperous camps such as Denver-Auraria and Central City with their popular newspapers, the *Rocky Mountain News* and the Central City *Register,* and decided that it could not qualify as a boom town without a weekly of its own. But how to get one was the question. None of the miners on hand was a journalist. After pondering the matter for a while, and with an ingenuity born of few scruples, they simply decided to steal one. On the pretext that the Boulder mines should be examined and a story written to verify their great richness, the miners enticed D. G. Scouten, the editor of the brisk little Valmont *Bulletin,* published downcreek in the tiny farming community of

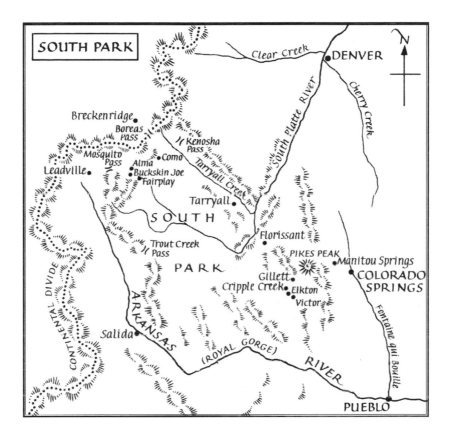

the same name, to climb aboard his mule and make the trip up to the diggings. When he arrived, the miners led him from one claim to another, showing him the high-grade ore samples and then cordially fed the inquiring editor glass after glass of 40-rod dynamite until he obligingly passed out. Then the miners hitched up a wagon, raced down the road to Valmont, loaded the town's newspress onto the wagon, and hauled it lickity-split back to Boulder before Scouten came to. They did not know it at the time, but the miners were adding one more story to the annals of the well-traveled Mormon handpress Jolly Jack Merrick used in his race with Byers to bring the first paper to Cherry Creek. After Merrick lost the race, he gave up publishing for mining and sold the venerable old press to Scouten. By now it was a rickety and cranky contraption, but it could still print and that was all that mattered to the miners of Boulder.

When Scouten woke up, he could scarcely believe his eyes. There, beside the cot on which he lay, was his press, all set up and ready to go. Surrounding him were his type cases and rolls of newsprint, neatly stacked. Beneath the window of the room was a desk. Sheaves of foolscap were carefully piled on one corner and half a dozen quill pens in a jar and a bottle of ink stood nearby. On top of the desk was a note that read "Welcome, Editor of the Boulder Valley *News!* Between the throbbing of his head and the enthusiastic persuasion of the miners who burst in on him as soon as they heard him stirring, Scouten surrendered. It was the eve of the day the paper was to come out, so Scouten quickly reset the name on the masthead, rewrote the front page feature to tell the diverting story of a shanghaied editor, and went to press. The next day, the miners proudly sent copies of the first edition of their prized weekly to Denver, over to Central City, and down to the astonished farmers of Valmont.

Other prospectors moved south and west in their search for a cure for what Byers liked to call Rocky Mountain yellow fever. One of the places that drew them was a spectacular alpine valley, lying ten thousand feet above sea level, surrounded by rugged peaks reaching above the valley floor to elevations of three thousand feet. South Park had been the scene of several waves of migrant populations long before the advent of the Pikes Peakers. The verdant grass and sparkling watercourses at the north end of the valley made it a favorite pasture for buffalo, and after the buffalo came the Ute, the Kiowa, and the Arapaho. The three tribes fought fiercely for the right of possession until at last the Ute won it for their own game preserve. In the eighteenth century, a few venturesome Spaniards crept into the area in search of gold, but they did not stay. The predatory Ute and the desolate salt marshes that lay across the southern end of this *bayo salado* drove them out. Later, mountain men, whose knack of living in peace with the Indians could well have been studied by frontier settlers and the US government, roamed at will over the seventy-five-mile-long park, trapping the highly prized beaver. It was here that Zebulon Pike's cellmate, James Purcell, had plucked gold flake from a source stream of the South Platte in 1806, and it was here that Frémont's friend "Parson" Bill Williams while on a trapping expedition stumbled onto color as late as 1849.

In the early summer of 1860, when the first prospecting parties from Clear Creek clambered up to the summits of 11,000- and 12,000-foot-high mountain passes, they were on the right track. There was surely gold in South Park, and seldom was it found in such beautiful surroundings. From the tops of the mountains where the tall pines dwindled to wind-stunted dwarfs, the miners paused to catch their breath in the thin air and to marvel at the carpet of wild iris stretched out below them. Around the long valley rose a circle of peaks, some of them snow-capped all year long, and through the valley floor gurgled sparkling streams that came together at the center to form the beginning of the South Platte River. It was to these streams that the gold-miners hurried helter-skelter down the mountainsides, gold pans in hand, past curious herds of bighorn sheep and villages of indignant prairie dogs whose shrill complaints strummed on the summer air. Once on the valley floor, the miners were not disappointed. On the first major tributary stream they came to they found sandbars that bore nearly a pound of gold a day for each man. The gold, in an unusually bright color, washed out in the form of scales "nearly as large as watermelon seeds" worth from 25 cents to $1.30 each, with an occasional nugget of heftier weight finding its way into the bottom of a pan. The miners were ecstatic as they quickly staked out claims and named their diggings Tarryall.

By August, there was a stampede to South Park. In Denver, Byers wrote in the *News* that "...A continual stream of miners have passed through our streets. Wagons, carts, pack animals, and footmen—all heading one way." Those on foot went over the mountains, finding the shortest routes possible. Those with teams left a dusty trail behind them as they scurried south for Pikes Peak and then turned west to scramble up and over Ute Pass and into South Park. But not all of the gold-hungry lived to reach the coveted diggings.

The park belonged to the Ute and, like the plains Indians, they resented the white interloper. The first word of trouble was a rumor passed around Denver that seven prospectors had been ambushed and murdered by a war party as they made their way over Ute Pass. The Indian danger was confirmed when Will Slaughter, postmaster Henry Allen's friend, straggled back to town with the harrowing story of his attempt to reach South Park. Slaughter, merchant Burt Kennedy, and

Doctor I. L. Shank set out for the headwaters of the South Platte from Gregory Gulch in early June. With a mule to carry their provisions and clothes, the three men were enjoying the warm sunlit days and mountain-cool nights on the trail, letting their thoughts occasionally rest on the prospect of uncovering a richly paying sandbar that would redeem their keen disappointment in the too-soon-exhausted claims on Clear Creek. Hiking along the mountainsides above tumultuous white water creeks, the three travelers edged upward until they reached the summit of a pass at the northeast corner of South Park. Descending on the other side, they had just made camp at the foot of a ridge of high hills skirting the valley floor when out of nowhere they were set upon by a screeching band of Ute. Slaughter was cutting wood for a fire in a grove of stubby piñon trees a few feet from the camp when he first heard the Indians' war whoop. Instinctively, he dropped his ax and dove into a shallow gulley behind the piñons, coming to a stop against the trunk of one of the scrubby trees at the bottom, hidden by the low-slung boughs. There he lay rigid with fear, listening to the sounds of death above him. When the Ute had done their bloody work and thundered away, Slaughter cautiously climbed back up to the campsite. His companions lay on the ground, Kennedy with an arrow in the stomach and Shank with one in the back—they were both scalped and the clothes were torn from their bodies. Gone was the mule and most of their provisions. Slaughter buried his friends in a shallow grave and headed back to Denver on terror-spurred feet.

Other stories of the same kind were on everyone's lips on Cherry Creek, some verified, and some the trumped-up tales of born liars. But true or not, they kept no one from beating a path to South Park.

When the vanguard of the stampede reached Tarryall they were enraged to find that the first prospectors to arrive there had changed the mining code to allow each man to claim 150 feet along the creek instead of the established 100 feet. Cursing the greed of the discoverers, the newcomers left, cautioning everyone they met to avoid that den of thieves called "Graball."

But derisive names affected the new camp not at all. In two months, Tarryall had two hundred miners' cabins, a branch of Denver's largest mercantile store, and a main street grandly named Broadway. The chief businesses of the town were keeping a tavern, dealing cards, and selling rotgut whiskey and unprotected mining claims to wide-eyed strangers. But come Sunday the miners put down their tools and shop owners closed their doors, and the camp took on the aspect of a model Sunday picnic:

> The miners are lying about loose. Some are washing and drying
> their shirts, some stretch in the sun for a snooze, some loading
> their rifles for a deer hunt, some singing snatches of Methodist

hymns, inter-larded with the chorus of Negro melodies, some are figuring on a pine log about their week's profit and loss, some are writing letters using the bottom of their gold pans for a desk, while others are deep in the mysteries of cooking.

Tarryall may have been despised by the other mining camps that sprang up in South Park, but at least she took care of her own. At the height of the boom, 150 miners got together and set aside one good-paying placer claim as town property. And they decreed that any man who was so low on funds that he could not buy a drink was welcome to come to the claim and pan out what he needed for a night on the town. For as long as the mine held out, the understanding was honored by all comers. There was no high-grading, and many a poor and thirsty man took a new lease on life after dipping into the Tarryall welfare system.

For three months in 1861 Tarryall boasted a newspaper, one they did not have to steal. William Byers, always a staunch Republican, sensed the need to do a little electioneering to bolster the ticket over in South Park in the upcoming elections, so he sent two men and a job press over the mountains to set up the *Miner's Record*, From July to the middle of September the little paper issued a weekly sheet that accommodatingly included, along with a series of eye-catching political diatribes, all of the local gossip of South Park. But after the election, in which Byers' candidates swept the field, the Denver-Auraria scribe shut down the *Record*, and South Park once again was dependent on a passing prospector, stage driver, trapper, or freighter for its news.

Even without a newspaper, it was not long until the miners of Tarryall heard about the handsomely paying gravels discovered farther up the South Platte. At the scene of the new strike was Charlie Nichols. He had given up neither gold-hunting nor townmaking after General Larimer and company did him out of "St. Charles Town." Also on hand were brash Jim Reynolds and his pal Sam Hardesty. The lines of battle between candidates or camp boss were soon drawn, but it was no contest. Nichols was in the habit of doing his persuading with words. Reynolds and his ally Hardesty favored bullets. Before Nichols could rally support for his cause, Reynolds called a midnight meeting of the miners. When the men were all assembled around the crackling campfire and Hardesty stood in the shadows, his hand on his fine-triggered Colt, coldly surveying the crowd, Reynolds stepped into the circle of the light and declared himself to be boss of the camp. Not one man protested. Charlie Nichols had lost out again. He might have wrested power from Reynolds later when Hardesty, aching to sample the pleasures of society that his first pile of gold dust could bring, slipped off south to Canyon City, got into a fight, killed a man, and was shot to death escaping from a posse. But still no one challenged Reynolds. What he said went, and when the time came to draw up rules for the new mining district and to give the camp a name, Reynolds laid down the law.

"This place ain't going to be no Graball," he said. "Every man gets a hundred-foot claim, not one hundred and fifty feet. That's fair play!"

For a man who would shortly give up mining for an easier way of making money, as leader of a gang of highwaymen, this was a memorable defense of morality, and for the first and probably the last time in his life, Reynolds was cheered.

The tent town became Fairplay and in the manner of boom camps everywhere it soon had a handful of rough-hewn log cabins, a false-fronted hotel, three newspapers, and twelve whorehouses. The residents were proud of their reputation for fairness and to emphasize their unique attribute they buttonholed anyone who would listen to tell them the story of "Banker" Hill. Tom Hill and three other men, fed up with the rebellious ores of Gregory Gulch, came along to South Park and latched onto a handsomely producing placer mine that in a very few weeks desposited hundreds of dollars in each man's pocket. With all this money came an uneasiness of soul to Hill's three companions. What if the temptations of drink and dice in the uproarious camp proved too much for them and they lost their hard-won pokes? The question gnawed at the miners until one of them got the bright idea of singling out Hill, who was a model of temperance, to hold their pouches of gold dust for safekeeping. The sober Hill agreed to be banker, and with the burden of potential moral turpitude combined with financial ruin lifted from their shoulders, Hill's three friends went back to work and play with a greater will than ever. Eventually, the time came to restock the larder of the foursome's cabin, but when the three depositors approached their banker he turned a deaf ear to their requests for a withdrawal, declaring he didn't believe in money-lending. Astounded and outraged, the miners demanded their money but Banker Hill was adamant. At length, when the coffee can was empty and the gunny sack of beans lay flat, the three miners hauled Hill up before a miners' court. During the proceedings, which were held at night around a campfire in the tradition of the town, Hill slipped away in the darkness and bee-lined it for the safety of their cabin. Once inside, he jumped into a bunk and drew the straw mattress over him, making himself as flat as possible. But the flight was futile. A few minutes later his partners burst in and quickly discovered his hiding place. They dragged him back to the campfire where he was made to produce the store of gold dust he held and count out four equal shares. When he finished, the three miners pocketed their bags of dust, but Hill turned down his share, claiming he wanted no part of such a violation of trust. Whereupon, two of his partners grabbed their erstwhile banker and held him down, while the third shoved Hill's poke of gold dust into his pocket. That bit of fair play taken care of, they set out to rustle up the first square meal they had eaten in days.

Despite the formidable range of mountains that lay between Cherry Creek and Fairplay, the gold camp was by no means isolated. The Denver-Auraria and Colorado Wagon Road Company made it easy to get there over the steep trails and passes if a man had $1.75 in toll charges. By 1870, some six thousand people had spread over the area, making the owner of the toll road a rich man.

With the passing of years, the placer mines grew from one-man panning operations to elaborate flume-fed extraction works. And as the camp grew, somehow the shining quality of fair play, of which the first residents were so proud, lost some of its brightness. The highly organized hydraulic mining operations demanded a sizable labor force, and over the vehement protests of many of the populace, Chinese "coolies" were imported to work the mines. People protested not because the Chinese would be exploited as cheap labor but because their ways were mysterious and different and they were therefore suspect. When the first groups arrived, they were compelled to raise their tents and cabins across the river away from town. The only exception was China Mary who was allowed to live in her laundry on the Fairplay side. Uneasy residents dreaded the early morning hours when the solemn, pigtailed coolies padded silently over the bridge to the tong house at the north end of town to receive their orders for the day. When two drunken Occidentals shot up a group of fifty unarmed Chinese miners, Fairplay saw fit to do nothing about it but to name the place where it happened "Chinaman Gulch."

With the creation of Colorado Territory in 1861, the region got its first formal, if sketchy, judicial system. Not that it made any difference to the remote mining camps. They had lived under their own rules for too long to bow suddenly to another authority, however tenuous that authority might be. The Washington-appointed territorial supreme court judges doubled in the thankless job as circuit judges, periodically making the perilous rounds of the gold camps to hear cases ranging from petty larceny to murder. The perils were many, some natural, some man-made: storms, predatory animals, Indians, highwaymen, vigilantes. The circuit judge could be sure of only one thing—his trip would be eventful. He might be sound asleep beneath a pine tree in the small hours of the morning when the *carcajou*, the mountain lion dreaded by the French *voyageurs* of an earlier day, sprang down from a branch overhead to shred him from limb to limb. If the judge was lucky and got to a settlement by nightfall, he might share a bed of straw in a windowless room of a saloon with several other travelers, all of whom reeked of rancid bacon and stale tobacco, to try and catch forty winks while the stamp-and-go of a raucous Mexican fandango went on full tilt in the next room. Court was anyplace the weary lawman hung his broadbrimmed hat: a saloon, an abandoned cabin, a stable, a vacant store. And the welcome he received could run the gamut from heartfelt cordiality to outright hostility.

In those early days, a judge soon discovered that the miners were their own law. And if they did not like the way a circuit judge handled a case, they had no compunction about taking matters into their own hands. Judge Bowen, on his regular rounds, came into Fairplay on the stage to find a full docket awaiting him, including a murder case. The alleged murderer, Hooper, had been rounded up by a posse and sat cooling his heels in a cell on the second floor of the town's spanking new sandstone courthouse. The trial was over in two days, and the prisoner was found guilty, but since Bowen felt there was some reasonable grounds to suspect the evidence, he gave the man a life sentence instead of the death penalty. There were murmurings of disapproval in the courtroom when sentence was pronounced, and the next morning when the judge came to the courtroom to try his next case, he found a hangman's noose pointedly placed on the bench. Bowen immediately took the hint and the afternoon stage back to Denver, leaving his clerk to bring his mule along later. As soon as Bowen was out of town, the vigilantes broke into the jail and hung Hooper from a second-story window of the courthouse in full view of the approving townspeople. In 1953, workmen repairing the floor of the old courthouse unearthed a hangman's noose and some other indications that pointed to the efficiency of the vigilantes, thus perpetuating the story, true or not.

Less fortunate was Judge Elias Dyer. In Lake County, across the mountains west of South Park, Elijah Gibbs and his neighbor Harrington squabbled bitterly over their rights to the water flowing through a nearby ditch in which they both had shares. The feud was the talk of the county and the settlers in the neighborhood took sides depending on which man they favored. One night in June, Harrington saw a flicker of flame through his cabin window, and he dashed outside to find his outhouse afire. As he ran to the ditch for water to douse the flames, a single shot rang out and he fell dead. Harrington's friends accused Gibbs of the murder. They succeeded in having him arrested and brought to trial but he was acquitted on the strength of insufficient evidence. Not content with the verdict, the Harrington faction recruited a gang of local roughnecks to form a "committee of safety" and using night-rider terrorist tactics set about to run the luckless Gibbs and all who supported him out of the county. Next to Gibbs himself, their chief target was Probate Judge Dyer who was convinced that Gibbs was innocent and said so. After Gibbs fled with his family in the face of the committee's intimidations, Dyer was seized in his home and dragged before the vigilantes gathered in an old stamp mill. The judge was roughed up and given three days to resign his office and leave town. Dyer was no coward, but neither was he a fool. He left and went to Denver. In his absence, the committee ruled the county with an iron hand, patrolling the roads, arresting anyone who displeased them, destroying public records, shooting cattle and firing barns, wrecking mines and machinery, bringing the populace cringing to their knees. In Denver, Dyer heard of the outrages in his circuit and his blood boiled. He hounded the chambers of his superiors until he finally got assurances of their backing and then boldly went back to Lake County to face down the committee. Once in his courthouse office he immediately issued warrants for the arrest of sixteen men he believed were the ringleaders, deputizing his friend Doc Dobbins to serve the summonses. Dobbins had rounded up eight of the men when he, Dyer, and the rest of the warrants were apprehended by the county sheriff obligingly working for the committee. Dyer was hauled up to face the committee who threatened him with death if he did not publicly dismiss the charges against the sixteen men named in the warrants. He was given the night to reach a decision and locked in a room across the way from the courthouse. The next morning, at eight o'clock, he was taken across the street and up the stairs to the courtroom where he called court into session and with grim reluctance dismissed the charges against the accused men for lack of evidence. The crowd who had followed the judge upstairs now moved out and down the stairs, congratulating themselves on another round won while Dyer sat in his chair, deep in thought and oblivious to those around him. The courtroom had been cleared but a moment when those on the street below saw five men steal up the back stairs of the building. Suddenly, the milling crowd in front of the courthouse heard two shots, a cry for mercy, and then another shot. The first man up the main stairs after the shots were fired found the courtroom empty except for the Judge who lay sprawled on the floor, conscious but unable to speak. A trickle of blood wound its way from behind his right ear along the line of his cheekbone. One arm lay outstretched, the hand grotesquely awry and the wrist a bloody pulp. Within fifteen minutes. Dyer was dead.

It was a few days later that Dyer's father received a letter his son had written during that long night before he was killed. "Dear Father," wrote the young lawman, "I don't know that the sun will ever rise and set for me again…The mob have me under guard…I die, if die I must, for law, order, and principle."

When the news of the crime reached Denver, the governor quickly ordered out a commission to investigate and offered a $200 reward for information leading to the judge's murderers. After weeks of inquiry no one came forward, no arrests were made, and the brutal killing remained unsolved. But out of Dyer's death some good came. The bloodletting apparently spent the violence that surged in the veins of the mob, and Lake County settled down to live within a semblance of the "law, order, and principle" the judge died for.

The fifty-niners, good and bad, were a tough lot. In the easy come, easy go atmosphere of the boom camps, life was cheap and men gambled with death with the same recklessness that they played lesser games of chance. Typical of the breed that threw in their lot with Lady Luck and ran after gold to the Rockies was Sam Porter. Boozey and bragadocious, Porter looked upon the activities of the Fairplay lynch committee, after the hasty departure of Judge Bowen, with drunken disdain and threatened to shoot the first man he saw, just for the hell of it. He did, and his bluff was promptly called by the vigilantes. They threw a rope over a pine branch, dropped a noose over Porter's head, and asked if he had any last requests.

"Yeah," said Sam, defiant to the end, "pull!" He was immediately obliged.

But if there was cold-blooded disregard for life in these roistering mountain camps, there could also be the gentle reverence of it.

In 1860, "Buckskin Joe" Higgenbotham, gophering north and west of Fairplay and its neighbor camp of Alma, found a good bed of placer gold that drew miners pell-mell from all the camps farther south. Higgenbotham's find was topped a short while later by a man named Harrison who was not looking for gold at all. On a deer-hunting expedition, Harrison spotted a handsome buck, took aim, and fired, only to see the creature pause then bound off into a thicket. Harrison followed his quarry looking for traces of blood. He never found the deer but he found the spot where the bullet struck a rock, making a small crater that glittered with bright flecks of gold. Harrison clandestinely worked his mine for weeks without telling of his discovery, stuffing bits of amalgam into every hiding place he could think of. After a while his friends thought it was mighty peculiar that a man went hunting every day and never bagged any game. So one day they followed him and his secret was out. Harrison finally sold his mine to a man called Phillips who found he had bought not just a claim of surface ore but a deep lode laden with gold. Within a year, the camp of Buckskin Joe was a rollicking settlement of two thousand people. By 1864, the mines on the Phillips lode had produced $3 million.

Besides a rich gold lode, Buckskin Joe had one other commodity that was famous from one end of South Park to the other. It was a dance hall girl whose beauty, grace, and winning nature earned her the name of "Silver Heels." A Saturday night in the swinging gold camp saw long lines of miners in freshly washed flannel shirts, with their faces scrubbed and their beards combed, waiting impatiently for a dance with their favorite girl. Then, in October of 1861, a calamity came to the booming little settlement. One of a party of drovers bringing in sheep from the south to sell to the meat-hungry miners came down with smallpox. In two weeks the disease had spread over the region, and those who could packed up and left town. Wagon after wagon loaded with settlers rumbled down the road to escape the plague. Every day saloons and stores banged closed their doors. Soon there was no one left to care for the sick and dying victims of the ravaging disease—no one except Silver Heels. Refusing the pleas of the rest of the dance hall girls to come with them before she caught the dreaded sickness, Silver Heels went from cabin to cabin nursing the feverish, delirious men who but a few weeks before had joyously whirled her around the dance floor in the bloom of life. In time, the inevitable happened and Silver Heels too fell desperately ill. Now the miners who had survived returned her kindness, taking turns caring for their idol and watching in helpless anguish as myriad pustules formed over her beautiful face. For weeks she lay close to death, and then one day the crisis was past and the disease left her and so, as a mirror showed, had her beauty. Silver Heels was disconsolate. To her benefactors however her scarred face changed nothing. What was important was that their favorite girl would live. As a token of their gratitude to the lovely creature who had risked her life to help them, the miners of Buckskin Joe got up a collection of $5,000. But on the day they went to her cabin to give her the money, Silver Heels had vanished. No one who could positively identify her ever saw her again, although there is a persistent story that for years afterward in October of every year a heavily veiled woman slipped into Buckskin Joe to appear briefly beside the graves of each of the miners who died in the epidemic, and slipped out of town again as mysteriously as she had come. The saddened men of Buckskin Joe returned the money to those who had donated it and went back to working their claims. But they were not to be done out of their tribute. To the east of town rose the ramparts of the Park range, and to the peak on which the rising sun turned to silver its snowy cap they gave the name of Mt. Silver Heels.

❧❧❧

The development of the Rocky Mountain gold rush was monitored by organized religious bodies as closely as it was by land speculators, gamblers, politicians, and Wall Street bankers. Alert to the westering tide of humanity, Methodist churchmen, from their Kansas conference, in early 1860 sent a mission to Cherry Creek headed by John M. Chivington, a great bull-throated moose of a man who could still a crowd of garrulous Pikes Peakers with one booming utterance. To carry the Word, Chivington had ten circuit preachers under him who penetrated every mountain pass ferreting out the godless with the same ardor as prospectors searching for gold. One of these itinerant carriers of the gospel was the Reverend John L. Dyer, father of murdered Judge Elias Dyer.

Hardy and fearless, Dyer was forty years old and a widower when he came west. He had spent his youth as a lead miner in Minnesota with an ear finely tuned for the call. When it came, he was ready. And so was Pikes Peak. The boisterous, wide-open mining camps rife with all the excesses so dear to the human spirit were fertile fields for the sowing of missionaries.

From Kansas to Denver, the strapping, 190-pound parson made his way on foot, stopping to preach and pass the hat whenever he found a gathering of settlers. Arriving in Denver in early 1860, footsore, broke (collections on the trail were scanty), and some thirty pounds lighter, he barely had time to catch his breath before the presiding elder sent him on his way west again to his circuit. The chunk of Rocky Mountain real estate that was to be Father Dyer's domain for an adventure-filled twenty years of itinerant preaching encompassed the wilderness of South Park, the headwaters of the Blue River to the north, and the high valley over the Mosquito range to the west where the Arkansas River rose.

Dyer's first stop on the circuit was Buckskin Joe. Here he set up shop in a store loaned to him on a Sunday by a generous and God-fearing merchant. At the last moment, Dyer spied a sign above the counter advertising "GOOD WHISKEY."

"But Mr. Hitchcock, I cannot preach under a sign like that," protested Dyer.

Hitchcock laughed heartily, "No trouble, Father; I'll just take it down. Whiskey's all sold out, anyway!"

Dyer's fire-and-brimstone style appealed to the miners and the preacher was gratified by the attendance at his Sunday meetings, and "in the face of every kind of opposition—at least two balls a week, a dancing school, a one-horse theater, two men shot," he fired up his congregations with fear and wonder, reclaiming from the fallen several repentant backsliders.

Sparing some of his hard-come-by collection cash for a miner's cabin, Dyer settled down to keep "bachelors hall." Hot bread baked in a frying pan, a sizzling venison steak, steaming coffee, and to read with his repast an issue of the *Northwest Christian Advocate* or the *Rocky Mountain News* before he stretched out on the hay-filled tick on his bunk was the treat Dyer set himself on his return from an exhausting two-week tour of his circuit, a tour made all on foot. No luxuries like a mule and wagon were his. Money was too dear.

A circuit preacher was on his own and could expect little financial help from his elders. As in most political bodies, it was easy to get an appropriation authorized in conference but the funds could be a long time coming, if they came at all. That first year, from early spring to fall. Dyer tramped nearly five hundred miles over Indian trails and over paths he cut for himself, in fair weather and foul, preaching three times a week, at nights and on Sundays, and working at any job he could find by day. At the end of his journeying, his Kansas-style city clothes were in rags, his hat was awkwardly patched with antelope skin, and his boots were half-soled in makeshift fashion with rawhide. For all his efforts, his collections came to $43.

Not that money was what Dyer prized. Poverty went with preaching. But a man had to keep himself somehow to carry forth the Word. So when winter came, the wily Dyer made himself a pair of skis and wangled the mail contract from Buckskin Joe over the Mosquito range to the boom camps on the Arkansas. For a once-a-week trip of thirty-seven miles, nearly straight up and straight down, carrying a thirty-pound pack, Dyer got $18, and managed to have time left over to do his preaching. The trip became considerably more worthwhile when the minister-mailman was asked to carry shipments of gold dust out of the mines destined for Denver and transshipment east. At this time, an ounce of dust brought $40 in greenbacks, the paper currency unbacked by gold or silver that Congress saw fit to print to raise ready money for the Civil War. This was double the price gold dust could be redeemed for in gold coin. Some miners jumped at the chance to make this kind of profit on their gold, despite the risk of capricious fluctuation of the greenback as the fortunes of the North rose and fell. And to them it was worth $5 a shipment to have the capable, trustworthy Dyer act as their express. To Dyer, $5, even in greenbacks, was a bonanza, and he was always ready to oblige.

One trip in the searing cold of the high mountain passes over a snow pack from five to twenty feet deep, in the face of sudden, knife-edged blizzards, and with every step the ever-present threat of a snow-slide, was generally enough for the average man. But Dyer was not an average man. And yet for all of his ability the elements sometimes came close to doing him in. It was on an unsettled day in March, with the sun now and then breaking through a light snowfall, that Dyer began one of his return trips to Buckskin Joe. He set out about one in the afternoon, climbing slowly upward through the sticky, thawing snow that clung tenaciously to his skis, creating an extra weight that soon sapped his strength. Pausing to rest, he looked back and saw above and behind him a menacing black cloud and with it came a blast of freezing wind. Dyer knew well the signs of an approaching storm and quickly checked his bearings against the mountain peaks on either side of him. But when the storm struck a few seconds later, the preacher was enveloped in a white mass of wind-driven snow that blotted out everything before his eyes. The wind whipped the breath from his throat and pinned him gasping against a boulder. Struggling to cover his mouth with a mittened hand so he could breathe, Dyer steadied himself and fervently prayed for help. Strengthened by his spiritual toddy, the parson carefully bent down and removed his skis, and holding them out in front of him as a wind shield he haltingly started back down what he thought was the trail, stopping to rest every hundred feet or so. He had gone but a few hundred feet when the ground abruptly clipped under his sodden boots and he found himself on a ledge of snow. In a split second his mind knew what lay below—space for a thousand feet and then a rocky canyon studded

with pine trees. His heart raced as he gingerly moved backward, wondering wildly if the ledge would hold his weight. At last he was on
firm ground and suddenly he was perspiring. He bumped up against
a wall of rock and sank to the ground on his rubbery knees, clinging
to his skis stuck upright in the snow to their bindings. After a few
minutes, Dyer got to work, digging himself a hole in the drifted snow
and climbed in to rest, sheltered partially from the lacerating wind.
He was numb from head to toe and he soon felt a warm and pleas-
ant drowsiness overtaking him as he feebly wiped away the swirling
snow from his face. He could not tell how long he fought the urge
to close his eyes and slip into beckoning sleep, but finally the wind
began to drop and he could see through the snowflakes to the dark-
ening shapes of the mountains around him. Dyer floundered his way
out of his cold cocoon and staggered upright, stamping his feet to
get the circulation going. Then he strapped on his skis and headed
for a familiar ridge off to the right. He knew there was an old Indian
lodge-pole trail on the other side that zigzagged down the mountain
to Four-Mile Creek just south of Buckskin Joe and Fairplay. After
slipping and sliding down among fallen trees and drifts of new snow,
Dyer came to the creek where he took off his skis and waded across.
On the other bank, he got onto his skis again and started north on
the gently sloping fan that led to Buckskin Joe, his feet now totally
without feeling. It was nearly midnight when Dyer dragged himself
into his little cabin, with his valuable pack of mail and gold dust still
intact. When he pulled off his boots, he saw the telltale whiteness of
his feet. He quickly plunged them into icy creek water but it was too
late. They were badly frostbitten and in the warmth of the firelit cabin,
the pain soon became excruciating. Dyer crawled into his bunk and
lay in agony through the rest of the night. The next morning one of
his congregation came by and when he saw the parson's predicament
he dashed out to cut a balsam sapling and strip it of its bark, making
a soothing poultice for the parson's throbbing feet. Three weeks later,
minus toenails and quantities of sloughed-off skin, Dyer was on his
feet once again carrying the mail over Mosquito Pass.

It was not the elements that were the bane of Dyer's long existence among the carefree gold-boom camps. It was drink and dancing.
Fighting valiantly but futilely against the indulgence of his congregations in these engaging ice-breakers, more often than not among the
generous miners, the parson came away richer in dollars, if poorer in
spirit, after an encounter on the battlefield.

On a snowy Christmas Eve, Dyer was invited to dinner and a
dance at the board-front hotel that graced the camp's main street.
Since a square meal was a treat even a man of the cloth could occasionally allow himself within the constraints of self-denial, Dyer
accepted. But he made sure to duck out before the dancing started.
His departure brought genial words of reprimand but the preacher

was unswayed and went off to his nearby cabin where he tried in vain to sleep as the vibrant strains of a fiddle and the thud of dancing feet hammered in his head. About daylight, there was a banging on his door and shouts of "Open up!" rocketed over the still crisp air. When the groggy Dyer opened the door, he saw the crowd of dancers, sniggeringly drunk and ready for devilment. One of the revelers lurched forward, grinning at the solemn, nightshirted figure of their preacher.

"We've decided you must come up to Walker's saloon and deliver us a temperance speech or treat us all to a drink!" crooned the speaker. The crowd roared its approval and Dyer, seeing "they were too drunk to be moderated," went along with them.

At the saloon, he climbed upon a box as the giggling, jostling miners gathered around him. Dyer let go his stentorian baritone and silenced them.

"Gentlemen, there is not a man here who would not be shamed to have his mother and father see him in this condition [cheers and stomping]. And if we were not so drunk we would not be here [more cheers and stomping]. And if we were any drunker we could not be here [even more cheers and stomping]. Therefore, let us go home while we can and save the shreds of our immortal souls!" (Wild cheering and ear-splitting stomping.)

While the gleeful miners swarmed around the preacher, slapping him on the back and pumping his hand up and down, someone passed the hat, and Dyer went home to breakfast with forty more dollars to carry on the work of the Lord.

Bad as drink was, of all the evils in a community in Dyer's opinion, dancing was the worst. Too many times the roving preacher had seen women persuaded to leave their children alone in order to be whirled around the dance floor a few giddy times, and too often had he seen a man run away with another's wife, all as a result of one night's fling at a cotillion. Venal men "leading silly women away from their families" to the dance floor was a sight that invariably prompted in the preacher an immediate and heartfelt text on damnation. And Dyer firmly believed in damnation: "...convictions are slight and conversions are not so clear as they ought to be when our preachers... 'snugly keep damnation out of sight,' " warned Dyer. "...[A] minister ought to present both sides, so that the sinner shall fear the torments of hell as much as he desires the glories of heaven. Men must believe there is danger, or they will not start."

In the early 1860s when war clouds drifted westward over the Rockies, it wasn't hard to get a man started over the danger. There was the pulsing danger of internal violence between southern and northern miners peopling the booming gold camps. And the imminent danger of a Confederate invasion of the gold-rich Rockies was more than idle rumor. Into the lives of the hard-working, hard-living miners, whose thoughts ran preferably to the price of gold, whiskey, and women, crept the question that stirred the blood in their veins with almost the same rapidity: slavery. Men who had fled its confrontation to the wilderness of the mountains now found that even in their remote, granite bastions the question had caught up with them. Old loyalties were rekindled, traditional prejudices bubbled to the surface, and the Civil War reached out its tentacles into the fingered fortress of the Rockies to suck up its victims.

Jim "the Bold" Reynolds was one of the first to succumb to its pressures. As camp boss of Fairplay, Jim aired his southern sympathies with little or no restraint. If he had bullied men with any other subject of conversation, he probably would have met no resistance. But war fever, like gold fever, supercharges men's sensibilities and Jim suddenly found himself in a Denver jail accused of sedition. With the help of his brother, John, and seven others, he escaped, making a beeline with his *compadres* for Texas. There, Reynolds and his gang reported to General Douglas H. Cooper, CSA, and on April 11, 1864, they left Texas augmented by fourteen additional men, with permission to raise troops for the Confederacy in Colorado. To Reynolds, it seemed only logical that while he was at it, he should raise some money for the South's cause. Their first raid was on a wagon train of freight at Cimarron Crossing on the Arkansas River. The take was $1,800 in gold specie and about the same amount in greenbacks. On June 12, the raiders split up, with Jim leading eight of the crew into his old stamping ground, South Park. On the road to Fairplay, Reynolds captured the manager of the fabulous Harris lode, relieving him gently but firmly of his strongbox. A few days later, just as stage driver Abe Williamson got his stubborn mules to break into a respectable trot on the daily run from Buckskin Joe to Alma and Fairplay, he saw a group of horsemen blocking the road ahead. Williamson jammed his foot on the brake, his curses lost in the jangle of harness and braying of mules as the coach skidded and corkscrewed to a stop in a cloud of powdery dust. Williamson was pondering why it was that the stage had to be held up the one time he carried stage-line-owner Billy McClelland as a passenger when he recognized the leader of the gunmen.

"Why, Jim Reynolds, you rebel bastard!" exploded Williamson.

"Why, Abe Williamson, you Copperhead lecher!" roared Jim the Bold as he came up to the stage. He peered into the windows of the coach. "Get down from there, Williamson, and bring that rascal McClelland with you!"

Fresh in Reynolds' mind was the sound-alike name of the Union general who a year before had stopped Lee's drive north at Antietam in one of the bloodiest battles of the war. Reynolds and his gang robbed both men. McClelland's pockets gave up $400 in currency and a gold watch. From the boot of the stage they got $3,000 in gold dust. Despite his fiery temperament, Reynolds was no killer. He kept his prisoners with him for a day or two until they cooled off and then he released them, unharmed.

The guerrilla leader was in his element, pouncing on wagon trains, stages, and lone riders with cheerful civility mixed with dire threats if his victims did not do as he instructed them. As their loot grew with each encounter, Reynolds decided raiding was a lot more fun and profitable than recruiting. He made no attempt to add to his followers and always released his prisoners within a few days of their capture. Invariably, every time Reynolds released a man, the captive scurried to the nearest camp to raise a posse to chase the rebel raiders. After a few weeks of uninterrupted success, the gang got its retribution. At dusk, one sultry day when smoke rose straight upward, they were surprised in camp at their cook-pots by a large posse of irate miners on horseback. There was a rapid exchange of fire, and the gang scattered. When the posse moved in they found only one of the raiders, and he was dead. The leaders of the posse cut off his head and bore it triumphantly into Fairplay where they got the town doctor out of bed to pickle their trophy in alcohol for the whole town to see in the morning.

Meanwhile, the Reynolds brothers had slipped away under cover of darkness and split up, planning to rendezvous after their pursuers gave up. But the posse did not give up. It was reinforced by two companies of cavalry, and within a few days one more of the rebels was caught. He was threatened with an immediate date with the noose unless he talked. He talked and led his captors to the gang's meeting place. The posse took up their hiding places and when the rebel band collected they captured four more, including Jim Reynolds. Two others escaped into New Mexico. The captured four were hustled into Denver and when the pockets of the guerrillas were searched, out came an oath of allegiance to the Confederacy. That made Reynolds and his bunch prisoners of war and they were promptly turned over to the Army.

The officer in whose custody they were placed was none other than John M. Chivington, the Methodist elder turned soldier. When the war broke out, Chivington was offered the post of chaplain of the First Regiment of Colorado Volunteers, but the bellicose preacher turned it down in favor of a combat role. When the Colorado First fought a campaign in New Mexico, he had proved himself to be as capable in battle as in the pulpit and was now a colonel. When the Reynolds gang was safely locked up in the guardhouse, Chivington petitioned Army headquarters in Kansas for authority to try the rebels and shoot them if they were convicted. Instead, he was ordered to conduct them under military guard to Fort Lyon for trial. The refusal of his request rankled the portly Chivington, but like a good soldier he carried out his orders—almost. On the first of September, a large detachment of cavalry set out from Denver with Jim Reynolds and his men in tow, headed eastward toward Fort Lyon, 240 miles away down on the Arkansas River near Bent's Fort. The soldiers drew rations for the nine-day trip. No rations, however, were drawn for the prisoners, and they never reached Fort Lyon. Not far out of Denver, near the head of Cherry Creek, so stated Chivington's official explanation, Reynolds and his men were shot and killed as they tried to make their escape. However, some time later, Uncle Dick Wootton, in traveling the lonely road taken by the soldiers escorting the rebels, saw a flock of ravens wheeling over a grove of cottonwoods not far from the trail. Trotting over to investigate, he found four skeletons tied to trees, their bones stripped of flesh, and free of clothes except for the boots they stood in. In each skull was a single bullet hole.

That was the end of the Reynolds gang's "reign of terror" in South Park. But the legend of their exploits lived on. It was and still is rumored that out of Fairplay, up Handcart Gulch, lies buried an immense cache of gold taken by the rebels in a raid on a wagon train in New Mexico while the gang was en route to South Park from Texas. To date, treasure-hunters have followed up clue after clue, including a dagger-marked tree, to no avail. If there is a fortune buried in Handcart Gulch, it is still safely hidden.

❦

North of South Park lies a region the Ute called *Nah-oon-kara*, translated roughly as "where the river of blue rises." Like South Park, the lush green upland valley with its north-flowing river was a favorite camping place of the Indians, and they objected strenuously as the antlike parade of gold-seekers spilled over Hoosier Pass from the placer camps in South Park to work the gulches of the Blue River. Ruben Spalding and thirty other prospectors were the first men to work the gravel beds, turning up 25 cents to the pan. This was enough to keep them there and without delay they whipsawed lumber from the stands of pine on the high sloping hills on either side of the wide river banks and set up sluices. As if drawn by a magnet, countless others cascaded into the scenic valley walled on the west by the jagged Ten-Mile range. Some came from the south, others from the site of Jackson's diggings on Clear Creek over a well-traveled Ute lodge-pole trail. By October of 1859 there were nearly a hundred miners on the Blue, living in tents and lean-to's and keeping a wary eye on the encampment of sullen Ute nearby. The uneasy whites left off placering long enough to build a log blockhouse for protection in case the owners of the land they mined decided to attack. Nautical fashion, the fort was named the "Mary B" for Mary Bigelow, the first woman in camp. The Ute however did not provide Fort Mary B with an opportunity to sail into history as the scene of a desperate struggle between whiteand red men, and by spring the settlers decided that it was safe to make the diggings a more permanent settlement. Ten log houses went up nearly overnight on a neatly platted townsite. The miners tacked the name Breckinridge on their camp in honor of the vice president of the United States in the fashion of the day that called for naming towns after prominent political figures with the hope of a little patronage. But the defection of their namesake to southern colors at the outbreak of the war was too much for the Union-favoring miners on the Blue so they changed the town's name to Breckenridge, not much of a change but enough to take the taint away. So rich in gold were the broad banks on which the camp grew up that a man could make a day's wages by panning the dust in the streets.

To Breckenridge in 1859 came diminutive, twinkling-eyed Judge Silverthorn to hold the first miner's court. He liked the camp so well that he and Mrs. Silverthorn built a huge log cabin and opened a hotel, covering the rough wooden walls with colored lithographs and newspapers that were avidly read by the news-starved miners. The feature of the hotel that made it a mecca for weary travelers was the downy soft feather beds covered with genuine sheets that were on every bunk.

Popular as it was, the Silverthorn Hotel had a flourishing competitor in the form of Alexander Sutherland's public house. Sutherland, a Scottish emigre and veteran of the Crimean War, set a table fit for royalty, with the mainstay a delectable oyster stew. Oysters, for some unaccountable reason, were the absolute favorite food of the rough and ready mining population. Later, as the settlements in the Rockies multiplied, eastern markets would be hard put to satisfy Colorado's voracious demand for the tasty mollusk. In the meantime, Breckenridge was one of the few places a man could satisfy his craving, and Sutherland's oyster stew was renowned from Denver to Salt Lake. Equally renowned was the way in which he announced dinner was served. Down the valley of the Blue River every evening at the appointed hour rolled the stirring notes of "Peas upon a Trencher," blown by Sutherland on the very same bugle he used to signal the charge of the Light Brigade at Balaclava on that fateful October 25, 1854.

Prospectors spilled out of Breckenridge, up French Gulch, up Black Man's Gulch, to the surrounding districts of Pollard, Independent, and Spalding, where a dozen or more mining camps mushroomed. The upstart town of Parkville set out to outshine Breckenridge as the major camp of the valley by setting up a mint, enticing Jack Langrishe to bring his gold camp repertory troupe to town on a monthly basis, persuading the politicos to name it the county seat, and inviting Parson Chivington over to dedicate the second Masonic lodge established in all of Colorado in a gala celebration that drew the elite of the pick and pan society on both sides of the Park range. But Parkville got its comeuppance, so the story goes, when the energetic

Mrs. Silverthorn suggested in 1862 that the people of Breckenridge act to make it the county seat by the simple expedient of stealing the county records from Parkville. That is exactly what they did, and from that time on, Breckenridge has gone unchallenged as the seat of the county.

One of the first men over the mountains to the Blue River Valley was Harry Farncomb, who poked around the banks of the river and turned up a half dozen or so pans of gold-bearing sand. But he wasn't satisfied with $2 pans of gold dust when there might be some pockets of pay dirt feeding the dust into the stream. So he started tracing his find. The trail led up French Gulch in a tortuous route of boulders and underbrush, but at the head of the gulch on a small rise Farncomb found what he suspected he would, and more. Lying just below the surface in one of its rare forms of occurrence was a lode of pure, crystallized gold looking for all the world like a tangled, twisted strand of glittering wire rope. Harry kept mum about his "Wire Patch" and quietly bought up the land adjacent his claim. But it wasn't long until his strike was discovered and miners rushed up Farncomb Hill to get a part of the fabulous ore for themselves. Harry ran them off at the point of a gun. The miners retreated, armed themselves, and tried again, but Harry had his own army by then and once more after an all-day pitched battle in which three miners were killed, they were driven off. For ten years the war to control Farncomb Hill kept the undertakers busy while Harry blithely took thousands of dollars out of his mine, successfully dodging all bullets aimed in his direction. Out of Farncomb's Wire Patch lode, so legend has it, came one of the largest nuggets found in the Rockies, weighing 33 pounds 7 ounces. Finally, in 1890, the mine was syndicated and Harry Farncomb, his pockets bulging with gold, happily retired from his embattled hill.

When the placer deposits that were worked by hand began to play out, things got a bit dull around Breckenridge. The nearby, smaller camps were abandoned one by one. Out-of-work miners and drifters, waiting for the next boom, hung around the Silverthorn Hotel, passing the time with a little poker, a little tanglefoot, and a lot of stories about the days when the Blue River diggings rivaled the richness of Gregory Gulch on the other side of the mountains. But soon even card-playing, whiskey-drinking, and story-swapping palled as amusements and boredom made its insidious way through the town. So it was that when "Captain" Sam Adams stepped down from the stage that rumbled to a stop at the porch of the hotel on a day early in July in 1869, he was attended by a devotedly curious crowd.

The unctuous, affluent-appearing Adams, who at first passed off questions about his business with a patronizing smile and winking eye, was quickly lionized by the town fathers of Breckenridge. He met everyone who was anyone, and that was his intention. Over cigars and a glass or two of whiskey in the homespun parlors of the town's most imposing residences, Adams skillfully wove the future of his hosts into the fabric of his reason for coming to Breckenridge. As he confidentially told his newfound friends, there was still on the books of the US government a sizable bounty to be paid to anyone who found the Northwest Passage. And, as Adams triumphantly told his hosts, he had found it—a negotiable waterway stretching all the way to California. All the people of Breckenridge had to do to share in the reward, according to Adams, was to finance the building of four boats and the purchase of supplies for the voyage. Adams was sure that he would have no trouble finding ten hardy volunteers among the miners of the Blue River diggings to come along as crew. He was right. He had no trouble finding the men or the money. He even found a dog to serve as a mascot. His extraordinary plan and persuasive manner caught up the gullible bonanza town in a fever of projected fame, and everyone dug deep into their diminished piles of gold dust in the hope of putting Breckenridge back in the black.

Carpenters, long idled by the lack of demand for houses and sluices, jumped at the chance to build the boats. The women of the town held a marathon sewing bee to appliqué a huge banner that read "WESTERN COLORADO TO CALIFORNIA—GREETINGS!" The backers of the project kept a constant vigil at the boatyard, noting with pride and excitement the record progress of the construction. In mid-July, the gala day came. With the gauze bunting flapping gently in the mountain breeze, and the quavery notes of the town band intermixed with rocketing cheers of the crowd drowning out the uplifting words of the send-off committee, Captain Sam's fleet set sail down the Blue, headed for the Colorado River and California. But only the start was auspicious. The rest of the voyage was a disaster. On the way down the Blue, the going was so rough that half of the landlubberly crew jumped ship, including the dog. The rest sailed on, determined to master the tricky channels and rapids. But on August 9, the entire expedition and the hopes of Breckenridge were dashed on the rocks in Glenwood Canyon, some 150 miles from the armada's launching site.

The bedraggled Adams came back to the mining town long enough to bemoan the cruel fates that had played havoc with the expedition and to console his depressed backers with the vow that he would go to Washington posthaste and demand the bounty for the courage Breckenridge had shown in joining him in this desperate but nationally important effort. That was the last Breckenridge saw of Captain Adams. He was however true to his word. He did try to collect the prize from Congress but whatever attention he might have gotten on the strength of the absurdity of his venture had disappeared with the completion of the nation's first transcontinental railroad that spring, and his petition was laughed into the wastebasket.

After four years of production, the placers of the Blue River were stripped of their surface gold and the mining camps swiftly became ghost towns. But Breckenridge rebounded to life in the seventies when lode gold in substantial quantities was found in the mountain slopes on either side of the valley. Shaft mining and hydraulic placering took the place of Long Toms and sluices. In seventy-seven years of operation one of these lode discoveries, the Wellington, produced $32 million, an average of about $415,000 a year.

While he prospected for souls on his travels among the peaks and valleys of the South Park and Blue River, Father Dyer, whose early experience in the lead mines of his home state came in handy, did not keep his eyes heavenward so long that he could not catch the glimmer of a piece of pay dirt beside his path. On one of his trips up Tarryall Gulch, over Boreas Pass, headed toward Breckenridge, he found a lode that assayed at six thousand ounces a ton. That helped the collection box some considerable amount and made possible the first church in Breckenridge.

⚜

A third boom sustained Breckenridge in the column of living towns when, in 1898, Ben Stanley Revett thought up a way to fish out the gold he knew was still buried in the beds of the Blue and its tributaries. He dammed the river to form a pool of water, and in this pool Revett floated his first "gold boat," a huge dredge whose chain of buckets scooped up tons of river bed, passing it over filters that separated the gold from the gravel. Gold dust was deposited in pontoon boats tied behind the dredge while the scoured gravel was disgorged in gravelike mounds along the banks of the river. When one part of the river was thoroughly dredged, the dam was moved and the dredge floated to the new pool to begin the process all over again. A dredge could eat its way through a segment of river bed one hundred yards wide and eight feet deep, producing $20,000 in gold a week. By 1910 five of these mechanical monsters worked the Blue and Swan rivers and French Creek. For thirty years afterward, the valley echoed to the constant clank of the buckets and the rattle of the falling gravel. Before the leviathans were finished, $35 million was recovered from the streams that were once thought to be entirely played out.

⚜

In the fall of 1859, A. G. Kelley, prospecting westward out of Denver-Auraria, tarried not at Tarryall, nor did he turn north to follow the South Platte to Fairplay, but instead hurried on to the west end of South Park, clambering over the 9,300-foot Trout Creek Pass in an early snowstorm, and dropping down into the narrow north-south valley formed by the headwaters of the Arkansas River. Moving northward along the river, he turned up a good sample of float gold about fifteen miles above the exit of the pass. He marked the location and scuttled back to Denver before he became snowbound for the winter. In late February he could wait no longer so he rounded up twenty-four other eager argonauts and trekked back to his claim on the Arkansas. The others staked out claims and the party formed a mining district they called Kelley's Bar. When another party happened on the scene, the two groups joined forces and started moving upriver still farther to the north. When they came to a series of gulches and streams fanning out on the east side of the river some twelve miles north of Kelley's Bar, Abe Lee, a gold-wise Georgian like Gregory and Green Russell, decided this was a likely spot to look for color. Lee and the others started up one ravine, struggling through snow drifts four feet deep, stopping every now and then ostensibly to pan the creek but actually to rest. At an altitude of ten thousand feet the cold air was so thin it seemed as if they could not get enough of it into their lungs to breathe. Their chests hurt and their heads ached, some were dizzy and nauseated, but slowly, doggedly they pushed on up the gulch. On the morning of April 26, when Abe Lee wearily sat up in his buffalo robe bedroll, the sun was already up, its bright, warm rays melting the snow-shelf overhanging the clear mountain stream that coursed its way down the gulch. Lee blinked his eyes and sleepily watched the drops of melting ice plop into the water. And then he opened his eyes wide and scrambled for his gold pan. Winking up at him in a still pool of water set apart from the main stream by a beaver dam was a telltale glitter. He scooped up a pan full of gravel and washed it quickly but carefully in the stream. As he poured away the water, bits of gold flake covered the bottom of the pan.

"Hey, boys, come up here!" Lee's excited shout bounced from rock to rock down the gulch. "I've got all of California right here in this pan!"

The others came as fast as they could up and over the snow-covered terrain to see Lee's discovery. There was no doubt about it. It was a rich sample and there was more where that came from. That afternoon, the miners shot off their guns to signal the stragglers farther down the gulch to hurry on up. Then they built a gigantic bonfire, the damp pine logs streaming blinding white smoke, and hurriedly made a camp of bough-thatched lean-to's. The next day "California Gulch" became a mining district with Abe Lee as its recorder. By midsummer the camp had spread for five miles up and down the gulch and was overrun by five thousand people—a thousand people per mile, most of them frenetically sluicing out the yellow metal from the pay dirt that generally lay under an eighteen-inch-thick layer of loosely cemented gravel. The least a man made was $10 a day, and the luckiest took out $100 a day. In the extravagant terms people use to characterize gold booms and that carry over magnified even more into legend, the figures for that first season's production are given as $2 million total, with some especially rich claims giving up $80,000 to $100,000 each. People in the gulch put in a giddy summer, and it was not just a result of the thin air. Even if the production figures are discounted, California Gulch turned out to be the richest placer deposit ever found in Colorado.

Even hard-luck H. A. W. Tabor latched onto a paying claim in the rich gulch. After the indignity of having his claim jumped at Idaho Springs by the friendly old mountaineer, Tabor packed Augusta and young Nat into the well-traveled wagon and followed the stream of prospectors into South Park in the early spring of 1860. The Tabors stopped first at Buckskin Joe, then at Breckenridge where Tabor went gold-hunting and Augusta took in boarders and baked bread and pies to sell to the local miners in order to keep the family solvent. But as soon as Abe Lee's discovery filtered over the mountains, the gold-itchy Tabor uprooted his brood and with two other men set out for California Gulch. Although he was unaware of it at the time, each westward step—from their homestead in Kansas, to Denver, to Idaho Springs, to South Park—was taking Tabor closer to a bonanza that would make him a legend in his own time. To his twenty-seven-year-old wife, this latest move was but one more bone-jarring goose chase. But Augusta was steadfast and loyal, and she never let Tabor hear one complaint. They tramped for miles across South Park, looking for the pass Father Dyer described that would take them over the Mosquito range down into the new diggings. But search as they might, the entrance to the pass eluded them. Augusta poured out her misery in her diary:

> The fourth day in the park we came late at night to Salt Creek. Tried the water and found that we could not let the cattle drink it, neither could we drink it. We tied the oxen to the wagon and went supperless to bed. The night was very cold and a jack came to our tent and stood in the hot embers until he burned his fetlocks off.

The next day the men left Augusta to guard the camp as usual and set out to look for the trail over the mountains. A spring wind whined around the tent all day, gusting funnels of powdered alkali through the camp. The cattle lowed mournfully in their thirst and Baby Nat fretted at being cooped up in the dust-ridden tent. No one passed by and for as far as Augusta could see there was only the emptiness of the valley and the silent mountains, impersonal and coldly blue in the distance. Darkness came and still there was no sign of Tabor. "The men had not arrived when night's shadows gathered round," wrote Augusta plaintively, "and I felt desolate indeed. The little jack came into the tent, and I bowed my head upon him and wept in loneliness of soul."

It was well into the night when the men returned, still without having found the trail. After a day or so more of dispirited searching, the Tabors gave up and were packing to head back across South Park when a prospector came into camp and gave them precise directions on how to reach the new diggings. Once more the mobile Tabors were on the road, unloading and loading their wagon, as they pushed and hauled the reluctant oxen and their cargo over the narrow rocky ledge that was Mosquito Pass. Many times Augusta, carrying her baby, had to stop, gasping for air, to let her racing heart settle down, but not daring to look over the side of the trail at the pine-covered floor a thousand feet below. At last they were over the mountains and at the Arkansas River where they made camp. Taking a day to recover their strength, the Tabors then pushed north to California Gulch to be among the first arrivals after Lee's lucky strike.

As they made their way up the crooked trail through the gulch, Augusta saw the familiar sights of settlement she had seen on Clear Creek and in South Park—wagon boxes converted to makeshift shelters, pine bough lean-to's, one or two canvas tents, and everywhere on the stream men busy at their pans and rockers. At the sight of Augusta, a miner let go a cheer that brought a dozen others running to see the new strike.

"Did you raise color?" shouted one man.

"Hell, no, better'n that. I raised me a cook!"

"Where?"

"Right there," said the man who had first cried out, pointing at the Tabor wagon. Augusta beamed, the blush on her thin cheeks reflected by the faded pink sunbonnet shading her rather weak eyes.

"Why, Miz Tabor," spoke up one miner who had sampled Augusta's table in Idaho Springs, "welcome to Boughtown!"

"Boughtown!" croaked the first man. "That ain't no name for a diggings as rich as we got. This here is Oro City, ma'm!"

The news that there was a woman in Oro City was quickly passed up and down the creek, and in their delight at having a female in their midst, especially one who could cook, the men of the camp dropped their mining tools and spread out over the pine-covered sides of the gulch cutting green logs to build their new residents a proper cabin. The 16 by 32 foot structure went up in one day. While half the miners laid up the building, chinking the cracks with mud and laying a roof of poles covered with bark and sod, the rest fashioned a table, a rough sideboard, and a set of three-legged stools.

The amiable Tabor basked smilingly in the light of the attention the doting miners showered on his young wife, and to return their favors he slaughtered the oxen that had brought them all the tortuous way from Kansas and declared Tabor's boarding house open for business. Augusta tied on her gingham apron and turned out the first decent meals the camp had tasted in weeks.

Once Augusta was set up in business with hungry miners paying $8 a week for vittles, Tabor went prospecting. A few days later, about twelve hundred feet up from Lee's strike, he turned up a considerable pan of gold flake. Tabor promptly recorded the claim and set to work. At night, after her kitchen chores were done, Augusta pitched in to help her husband clean out the sluice box where they found "fine gold in an abundance of black sand." All along the gulch, everyone was complaining bitterly about the black sand. It was nearly as heavy as gold and would not wash out readily, so that to get the gold out of the sand there was the long and tedious process of picking the grains of gold out with a pair of tiny iron tweezers. Nevertheless, there was enough money between the profits of the boarding house and the mine to send Augusta east in September to buy the Kansas homestead, a parcel of new clothes, a team of mules, and a couple of wagonloads of supplies to stock the store Tabor decided to open in Oro City.

After Lee's bonanza it was only a matter of weeks before the winding gulch seethed with thousands of newcomers. Web Anthony, who came over from Central City to see for himself what truth there was to his brother's extravagant pictures of the richness of the diggings, left his team at the bottom of the gulch and walked up its crooked street "walled up with log Palaces." He found Hinckly & Company's express office, numerous gambling houses, and Tabor's store (with real glass windows) where the popular Augusta ruled as postmistress and keeper of the scales, weighing out each man's daily take for him. As one who knew the virtue of frugality from having had to practice it all of her life, it pained the postmistress to see the miners get their gold weighed and promptly go out on the town and "spree all night and return dead broke in the morning to commence again." She would never be able to get used to the idea of sudden and seemingly unlimited wealth accruing without the dint of hard work.

Tabor's store did well, but that was not surprising. Supplies were hard to come by and there was a constant demand. Every item of clothing and food (except for wild game), of tools and machinery, had to be freighted into California Gulch over mountain roads that climbed nearly straight upward along the edge of precipices and down nearly vertical grades. In summer, Tom Truitt's wagons carted in what the storekeepers needed. In winter, men on showshoes, traveling mainly after dark so they could walk on top of the crusted snow and not sink into it, brought in whatever supplies they could carry on their backs. The freight charge was 20 cents a pound. Tabor, selling flour for $1 a pound, bacon for 50 cents a pound, and sugar and rice for 40 cents a pound, made a comfortable profit. And it was a sight easier work keeping a store than bending over a Long Tom all day, especially since his mine showed daily signs of pinching out.

Day and night the log palace saloons and sporting houses were clogged with hardworking, hard-drinking, and hard-gambling miners. It was a simple life with simple pleasures, and when the money ran out all a man had to do was to go wash out some more. So rich was the mine opened by Ferguson and Wells, two hardbitten gold gophers from California days, that an enterprising saloon owner built his spa right next to their claim so the men would not have to go but one step off the property to drop their daily poke.

Curiously enough, there was little lawlessness in Oro City. Mining claims were the only pieces of property considered worth recording, and every other transaction was governed by an unwritten code of fair play. As long as a man minded his own business and did not infringe on another's, he was welcome and treated as a friend. The only killing during those first boom days in Oro revolved around a disputed claim. One of the laws of California Gulch mining district was that no man could hold a claim by proxy. A man named Kennedy defied the ruling and recorded a claim in the name of his son-in-law in the East. Before the son-in-law arrived at the gulch to assume ownership, two men jumped the claim. Kennedy loaded his shotgun and waited for the interlopers to start mining. When they appeared, he raised his gun but one of the two miners was quicker and Kennedy fell dead with a rifle bullet in his heart. The killer was acquitted by a miner's court, and that was lesson enough for the rest of the populace. There were no more shootings in the gulch.

By 1867, Oro was nearly a ghost town. The placers had played out, leaving only a muddy slurry in the well-worked stream bed. For a time in 1868 it looked as if Oro City would come alive again when lode gold was uncovered in the Printer Boy mine, but this small-scale boom fizzled out, too, and by 1870 the gulch was deserted, except for a few spiritless hangers-on like H. A. W. Tabor who had nothing to look back on but a series of flash-in-the-pan claims and nothing to look forward to. Sage overran the crooked street. Pale shoots of runnergrass wound their way into the dank earthen floors of abandoned, rotting cabins. Sage hens nested in weathered sluice boxes, and only the wind whistling down the gulch through the windowless, doorless cabins and over rusting machinery interrupted the slumber of the once whooping, sleepless, gold-drunk camp.

Bucksin, Broadcloth, and Bullets

WITH EVERY NEW STRIKE AND BOOM CAMP THAT BYERS HEADLINED IN THE *Rocky Mountain News*, the rivalry between the twin towns on Cherry Creek waned. The new camps needed food, tools, lumber, hardware, clothing, booze, commercial banks, draft animals, and wagons. To supply all of these things, the thousands of miners scattered among the gulches and bottomlands of the new Eldorado looked to Auraria and Denver. Business boomed. Five months after the two towns were founded, merchants grossed receipts of $250,000. There were so many stores that peddlers were required to take out licenses to ply their trade. So brisk was business in one shop that the owner, unwilling to lose a penny, paused on his way to a funeral to hang out a sign on his door: "Gone to bury my wife. Back in half an hour."

In the opening months of 1860, journalist Henry Villard carefully noted there were now thirty-nine wholesale and retail houses, fifteen hotels and boarding houses, twenty-three bars, eleven restaurants, two schools, two theaters, and one newspaper between the two towns. Neither the professions nor the trades were lacking—lawyers, doctors, carpenters, blacksmiths, watchmakers, shoemakers, tailors, and barbers all flocked into the bustling towns. The traffic between the mountain mining camps and Cherry Creek increased every day. Drummers, miners, realtors, tinsmiths, wheelwrights, itinerant preachers, and merchants swarmed over the rugged roads. There was even a traveling saleslady. All by herself, Mrs. L. E. Miller drove a wagon full of sewing machines over the plains from Wisconsin to the Rockies. Among the flock of newly arrived home-makers who eagerly watched her demonstrate the new iron treadle machines and then promptly forked over $160 appropriated from husbands' buckskin bags of gold dust the formidable saleslady did a tidy business.

Tom Wildman, a college-bred young adventurer from Danbury, Connecticut, come to strike it rich in the new diggings, arrived on Cherry Creek to find two glaring shortages of materials and services. "There is not a single hat in the town," he wrote to his banker father in Danbury, the hat-manufacturing center of the country, "...send us a few cases...good broad brims, planters black and brown brush...[and] we can make a dollar and a half on every hat..."

Bachelor Wildman paid $7 a week for board. The food was plain and filling. Although flour, cornmeal, and some of the root vegetables were making their appearance in town from mills and farms down in Taos and on the Huerfano River to the south, fresh meat was still scarce. And if Wildman got a toothache from chewing on the daily fare of jerked venison and bear meat he discovered, to compound his pain, another more urgent shortage in town. "A good dentist's office is wanted here; there are none...and $5 is freely offered to have a tooth extracted with the customary instruments, to wit—*a pair of bullet molds.*"

As uncomfortable as a toothache (or its extraction) might be, there was another physical ailment that sent many a resident of Denver and Auraria running to a saloon for a jolt of tanglefoot to dull the nerve ends or to the doctor for a pint of calamine lotion. The lack of green leafy vegetables and dairy products brought on a vitamin deficiency disease that could make a man's life miserable. Wildman was one of those afflicted and, like scores of others, his skin itched and burned unmercifully. "It keeps me awake half the night...The common salutation is, instead of 'How are you this morning?' 'How is your itch today?' "

Despite these discomforts, in the itch to bring an air of permanence to the twin towns, the inhabitants sent the business of settlement catapaulting forward at a brisk pace. From tens of dollars the cost of town lots jumped to hundreds of dollars. Lots 20 by 120 feet went for prices between $250 and $500 depending on their location. To fill these lots, some of the more affluent arrivals had their eastern homes dismantled, stone by stone, and carted out to be rebuilt on Cherry Creek. Prosperity argued for union of the two rival towns. The argument was bolstered by a growing threat of competition presented by a little upstart town a few miles to the west. Tom Golden, George Jackson's hunting companion, turned up a modest placer on the lower part of Clear Creek shortly after Jackson made his lucky find on the South Fork of the creek near the hot springs. Others followed Golden to his sandbar, and in the usual pattern a town sprang up around the new diggings that soon became noteworthy because of a peculiar case of justice, one that might be called "preventive lynching." Miner Edgar Vanover drank more than he mined and when he was drunk he subjected the whole camp to trigger-fingered threats and wild shots of his Navy revolver. The citizens at last called a mass meeting to decide what to do about the thorn in their collective side. Put before the assembled miners was one question: Deport him or hang him? The verdict was hang him, and so they did on a bright, calm, and peaceful September morning.

Boss of the camp, David Wall, named the new settlement Golden in honor of Tom and the town's chief product. But Golden's auriferous sands did not fulfill their early promise and the townspeople turned to another more reliable source of revenue. Situated as it was fifteen miles closer to Gregory, Jackson, and Russell gulches than were Denver and Auraria, Golden set its cap to steal away from the twin towns on Cherry Creek the lucrative commerce they enjoyed with those booming mountain mining districts.

To the shrewd Larimer and Byers, it was clear that consolidation would best serve the common good of both Denver and Auraria. All through the fall of 1859 Byers editorialized on the virtue of merger, pointing out that one of the few good things the ill-founded Jefferson legislature did was to approve the joining of Auraria and Denver. For his part, General Larimer stumped the two towns week after week, harping on consolidation as "an absolute necessity for the survival of all." At a mass meeting on April 3, 1860, the question was put to a vote. Unification carried by 146 to 39 votes. Three nights later, in the light of the full moon, the event was heralded by a parade of Aurarians and Denverites, each marching from his own side of the creek to the center of the Lawrence Street Bridge. Bursts of enthusiastic gunfire and skyrocketing cheers detracted from what Larimer had hoped would be a solemn reading of the proclamation:

> Whereas, the towns at and near the mouth of Cherry Creek are and ought to be one; therefore, be it resolved that from this time Auraria proper shall be known as Denver City west division...

To prove that politics makes strange bedfellows, the Auraria Town Company promptly presented their recent arch foe Larimer with four town lots in West Denver. Wrote the nonplussed townmaker to his wife, "They and I are now cheek-by-jowl good friends. They say...I have beaten them fairly."

The unified Denver now got a bona fide mayor, John Moore, and a charter that gave the city council authority to "restrain, suppress, and prohibit tippling shops, billiard tables, tenpin alleys, houses of prostitution, and all disorderly houses and restrain gaming and gambling houses and all kinds of public indecencies." Authority was one thing. Power was quite another. The new government, to say the least, did not enjoy the full support of the population, and for the next year the newly united towns on either side of Cherry Creek were beset by alternate waves of lawlessness and retribution that staggered the inhabitants.

It started with what has come down in Denver history as the Turkey War. On February 1, 1860, a rancher from down south near Bent's Fort hauled a wagon load of wild turkeys into town. He pulled up near Cibola Hall across the street from Uncle Dick Wootton's Western Saloon and left his wagon to find a buyer for his raucous, wing-beating cargo. It was late afternoon and Ferry Street was crowded with people. In the dark game room of Cibola Hall four drifters, headed by Chuck-a-Luck Bill Harvey, sat near the window and watched the rancher walk away from his wagon. In each of their minds was the same thought. A passel of turkeys in a town starved for fresh meat would bring a tidy poker stake for all four of them. Within seconds, after a whispered conference, Harvey and his pals got up from their card table and moved out onto the street. Harvey untied the droop-eared mules, while the others leaped aboard the wagon. With the mules free, Harvey jumped atop the wagon, picked up the reins, and gave a yell that started the animals forward with a sudden lurch, setting the turkeys to squawking frantically. Men on the boardwalks stopped to stare at the noisy, careening wagon as it flew down the crowded street toward the river crossing, scattering pedestrians before it like tenpins. Just as the thieves reached the corner of the first block, William Middaugh stepped out from a store in time to catch sight of the hellbent wagon and the wildly gobbling turkeys. Recognizing Harvey and his no-account friends as the vehicle rattled by, Middaugh rightly guessed what was happening. He streaked over the three blocks to the office of the *Rocky Mountain News* to give the alarm. Byers, as a major in the Jefferson Rangers, a more for show than blow local militia got up by the paper legislature, sent a runner from store to store to gather up the men to give chase.

When the word of the theft and the fact that there was an eyewitness got to the rest of Denver's arrogant clan of bummers, they went to work. As darkness came, all the riffraff in town roamed the streets, stopping citizens and hissing threats of death reinforced by ominous cockings of pistols if any harm came to Harvey and his gang. Blacksmith Tom Pollock, who was town marshal at the time, raced home to get his gun, but on the way he was attacked by a knife-wielding, drunken bummer. Pollock managed to escape only to round a corner and come face to face with Harvey, who stuck a pistol in the marshal's belly and demanded the Rangers be called off the hunt. Pollock was a big man with a renowned bull horn voice. He used it now to lambaste the turkey thief. Harvey's courage wilted and he scuttled off into the crisp night to look for his real target, eyewitness Middaugh. Someone said the witness was having his dinner at the Vasquez House. Harvey took up his vigil on the second story of the hotel and when Middaugh appeared he took aim and fired. The shot missed. Another ball whizzed by Middaugh's head from the opposite direction and he turned back into the hotel as a company of Rangers charged into the street. At the sight of the mounted militia, Harvey decided a load of turkeys was not worth dying for so he ducked down an alley through a fence and was gone. His henchmen had long since made the same decision and fled also. The four of them were tried, convicted, and sentenced to exile, in absentia, but neither the turkeys nor the thieves were found.

The town quieted down briefly, but the bummers were not finished with Denver, not by a long shot. A few weeks later, Harvey regained his bravado, and with George Steele, another cut-throat who had also been exiled under penalty of hanging, boldly rode into Denver. At the bar of the Criterion, the bummers' hangout, the pair swore to get Middaugh, Pollock, and all the citizens who had helped convict the turkey thieves. Someone who overheard the drifters' boasts slipped out to tell the marshal, but by the time the hastily mustered Rangers arrived to flush out the bummers, Harvey and Steele were nowhere to be found. The word was quickly passed around town that the two were on the prowl, drunk and dangerous. The townspeople need not have worried. After Harvey and Steele left the saloon, they caught sight of a large party of Arapaho and Cheyenne chieftains and braves riding in from their encampment downriver headed toward D'Aubrey's trading post on Blake Street. Saturated to the gills with tanglefoot whiskey and thirsting for trouble, the two renegades were still clear-headed enough to figure out that with the Indian men in town the Arapaho and Cheyenne squaws were without protection. Harvey and Steele hit the Indian village just at nightfall, making their way from one tent to another, raping, beating, and stealing. For a finale they cut out two swift ponies from the Indians' herd and hurtled off into the night.

When the Arapaho and Cheyenne men returned to find their village in shambles and their women ravished, the people of Denver very nearly got the vengeance Harvey and Steele swore to give them. Outraged at the white man's pillage, Chiefs Little Raven and Left Hand vowed angrily to burn Denver and slaughter its inhabitants. The town quaked. The nearest cavalry troop was two hundred miles away at Fort Laramie, and everyone knew that the merchant militia, the Jefferson Rangers, was scarcely a match for hundreds of Indians on the warpath. Mayor Moore called a town meeting. The embarrassed and scared whites decided unanimously that the only man who could make their peace with the wronged tribesmen was the sagacious trapper and squaw man Jim Beckworth, whose unbroken forty-year association with the plains Indians, sometimes as a chief of the Crow, was a mark of their respect for him.

When Beckworth came to Cherry Creek in November of 1859, he scarcely recognized his and the rest of the trappers' sometime camp. On the night of his arrival he poked around Auraria and Denver and seeing not one familiar face he proceeded to get memorably drunk. He recovered and decided miners were not such a bad lot after all and proceeded to become a fond and familiar figure in the saloons and streets of Denver. He was nearly always surrounded by a knot of men who hung on every word of his incredibly tall tales. Occasionally, the captivating yarn-spinner was asked into the homes of Denver's newly evolved social set. When Beckworth, who was a mulatto, came to dinner at William Byers' house with his Indian wife, "Lady Beckworth," Byers' wife Lib swallowed hard on her Virginia prejudice and then "... instantly decided that there was but one thing to do and that was to ignore it altogether and treat them all alike, which I did."

When Denver's Indian problem was put to him, Beckworth agreeably undertook to help the whites out of their predicament. He trotted out to the Indians' camp and persuaded them to settle for full reimbursement of all stolen and damaged property. Restitution of virtue was not discussed. Although he pleaded the cause of the white man, Beckworth at the same time wrote a letter to the *News* denouncing the wanton acts of "drunken devils and bummers" against his Indian friends, and warned that the Indians were "tenacious of revenge."

Denver was relieved that the crisis was over and once more settled down to a cautious coexistence with the Indians. The perpetrators of the trouble, Harvey and his cohort Steele, had again gotten safely beyond frontier justice.

No sooner had Denver relaxed from its most recent scare when trouble came again. John Rooker, a son of one of the original founders of the Auraria Town Company, had fallen under the spell of C. H. "Charley" Harrison's dissolute crowd at the Criterion. He admired the supple-fingered gamblers and the two-gun drifters who never worked and yet had all the money they needed to play the games of chance and to buy the favors of women. Young John soon forgot his strict religious upbringing and ran with the bummers, drinking up his days and spending one night plucking the poke from a whiskey-sodden miner so he could pluck the flower of one of Ada Lamont's maidens the next. Between drinks he played poker. On the night of March 29, 1860, he sat in Uncle Dick Wootton's saloon nervously hunched over his hand waiting for affable Jack O'Neil to ante up. O'Neil had been one of the first gamblers to hit Cherry Creek and before more than half a dozen cabins were put up, his shanty saloon, the Hote de Dunk, was open for business. Easy-going O'Neil never wore a gun. There was no need to—he ran a peaceable gambling hall, presided over by his stunning mistress, Salt Lake Kate. Now and again, O'Neil liked to try his luck at another's table. Not a man to be hurried, the gambler now took his time sizing up his cards. Rooker watched him in glowing impatience. He had been losing steadily, and he desperately needed a win off the genial O'Neil. The pot already bulged. The proprietor of the Hote de Dunk cast a quick look at his opponent and then threw another twenty-dollar gold piece on the pile, calling Rooker's hand. The boy called out two pair. O'Neil, holding three jacks, thought he had won, but as he reached out to rake in the stake, Rooker flashed his hand under his nose.

"I've got four kings, O'Neil," snarled young Rooker. "The pot is mine."

O'Neil stared at the boy's cards. "Why, you tinhorn kid, you called it wrong!"

Rooker stood up. "I ain't no tinhorn kid, and the money is mine!" he shouted.

O'Neil was incredulous. "What kind of a whore's son are you, anyway! Can't you count four of a kind?"

Rooker stiffened. "You can call me names, O'Neil, but not my mother. Put on your gun because the next time we meet it's you or me." Rooker then stalked out of the saloon, his face set in fury.

O'Neil laughed nervously and shook his head. "Tinhorn bluffer," he said to the curious crowd who had gathered at the first sign of a squabble. But the next morning when the gambler showed up at the Hote de Dunk he was wearing a gun. It was a beautiful morning, one of those still, bright days of early spring on the edge of the plains when the sky is a phenomenal blue and a light warmth pervades the air, beguilingly promising tranquillity.

In an alley down the street John Rooker was setting the stage for a far from tranquil morning. Aware that Uncle Dick Wootton never opened his saloon before noon, Rooker forced open a window and hoisted himself and his shotgun inside. Creeping to the front of the building, he took up station just behind the front door where he could get a good view of the street above the half-curtain on the door window. Minutes later, O'Neil, his gun in his holster, walked up Ferry Street. When he drew opposite Uncle Dick's saloon, Rooker fired both barrels at the gambler. The sound bounced back and forth between the buildings as O'Neil, spurting blood bright red in the sunlight from thirteen holes in his gut, crumpled into the street. He died within minutes, cradled in the arms of a distraught Salt Lake Kate whom a passerby fetched to the scene when O'Neil first fell.

The day after O'Neil was buried, Rooker's father held a mock trial in Cibola Hall, corraling a jury of townspeople and pleading his son's action as justifiable homicide in the light of the gambler's provocation and vilification of the boy's mother. Young John went free.

Like John Rooker, twenty-three-year-old James Gordon quickly lapsed from a life of industry and promise as soon as he hit the frontier. Whiskey, women, and weapons offered more gratification to his newborn restlessness than did his diploma in civil engineering. On a blistering July afternoon, after a morning at the bar, Gordon lurched into Ada Lamont's fancy house, and in the cool, shade-drawn parlor he shot the bartender who tried to protect one of the terrified girls from his drunken advances. From Ada's, Gordon moved on to the Elephant Corral, the new name for the Denver House, where he tried to pick a fight with Big Phil the Cannibal, an unsavory brute who boasted he had eaten two of his wives, and one Indian, and one Frenchman.

But for once, Big Phil declined the challenge, brushed Gordon aside as he would a gnat, and left the saloon. Gordon pulled out his six-shooter and blazed away at the departing giant, his aim sufficiently skewed by alcohol that Big Phil went unscathed. After another drink, Gordon staggered out into the heat again, pausing only long enough to put two bullets into the head of a dog trotting dutifully at the side of his owner. By now, Gordon's rampage was known all over town. The streets emptied. Doors on houses were slammed shut and bolted. Town marshal Tom Pollock rounded up some men to help him track down the drink-crazed bummer, but Gordon managed to stay one step ahead of the posse.

Gordon singled out the Louisiana Saloon for his next port of call. At the door he met John Gantz, a German settler just arrived from Kansas. With a leer, Gordon offered to buy him a drink. In broken English, Gantz declined. Gordon then knocked him down and sat on him. And as the terrified German pleaded for his life, Gordon held him by the hair and fired his gun at the man's head. Four times the Colt misfired but on the fifth try a bullet smashed into Gantz' brain, killing him instantly. Gordon got to his feet laughing insanely and stepped over the body and out into the street. After one last stop at Cibola Hall where he contented himself with a drink instead of a killing, he disappeared. Throughout the whole bloody afternoon, not one soul who was present when Gordon started shooting moved to stop him.

Late in the day news of Gordon's orgy reached Charley Harrison and he, too, set out to look for the culprit. Since Gordon was one of his boys, Harrison was determined to take care of his own. Shortly before dawn the next morning Harrison spotted his boy dead asleep in a thicket of brush near the Cibola. Quickly, and without being observed, Harrison got the woozy Gordon on a horse and safely out of town.

Ever since he had come to Denver the year before, Harrison had quietly waged a power struggle against the shaky civil government of the brawling, bustling twin towns on Cherry Creek. He had come in on a trail-tired horse as an escapee from a posse in Utah, but no one in Denver suspected he was fleeing from the law. Suave, gracious, quietly forceful if the occasion arose, and always modestly attired in black broadcloth, Harrison was the personification of a southern gentleman in appearance and public behavior. With his bent toward elegance, Harrison naturally gravitated toward the Criterion. Its owner, Ed Jump, had made it uniquely attractive among its kind in town. For its decor he had freighted across the plains ornately carved wooden columns and cornices, and frosted glass etched with designs of trailing vines to grace the front entrance of the establishment. The saloon's menu "rivaled Delmonico's," and its cellar was crammed full of the finest Kentucky bourbon and French champagne. When the

Criterion opened, even Editor Byers was impressed. "We have only to say it is just what its name indicates," applauded the *News*. Harrison thought so, too, and he wasted no time renting a faro table in the popular watering place. He ran an honest but profitable table, and before the year was out he had made enough to buy into the operation.

With a liberal profit from the saloon to play with, handsome Charley amused himself by financing a couple of bordellos, and by dabbling, under the table, in politics. His money saw to it that candidates he backed were elected to city offices. The renegade bummers, most of them from the South, took their orders from him. No man joined the sporting fraternity in town until or unless Harrison approved of him. His influence kept saloons and bawdy houses free of taxation, and it kept his lawbreaking friends out of jail. As his power grew, so did his circle of prominent friends as people sought his protection against the very lawless elements he controlled. This spiral of prestige added arrogance to his other qualities and no one doubted for one moment that he meant what he said when he vowed to kill twelve men in his lifetime to assure himself a jury of his contemporaries in hell. It came as no surprise to Denver therefore when on July 12 the celebrated gambler pumped not one but three bullets into Charles Stark. Stark, a blacksmith with aspirations toward a higher station in life, spent much of his money on fancy clothes and much of his time around the higher-toned saloons in town hoping to be noticed. One afternoon he was playing cards in Cibola Hall when Harrison and two of his prominent southern friends about town, Park McClure and Judge Seymour Waggoner, came in for a drink. Stark left his game and offered to join Harrison and his party. Harrison looked disdainfully over his shoulder at Stark's eager face and coldly told the blacksmith to move on. Embarrassed and enraged, the blacksmith whipped out a knife and lunged at Harrison. The gambler nimbly sidestepped the flashing blade for several passes, and then he pulled out one of his pearl-handled Colts and fired at Stark. Two bullets lodged in Stark's chest, the other in his thigh. For a time the doctor thought he would live, but on July 21 the blacksmith died.

In his office on the south bank of Cherry Creek, William Byers had watched with growing rage the headlong degeneration of law and order in his adopted town. Neither the Jefferson Rangers, Mayor Moore, nor Marshal Pollock and his aides were able to cope with Harrison and his gang. After this latest murder, Byers contained himself for three days but on July 25 he lashed out against the czar of the Criterion. Referring to Stark's death, he wrote in the *News* "...the act was wanton and unprovoked, in short, a cold-blooded murder..." And then he followed it up with an unmistakable message: "The rowdies, ruffians, and bullies generally that infest our city had better be warned in time and at once desist from their outrages upon the people...One more act of violence will at once precipitate the inevitable fate; and the terrors that swept over the fields of California at various times, and first purified its society, will be reenacted here with terrible results to outlaws and villains…"

Vigilantes. That was what Byers was driving at. Those who had lived through the days of the California gold rush felt a chill as they read Byers' words. Those who shared his frustration at the high-handed and unbridled outlawry welcomed the verbalization of their own secret inclinations. The Denver Committee of Safety met clandestinely over Graham's drug store on the night of July 28, only three days after Byers' clarion call appeared in the *News*. That was one result of the editor's thinly disguised call to arms. The other result involved the formation of another vigilance committee, of sorts.

Hot-headed, arrogant, and defiant, the bummers responded to the editorial in their own inimitable fashion of vigilance. They were boiling mad and they decided that if the outspoken editor was out to get Harrison, they would get Byers first. Carl Woods, a brash fortune-hunter of thirty, goaded the bummers hanging around the Criterion until he got three of them to join him. Rallying to his cause were the now familiar troublemakers Chuck-a-Luck Harvey, George Steele, and John Rooker, all of whom felt safe enough under Charley Harrison's protection to return to Denver. Fortified by a morning at the bar, the four burst into the *News* office and at gunpoint kidnapped the not exactly surprised Byers. The bummers pushed and shoved their captive across the bridge on McGaa Street and into the back door of the Criterion, promising him that "Charley will stuff your words down your rotten throat."

But for once, Charley was totally unaware of the plans of his headstrong band of young toughs, and when they sent Byers reeling into his presence his reaction was not precisely what the kidnappers expected. Harrison rose from his chair, solemnly and earnestly shook the hand of the beleaguered editor, and angrily turned on the bummers.

"What in hell do you think you're doing bringing this man here at the point of a gun?" Harrison was livid.

Woods blinked. "Listen, Charley, this bastard is out to get your hide and ours. We've got to settle his hash once and for all!" Woods declared, brandishing his pistol. The other three bummers loudly backed him up.

Harrison looked quickly at each of the gunmen. He knew full well that to do what Woods and the others expected, to pistol whip the popular editor of the *News*, would bring the wrath of all Denver down on his head and put an end forever to the comfortable sinecure he had created for himself. At the same time he could plainly see that the four desperadoes were in no mood to be trifled with. Harrison's features softened slightly as he gave his henchmen a crooked smile.

"Well, now that you've brought him here," he said, "leave him to me. I'll take care of him. Now get back to the bar before somebody gets suspicious."

Woods hesitated, then he shoved his gun into its holster. "See you fix him good," he growled and walked heavily and unevenly out of the room, followed by the other three men. As soon as the door closed behind them, Harrison propelled the compliant newsman through another door and into the kitchen. The owner of the Criterion prodded Byers with his pistol. "Now get back to your office," he said urgently, "and be ready for the sons of bitches. They're looking for trouble and they're too likkered up to stop now. This time they'll come shooting."

Byers sprinted back to the *News* office. Once inside, he barricaded the door and ordered his astonished but relieved printers to break out the rifles and shotguns kept handy beside the cases of type. The four printers, the two editors, and the printer's devil took up positions behind windows and desks and waited in tense silence for the onslaught of the bummers. Already the word of Byers' kidnapping had burst over the town and the usual curious throng crowded onto side streets on both sides of the creek to watch the show. They got their money's worth. When the bummers discovered Harrison's duplicity, they went into a rage, just as the gambler warned they would, and stormed after the editor. It was Steele who first rode past the *News* office, aiming an ear-splitting yell and a series of obscene gestures at the besieged journalists. Then he wheeled his horse and clattered past again, this time firing two shots at the windows of the *News*. Neither bullet found its human mark. As Steele dashed by, however, Byers opened fire and his aim was truer; the outlaw took a round of buckshot in the back. Steele stayed on his horse, heading toward the safety of the Criterion when he met Marshal Tom Pollock running down the street, gun in hand, to halt the attack. Pollock raised his gun and fired. Steele, with a part of his head blown off, spun off his horse and lay in a pool of blood at the marshal's feet. That was enough for Woods, Harvey, and Rooker. They threw down their guns and were more than glad to take exile for their punishment.

For the next few weeks, Denver was treated to an unusual period of peace, if peace it could be called. There were no more public bloodlettings, but the work of the Committee of Safety turned up in the form of masked intruders swiftly and ruthlessly searching homes for stolen articles, sudden dead-of-night inquisitions accompanied by thinly veiled threats to loosen tongues, and the occasional man mysteriously missing. Who belonged to the one-hundred-man group was a well-kept secret. A few rumored members were the marshal himself, his deputy William Middaugh, busybody D. C. Oakes, the militant Methodist elder John Chivington, General Larimer, and the man who had called publicly for the formation of a vigilance group in the first place, the editor of the *News*. But if Byers knew who the one hundred were, he wasn't telling. So expert was the committee that no one not a member knew when or where they carried out their operations until after the fact. So effective was the group that the lawyers in town announced in the *Rocky Mountain Herald* that they were closing their law offices for lack of business and until "regular and constitutional tribunals of justice are established in our midst."

When Byers published his broadside that opened the door to the formation of the Committee of Safety, he got more than he bargained for. Shortly after its formation, wholesale horse-stealing broke out in the countryside. Time and time again, the marauders, poorly disguised as Indians, raided the corrals of Denver and the surrounding ranches. The raids were too well organized to be the work of a two-bit thief. They had to be the work of a well-run ring of rustlers. On September 2, the vigilantes captured a known rustler who called himself Black Hawk. At the threat of being lynched if he did not reveal the names of the rest of the gang, Black Hawk agreed to the demands of his captors, but asked that he be allowed to leave town unharmed after he confessed. The committee agreed, and they locked the rustler in a cellar for the night with a pencil and a piece of paper so he could write down the names of the gang. The next morning the vigilantes found Black Hawk's body swinging from a cottonwood branch down on Cherry Creek. It was obvious to the committee that one or more of their group was serving two causes and did not want Black Hawk to talk. But in their hurry to do away with the key witness to their identity and activities, whoever murdered Black Hawk missed the incriminating list of names lying on the floor of the cellar. When the rest of the Committee of Safety found it, it touched off a reign of terror that matched the vigilante horrors of California gold rush days just as Byers predicted.

On the same afternoon of Black Hawk's lynching, gambler Jim Latty and five of his cronies were discovered hanging from tree limbs a few miles out of Denver. The next day, the body of John Shear, manager of the Vasquez House, was found dangling from a tree two miles out of town. On a stump beside the dead man was a playing card scribbled with the words, "This man hung for being a proven horse thief." Two days later, AC Ford, a prominent if promiscuous attorney, was abducted without protest from a stagecoach by four masked men. His partially buried body, riddled with bullet holes, turned up a mile from the road with a note pinned to his coat reading, "Executed by the Committee of Safety."

Whether or not the vigilantes were responsible for the deaths of these men was moot, but the impact of the ruthless killings on Denver was prompt. Letters of protest rained onto the desks of the editors of the rival *News* and *Herald*. Quoted the *Herald* from one letter, echoing the thoughts of all Denver, "Is the Inquisition revived in our midst?" Byers quickly responded with a plea for the public organization of a posse who would see to it that captured suspects were safeguarded from lynching and were handed over to the authorities for a fair trial. The reaction of Denver was immediate. In front of the new post office on a balmy September evening, the town's leading citizens stepped forward to offer their services for the establishment of law and order. The new Vigilance Committee was chartered then and there to back up Marshal Pollock's activities. Its chief was William Middangh, deputized as assistant marshal. On his staff were General Larimer, Postmaster Park McClure, and Judge Waggoner. The mayor, John Moore, the editor of the *Herald,* and six august merchants completed the roster. William Byers was not a member. If, as it had been rumored, some of these men had been members of the original, secret tribunal, once they were openly acknowledged as members of a town-authorized vigilance group, they became models of legitimacy. The systematic lynching abruptly ceased and so did the horse-stealing. With some relief, Denver settled down to its earlier and considerably more palatable diet of drunken brawls laced with an occasional shoot-out.

On the frontier where a gun was as much an article of clothing as were boots, and was worn for the same reason—protection—the definition of murder became very precise. As Uncle Dick Wootton put it, "...the killing of persons for the purpose of robbery or of a man not prepared to defend himself...we looked upon as murder..." If, on the other hand, two mean drunk men shot it out and one killed the other, that didn't count as murder. The chances were that one party was as much at fault as the other, recalled Uncle Dick, "...and we allowed them to...settle their own difficulties in their own way."

Despite the presence of the publicly acclaimed Vigilance Committee, the matter of law enforcement, or the lack of it, was not a dead issue. The irascible Park McClure saw to that. The restless energy he built up while waiting for the bullet hole Dick Whitsett had put in his thigh during their duel to heal now centered on the popular schoolmaster-turned-journalist, Owen Goldrick. The professor's prose could be as incendiary as the postmaster's personality. Most subscribers of the *News,* for whom Goldrick wrote, read with amused tolerance his accusations and exaggerations. But when Goldrick's virulent pen notified the world that "Of all the low cunning, small dealing, petty stealing, downright robbery and miserable 'skullduggery' carried on in the City of Denver that in the postal arrangement beats the whole..." the short-fused McClure erupted. It mattered not that the reference was to the express company which supposedly had a contract to carry the US mail at 3 cents a letter, but which stubbornly persisted in charging each recipient of a letter the express rate of 25 cents before handing over the missive. Postmaster McClure convinced himself that his honor had been gravely and publicly compromised. He puffed out his chest and stomped around to the *News* office to beard Goldrick in his den of insult. At gunpoint, he forced the frightened Irishman to sign a statement of apology clearing the postmaster of any charges of mishandling the mail. When that was done, The Thunderer called on Tom Gibson of the *Herald* and asked him to publish the apologia.

After a falling-out with William Byers, Gibson had left the *News* and Denver to go up to Mountain City on Clear Creek and start his own weekly. When that sheet failed after a few issues, he came back to Denver to challenge Byers in open competition. The *News* took up the gauntlet and the two papers carried on a lively duel of words. For subscribers on the frontier where life was more often hard and humorless than otherwise, the rivalry between competing newspapers had high entertainment value. And rival editors were delighted to find that the noisier their feuds, the more papers they sold. Consequently, the *Herald* was more than glad to print any deprecations aimed at the editors of the *News*. After the apologia came out in the *Herald,* Goldrick retaliated in the *News* by demanding that the postmaster retract the falsely obtained "confession." McClure's answer followed to the letter the rules of personal journalism: when he happened to meet Goldrick at a party two days later, he pulled out a gun and pointed it squarely at the heart of the word-mongering schoolmaster. Before McClure could fire, Goldrick's friends rescued him and he swore out a warrant on the spot for the arrest of the truculent postmaster. Tom Pollock gave chase and found the fuming McClure in short order, still brandishing his gun, and carried him off to face Judge Jacob Downing, a merchant who doubled in brass as judge of the common pleas court. Downing heard the evidence and deliberated only a few minutes before pronouncing his verdict.

"Park McClure, this Court finds you guilty of attempted murder," said Downing, "and we sentence you to one year in prison. But we will suspend sentence if you put up a two-thousand-dollar bond as a guarantee of good behavior."

McClure threw the judge a contemptuous look. "Go to hell, you thread-peddling reprobate," he said loudly. "Neither you nor anyone else in this one-horse town can tell me what to do!"

The judge's gavel came down on the tabletop that served as a bench with the sound of a battering ram. "Jail him!" he cried to Marshal Pollock.

In a town where punishment for crime took three forms—hanging, flogging, or exile—the order was easier given than executed. Denver, owing to a lack of demand, had no jail. But the marshal proceeded to carry out the spirit if not the letter of the law, and locked McClure upstairs in the judge's office under guard until he could be conducted to Leavenworth.

Down at the Criterion, there were dark words of disapproval when McClure's fate was known. And that night a gang of bummers rushed the guard and freed their fiery friend who immediately invited his rescuers down to the Criterion where he set up drinks for the house to celebrate his triumph over local authority. But The Thunderer's audacity backfired. Denver was at last thoroughly tired of the tyranny of the lawless. The townspeople vehemently called for prompt action. In response, Middaugh's posse marched resolutely on the Criterion. Warned of their coming, McClure made a dash for the post office and barricaded himself inside as grim-faced citizenry armed with pistols, shotguns, rifles, and buffalo guns crowded around the building. Marshal Pollock stepped out from the ranks of the posse and ordered McClure to come out. McClure refused. The crowd surged toward the building. Pollock warned the postmaster that he could not guarantee his safety if he did not surrender to the local authorities. After a few minutes, McClure sent word he would surrender to Goldrick or Judge Downing but not to any law officers. Again the crowd pushed nearer, a low rumble of indignation audible over the shuffling of hundreds of feet. McClure called out that he would talk with the marshal, and Pollock went into the post office. When he came out a few minutes later, he announced that McClure had capitulated. The mob broke into wild cheers, breaking up and moving away in twos and threes, and in larger groups, congratulating themselves on their victory over the arrogant postmaster.

The Goldrick-McClure feud appeared to be the catalyst that was needed to bring stable government to the embroiled town. In theory, Denver was under three jurisdictions: the Arapaho County government imposed by the legislature of Kansas, the wishy-washy extralegal Jefferson territorial government, and a pseudo-city government headed by Mayor Moore. In the midst of varying loyalties to these conflicting bodies, all of which claimed ultimate authority, the cutthroats, toughs, and hotheads had found a soft berth for their operations. But on September 21, 1860, a group of responsible men in Denver united to dump Moore's do-nothing regime. They met to write up a constitution and to elect six councilors to govern the town instead of a mayor. This People's Government of Denver, as they called themselves, held a tight rein on the outlaw element for the next eight months until Congress formally created the Territory of Colorado.

Meanwhile, Jefferson Territory refused to lie down and die a graceful death. As directed by the resolution passed by the paper legislature, Bev Williams, who had replaced H. J. Graham as the territory's spokesman in Washington, kept to the road and rails between Denver and the nation's capital, haranguing Congress with petition after petition for recognitionof the new Eldorado as a territory. For Williams, it was a happy combination of work and lobbying since he was superintendent of the Leavenworth and Pikes Peak Express Company and had to make the trip across country often, anyway. But his efforts were getting nowhere, with Congress too enmeshed at the time in the rising crisis between the North and South to pay much attention to a faraway gold-mining region's demand for self-government. Home-rule advocates in Denver despaired, and to top off their gloomy outlook they felt betrayed when delegates of the Jefferson territorial legislature chose to meet for their second session in Golden, Denver's rival, where living was cheaper (board "6 dollars a week—wood, lights, and hall rent free"). Cut-rate accommodations were an important factor to an organization which no one would support financially. But even reduced government spending did not rally any backers. It was a sign of the fatal illness of Jefferson Territory that hardly a quorum could be mustered for the second session no matter how cheap the cost of attendance.

While Golden watched the political flounderings of the inept extralegal legislature of Jefferson Territory, Denver was treated to a devastating burlesque of the entire proceedings. Billing themselves as the "Third House," a group of town wags, including Tom Wildman, took to meeting every night in the upstairs room of Libeus Barney's Apollo Hall to play at politicking. Before long their monkeyshines drew a packed house, and Wildman and his friends, struggling manfully for solemnity, heard reports from standing committees on Morality, Skullduggery, Mountain Naming, and Election Rigging; passed a resolution calling for negotiations with the governor of Massachusetts for the import of 999 comely school marms for the purpose of teaching the population of the territory (motion from the floor to strike out "teaching," seconded and carried); authorized Bill No. 323 chartering a circle route railroad from Denver to Taos (to take on a tankful of Lightning), to the top of Pikes Peak, to Laramie, and return, no passenger to be unaccompanied by ladies; and debated and passed a bill establishing the free use of jackasses by judges of the Supreme Court.

Then in December, 1860, in a move that surprised the jocular members of the Third House no more than it did Delegate Williams, the 36th Congress, meeting in short session, heard a bill to provide a government for the "Territory of Idaho." By this time the territorially-inclined of the new Eldorado didn't care what name they got just as long as it went along with an authorization for self-government. They had run through a dozen or more names in their various petitions, including Shoshone, Tahosa ("dweller atop mountains"), Cibola, Colona, San Juan, Arapaho, and Pikes Peak. The Idaho bill was shelved for two months while Congress decided whether to admit Kansas and Nebraska as free or slave states. When the bill did come up before the House on February 4, its name was now Colorado. One of the congressmen had suggested the namechange since the mighty Colorado River rose within the boundaries of the proposed territory. On February 6 the House passed the bill, and on February 26, the Senate followed suit. President Buchanan, anxious to get out of office before the war clouds broke around him, signed the measure but made no appointees to the new territory that would add to the already fulminating controversy between the northern and southern states. He left to Lincoln the delicate job of staffing the administration of Colorado Territory.

The moment Buchanan put his signature to the act, Delegate Williams dashed to the nearest telegraph office. But the people of Pikes Peak had to wait six days before leading the news in Byers' paper. The well-handled dispatch came by telegraph, Pony Express, and finally stagecoach to waiting Denver. For some, like William Larimer, the news was the satisfying culmination of a long and taxing effort that brought Colorado recognition as a political reality. For others, whose interests lay in not having the gold-rich region organized, it was a bitter defeat.

Despite their preoccupation with prospecting, gold-mining, and other adventures in the new Eldorado, a number of Colorado settlers of southern heritage or sympathy kept an eye and ear turned toward the fateful events that in the early part of 1861 were dividing the nation and which culminated in the organization of the Confederacy on February 4, the very day the House passed the Colorado bill. The bill that passed the Congress was a compromise put together by senators Green of Missouri and Wade of Ohio. In the hope that new territories admitted by bills that made no reference to slavery would pacify and return the rebellious states to the Union, they adroitly convinced their colleagues they should go along with what a contemporary called "the conspiracy of silence." The scheme fell far and away short of persuading the South to return to the fold, and in Colorado it encouraged southerners to believe that the new territory might yet be won for the Confederate cause. With its money-metal resources that sent $7 million in new gold to the Philadelphia mint in 1861, Colorado would be a fat prize for loyal southerners to deliver into the hands of their new government.

As the situation between the North and the South grew more tense in the East, this tension was reflected six hundred miles west on Cherry Creek. There was nothing secret about the rebel movement in Colorado. Ex-Mayor John Moore and his fellow southerner, James T. Coleman, bought up a weekly paper in Denver, renamed it the *Daily Mountaineer,* and blasted away at the North with their own editorials and those reprinted from fire-eating southern papers. To whip up the emotions of the Dixie-leaning populace, Moore and Coleman used the most inflammatory statements they could find:

> It may be concluded that the policy of the incoming administration [Lincoln's] will enforce the Southern Confederacy to maintain "the right of Revolution" with the flaming sword and the thunders of cannon...Mr. Lincoln cannot rate higher than a political cheat who has plunged his country into revolution for his own aggrandizement...he is trying to lure the South to its destruction by diplomatic tricks and words...

When these seditious words appeared in the *Mountaineer,* no bricks smashed the windows of the newsroom, no one set fire to the building or stormed its doors to tar and feather the editors. Even the usually astute Byers seemed not to see the picture clearly. His editorials were for the most part curiously optimistic. "The fears of the South are in great measure groundless," he wrote. "The country is as safe as ever; the good ship of state will glide onward with colors as bright and beautiful as ever."

But he was wrong. The good ship of state glided smack onto the rocks. Fort Sumter fell on April 13, 1861. And on the morning of April 24, Denver awoke to see the Stars and Bars of the Confederacy flapping in the spring wind atop the two-story Wallingford and Murphy store next door to Charley Harrison's Criterion. The word of the flag's presence rippled over the town on the wind. Coveys of citizens gathered on Larimer Street in front of the store, some calling angrily for the odious flag to be torn down, others demanding that it stay up. In the crowd was brawny Sam Logan who had been a member of the constitutional convention for the ill-fated Jefferson Territory. A strong Union man, Logan now decided he should take an authoritative hand in putting things right. He called for a few men to help him charge the porch of the store and snatch down the Confederate banner. As Logan led his handful of volunteers up the stairs, Harrison's rebel blacklegs, with pistol hands hovering over holsters, promptly blocked Logan's way. Face to face with Harrison, Logan inched forward, testing his opponent. Steely-eyed, the gambler let Logan come so close their coats touched. While the two antagonists eyed each other menacingly, Logan's men spat out threats of a pitched battle, but the rebels did not budge. The flag, said Harrison loudly and firmly, would stay where it was. Insults grew nastier and necks redder. The crowd moved uneasily, jostling and goading one another.

Down the street, in his law office doorway, talking with fellow attorney Bela Hughes, young Henry Teller heard the commotion and went round the corner to find its source. As soon as he saw what was happening, Teller beckoned to Hughes, and the two men without a word to each other ran down the street and pushed through the mob to stand on the steps of the porch beside Logan. Turning to face the jeering faces of the crowd, Teller, an ardent northerner, and Hughes a Democrat whose sympathies lay not with the South, each in turn used his best barrister's oratory to appeal to the fast dissolving reason of the mob. Tension crackled over the crowd, gun hammers clicked, knives came into hand, but when the two lawyers finished speaking, the mass of men were silent, with eyes shifting from one group to another. In this pause, Logan seized the moment to act. With a boost from Teller, he climbed hand over hand up the front of the building and brought down the flag. The crowd, as fond of bravado as a battle, let go with cheers and catcalls, depending on each man's bias, while Charley Harrison and his secessionist pals withdrew to the bar of the Criterion to bathe their wounded pride in rotgut.

Ludicrous as the flag-raising episode was, it proved the presence in Colorado of active southern sympathizers. Moore and Coleman's *Mountaineer* continued to snipe at the northerners in the territory, carrying a steady dose of Confederate propaganda to the miners in the gulches and the parks:

> The Lincoln administration now openly avows...[that] the Slave states must be annihilated by an armed force...You who are in the mountains are as yet free from this turmoil but sooner or later you will be made to know where you stand...

More important than its propaganda was its role as a clearing-house for the announcement of "secesh" meetings and drives to collect men and money for the southern treasury.

The harangue among abolitionists and proslavery factions clouded issues but sharpened sensibilities in the days that followed until the chips on the shoulders of Coloradans were as numerous as buffalo chips on the plains. The slightest slight, the barest innuendo, was enough to ignite explosive emotions. Ascetic, Dixie-bred Dr. James Stone, judge of the miners' court up in Mountain City, journeyed down to Denver on a blustery March day, stopping at Broadwell House for luncheon. Shortly after he was seated, a party of Denver's civil leaders were ushered in and seated at a nearby table. Included in the party were William Byers, Professor Goldrick, ex-Mayor Moore, and the secretary of Jefferson Territory, Lucien W. Bliss. Bliss, who had met Stone when they were both in the first Jefferson legislature, nodded to the judge as he passed his table. Once seated, the party of Denver men soon got into a heated discussion over their oysters and champagne on the matter of the freeing of slaves, a subject that seemed to crop up with invariable regularity at every gathering in those troubled days. During the interchange of words, the voice of Bliss, an adamant abolitionist, carried above all the others. To the pro-slavery judge seated within easy earshot, every word Bliss uttered appeared to be aimed at him, and at last Stone could stand it no longer. He jumped to his feet and burst in upon the conversation.

"Are those words meant for me, Bliss?" demanded Stone.

Annoyed at the interruption of his peroration, Bliss eyed the pale-figure standing over him and shrugged. "If the shoe fits, wear it," he said.

"You damn liar!" cried the judge.

Now it was Bliss's turn to jump up, and as he did so, he pitched his glass of champagne into Stone's white face.

The judge, with trickles of champagne coursing down his cheeks, trembled with rage. "My seconds will call on yours," he whispered, and left the room.

Bliss nodded absently and watched Stone depart. Then he walked slowly out of the dining room followed by his silent, grim-visaged luncheon partners.

Two days later, Dick Whitsett, playing the role of a second and not a principal in this affair of honor, and Stone's second, Bill Bates, stood aside as the duelists lined up with shotguns at thirty paces. At the signal to fire, two blasts shattered the silence of the afternoon, and Stone fell to the ground with a mortal wound in the bladder. Lucien Bliss was unhurt, but the death of Stone preyed on his mind. In conversation he was distracted, in business he seemed bewildered, and at last he resigned his post, sold his holdings, and left Colorado, a withered and directionless shell of a man.

None of these events was lost on the new president in Washington. Not only were there Confederate activists at work in the faraway fledgling territory, but there were also a great number of people among the settlers who, although not pro-South, had no particular affection for the remote, impersonal federal government. It was very plain to Lincoln that Colorado's tie with the Union was as tenuous as the thread of a spider, and it was also plain to him how important it was to strengthen and solidify that tie. To accomplish this amid what he called the "leaven of treason" Lincoln chose William Gilpin to be Colorado's first territorial governor.

The choice was a logical one. The keen-minded Gilpin had an impeccable background: sound Quaker rearing, English public school, West Point, distinguished military service in the Seminole and Mexican wars. Already, at thirty-eight, he had had a career in law and in journalism to top off his army experience; he was close to the president, having been one of the one hundred handpicked men to accompany the new chief executive to Washington from Illinois; and perhaps most persuasive of all his qualifications, he knew the Rocky Mountain region well.

It was as a member of Frémont's 1843 expedition that Gilpin first came under the spell of the spectacular mountain wilderness of the Rockies. Later, as a major in Doniphan's Volunteer Missouri Cavalry, it fell to him to pacify the rebellious Ute of southern Colorado. He had no sooner extracted a promise of peace from the Ute when the Navajo kicked up a fuss and he was sent out to bring them to the peace table. It was November, and the fury of winter caught up with Gilpin's little army as they struggled northward along the banks of the Chama River and over the snow-swept continental divide to the San Juan Mountains. The expedition produced no Navajo, several cases of pneumonia, and a characteristic memoir by Gilpin, whose enthusiasm often carried him into nights of fancy, comparing his campaign to Hannibal's march over the Appennines and Napoleon's over the

Alps. But Gilpin proved himself as an Indian fighter a few months later when the marauding Comanche and Kiowa made a bloody, fire-blackened gauntlet of the Santa Fe Trail, and the governor of Missouri ordered out a punitive force to tame the troublemakers. Gilpin led 850 men onto the plains and into nine battles in the summer of 1847, and according to his biographer "253 scalps of warriors were taken from first to last."

Retiring from the Army a lieutenant colonel, Gilpin took up with his friend Senator Thomas Hart Benton of Missouri, in whose company his growing belief in the theory of Manifest Destiny blossomed. For fascinated audiences Gilpin drew glorious word pictures of a future civilization centered in the Mississippi Valley and the Rocky Mountains, with Denver as the hub. His visions carried him upward and onward to describe the practicability of the United States uniting the world under a system of railways, one of which would reach the European continent via the Bering Strait. But fanciful as some of his visions were, Gilpin now and then came down to earth. As early as 1859, he spoke to a group of Kansans on the "New Gold Fields" of Pikes Peak, saying, "…[T]he facts…collected by me are so numerous and so positive that I entertain an absolute conviction…that gold in mass and in position and infinite quantity will within the next three years reveal itself to the energy of our pioneers."

And now, three years later, he was riding in a hard-sprung stagecoach across the plains to take up his job as governor of a new political entity made necessary because of the truth of his prophecy.

The Denver toward which Gilpin rode could by no standards be compared to capitals of the Midwest and East. Its treeless, dusty streets, lined with false-fronted buildings, its occasional brick edifice strangely incongruous in the incivility of its surroundings, and its motley-dressed population of miners, merchants, Indians, town boomers, drifters and drovers, lawyers, and gamblers were marks of its individuality. Yet progress toward becoming a conventional city had steadily been made. Denver had come a long way from the first crude log huts built by McGaa and Green Russell in 1858. Construction in 1861 represented an investment of nearly $700,000. Among the business buildings, Denverites pointed with pride to the bright new brick structure on McGaa Street housing Clark and Gruber's banking house. When Austen and Milton Clark and their partner, E. H. Gruber, saw the first shipment of Clear Creek gold dust pass through their Leavenworth counting house, they wasted no time in becoming buyers of the new Eldorado's precious commodity. In June of 1860, they opened their Denver bank and soon they began minting $2, $5, $10, and $20 gold coins from the amalgam and dust they bought from the miners in the gulches. The settlers were delighted to trade their buckskin pouches of gold dust for the sparkling coins that on one side carried a rough image of Pikes Peak and on the other, the name of the mint. Gone now was much of the tidy profit to be made by gathering up drifted dust from the floor around a store's gold scales, but gone also was the nuisance of weighing out the dust for each purchase.

Down the street from Clark and Gruber's mint, the ladies in town could shop for the latest Paris style in hats at the Bon Ton. Other streets were crammed with offices of physicians, lawyers, surveyors, and claim agents. Young Dave Moffat's book and stationery store did a thriving business in mining texts and dime novels. But the busiest place in town was the post office, next door to Moffat's store, where three times a week, long lines of eager men and women waited for McClure's staff to put up mail from the East. Brisk and bustling as life was in the new Eldorado, to be denied word from home dulled a person's enthusiasm for the adventurous West.

To take one's mind off not hearing from the folks back home, there were scores of diversions available. The genesis of spare time activities, cultural and otherwise, had been as swift as the development of commerce. Professor Goldrick was pleased to announce in the *News* that Denver had a thriving chess club and a brand new Masonic hall. A course of lectures was scheduled for Apollo Hall, with Kit Carson as the opening speaker. Madame Carolista, a curvaceous tightrope walker, set mouths agaping as she tiptoed her way across Larimer Street on a highwire stretched between the Criterion and the New York Store fifty feet above the ground. A clerk of the Jefferson Territory senate opened a night school in which he taught Spanish and French, which, according to the peripatetic professor, "were much spoken in business." With E. A. Pierce, Goldrick himself founded a circulating library, and after taking a good look at the goings-on down on Indian Row and at the all-day, all-night saloons and variety halls, he promoted the first Sunday school. Church attendance, however, could not match saloon attendance. When the Reverend John H. Kehler arrived to give his first service in the log cabin that bore a hastily painted sign reading "ST. JOHN'S EPISCOPAL CHURCH," he found two people in his congregation. Undaunted, he began his text with "Where two or three are gathered together in My name, there will I be, in the midst of them." For a time, the Reverend Mr. Kehler fought a losing battle against poor attendance. It wasn't because of disinterest among his congregation; it was just that they were suddenly called elsewhere. Out of twelve early parishioners, two were executed for murder, five were shot, one committed suicide, and one died of the d.t.'s.

Not all of Denver's clergy were fortunate enough to have a roof, however rude, over their heads. But the fact that he was churchless bothered the Reverend George W. Fisher not at all. He carefully inspected all the saloons in town, picked the largest, and with the proprietor's permission loudly held forth on Sunday morning with "everyone that thirsteth come ye to the waters, and he that hath no money, come ye to buy, and eat; yea, come buy wine and milk without money and without price." At that pulpit, his was a one-Sunday stand. The saloon owner decided that such talk was a demoralizing influence on his customers who, up to that point, had strictly adhered to the large printed admonishments behind the bar stating "NO CREDIT" and "PAY AS YOU GO."

"PAY AS YOU DANCE" was also the rule in the variety halls. A dollar bought one dance with a flirty-eyed, scantily clad, hurdy-gurdy girl and one bottle of beer. If the Last Chance mine or the Excelsior or the whatever mine were paying well, a man might favor a bottle of champagne with his dancing. For that he laid down one of Clark and Gruber's $5 gold pieces. If his mine were paying extraordinarily well, he might lay down several $5 gold pieces in the course of the evening, and as the heady wine titillated his tired body, the miner might find his hurdy-gurdy girl attractive enough to put out one more gold piece for the key to her room which she demurely sold him for a later rendezvous. And when he found there was no such room number as that scratched on the key, he might, if his sorrow were greater than his anger, visit Madame Mortimer, a "clairvoyant physician" by her own advertisement, who specialized in "diseases of the heart." But even her services were not to be found when she forgot the maxim "physician, heal thyself," and succumbed to her own specialty, retiring into domesticity, as one rueful observer put it, with the "brevet rank of lady and wife."

In another part of town on a Saturday evening, away from the rough, flannel-clad miner and the painted dance-hall girl, away from the clamorous brassy sounds of the rinky-dink saloon pianos, could be heard the measured notes of a waltz and the light rustle of crinoline. The legislators of defunct Jefferson Territory in shining linen and fine broadcloth were having their last fling. The supper table, decorated with sprays of spruce and pinecones and laden with delicacies imported over the plains, was as enticing as any to be found "in St. Louis or Baltimore." But one flaw marred the occasion. One or two of the bachelor lawmakers were accompanied by "theatrical ladies," and on this account William Byers' prissy wife Lib and several other of Denver's matrons saw fit to leave the room.

Into this potpourri of buckskin and broadcloth rode William Gilpin to take his place as governor of the brand new territory. It was late in the afternoon of May 27 when his coach rolled to a stop in front of the Tremont House, Denver's newest and fanciest hotel. On the way into town from the high plateau to the east, Gilpin had looked across the broad bottomland of the South Platte to the wall of uneven, snowcapped peaks to the west. Before him lay his capital, minute, expectant, assured, and so he hoped, as durable as the majestic Rockies stretching north and south as far as he could see. As he gazed on his charge, there formed in his mind the words he would deliver to his first legislative assembly: "The stern and delicate duty which is confided to you is to create and condense into system and order the elements of stable government for this commonwealth of the primeval mountains."

It would appear that William Gilpin had been very carefully briefed on the state of affairs in his infant territory.

The reception for the new governor that evening began promptly at "7½ PM" at the Tremont House. The good-looking, tall, full-bearded governor, a bachelor at this juncture, stood on the balcony of the hotel surrounded, as Goldrick painstakingly recorded, by "a large number of Denver's fairest ladies...like flowers on our mountain's brow...creating an ornament and an interest to the lookers up from below."

"The lookers up from below" liked what they saw, and at the end of Gilpin's somewhat impassioned speech on the past and future progress of Denver, the vast production of gold in the territory, and the need to hold fast to reason in the face of troubled times, they gave him "three loud cheers." The band played the national anthem, General Larimer got in his two cents' worth, and the crowd dispersed, most of them to file into the hotel to shake the hand of their new chief executive, the rest to move on to their homes or to saloons to speculate on what the presence of this hazel-eyed spellbinder meant to their futures.

They had not long to wait before finding out. On July 10, Gilpin admitted justices to the newly created supreme court. On July 12, he announced election districts and a general election to select a delegate to Congress and a territorial legislature. The election was held on August 19, and Republican Judge H. P. Bennett was chosen to succeed Beverly Williams whose secessionist leanings were a little too strong for the majority of the voters. Twenty-two members were elected to the legislature to represent the 25,331 people Gilpin's prompt census revealed were scattered throughout the territory.

By September 9, the new legislature was hard at work, sitting in the session day and night in a two-story building on Larimer Street across the street from the Criterion saloon. In his inaugural message to the lawmakers, Gilpin called for the establishment of counties, schools, and a civil and criminal code. The tyro legislators were quick to give the governor what he asked for. Seventeen counties were carved out of the bounds of the territory, a school system was devised, and the common law of England was made the basic law of the territory. Imprisonment for debt was prohibited, but the lawmakers, remembering the previous year's rash of rustling and its tragic consequences, decreed that horse-stealing would be punishable by a minimum of twenty years in jail and a maximum of death. On the other side of the coin, convicted kidnappers could look forward to an easy one to seven years in the pokey. The legislators also remembered that Jefferson Territory foundered on its $1 poll tax, so they made a new levy of 50 cents.

With all Denver caught up in the progress of the new government and its official business, Governor Steele, not to be left out of the political arena, formally dissolved the inoperative Jefferson Territory. With its public demise went Auraria's blue-uniformed, white-plumed Jefferson Rangers and Denver's smartly turned-out Guards, to the consternation of Major Byers of the Rangers and Captain Park McClure of the Guards. Byers thought it exceedingly importune at this time of divided loyalties to deny the new capital the benefit of an organized militia. McClure, as leader of the Guards, regretted its passing as a loss of a chink of his prestige, but he would soon find another uniform that would suit him just as well.

When the first joint resolution of the little band of territorial legislators proclaimed Colorado loyal to the Union there were grim faces to be seen at the Criterion. After the flag fracas in front of the Wallingford and Murphy store, it was no secret that Harrison's spa was the headquarters of the southern sympathizers in Colorado. Avowed local and very vocal secessionists including Park McClure, Beverly Williams, John Moore, Jim Coleman, and their cronies commandeered the same tables every night to read the war news in the six-day-old Kansas papers and vociferously argue the cause of the Confederacy with any who would listen. While the vocal champions of slavery held the attention of the patrons, only a few noticed sunburned, trail-tired newcomers with pronounced southern drawls who bellied up to the bar and at a high sign from Harrison quietly followed him into a back room. Later, in the dead of night, grim-faced and thoughtful, the newcomers slipped out the side door and onto awaiting ponies. Coincidentally, robberies of mine strongboxes and stages taking gold shipments to Denver and eastward took a sudden jump upward. And horse-stealing once more plagued the territory. Denver's two thousand residents grew edgy. To add to their anxiety, word went around town in the middle of May that Confederate agents would pay $40 for every Navy Colt, $50 for every Dragoon pistol, and a lot more for every rifle delivered to the Criterion. Despite public warnings that without an organized militia Denverites might well need their guns, arms of all descriptions poured into Harrison's emporium.

The unrest that spread over Denver as a result of the war received new fuel in the foreshadowment of a growing danger from the plains Indians. From some source, and few were convinced it was not the rebels, the local Arapaho and Cheyenne were getting enough whiskey to turn them into brazen thieves who bullied their way into homes to take what they wanted. When new contingents of the two tribes moved into camp downstream from Denver on the Platte swelling the Indian population there to several hundred, rumors raced through town. The Indians, said confidential informants, had wiped out Fort Wise and soon it was to be Denver's turn to bear the wrath of the red man. These stories were followed by tales of a rebel plot to sell guns to the Indians and then get them hopped up on rotgut so they would sweep down on the settlers in Denver. About this time a message from Chief Little Raven on the subject of Indian thefts did not help to calm the populace. He explained that he could control his Arapaho braves, but he could not guarantee Black Kettle's fractious Cheyenne, and he went on to say that he had heard via the Indian grapevine that the Comanche were on their way to raid Denver. He urged caution and restraint on the part of the white man.

For an Indian to be cautioning the whites to use restraint was almost more than an already distracted Denver could bear. A plea for federal troops was fired off at once to Fort Wise. The fort's commander, Colonel Sedgwick, whose opinion of Colorado miners was succinct if sour ("there never was a viler set of men in the world") was totally unsympathetic to Denver's request. He answered that he had no troops to spare and suggested that Denver try to persuade chiefs Little Raven and Black Kettle to remove the danger by taking their tribes away onto the plains. Despite its bad news for Denver, the message allayed one rumor: Fort Wise was still in the hands of the Army.

Meanwhile, small bands of drunken Indians roamed the countryside around Denver stealing from the settlers whenever they got a chance. Every new incident strung the town tighter with tension. But suddenly the tension eased when three Mexicans were arrested and charged with selling firewater to the marauding braves. The culprits got off with a suspended sentence and were exiled. The Indians calmed down once their source of whiskey was removed, and the raids stopped. With the crisis past, Denver's equilibrium returned and people laughed off as preposterous the notion that the Confederates were plotting to incite the Indians against the whites and then, with the white settlers occupied in battling the red man, move in to annex the territory for the South. But on May 23, when Park McClure suddenly left town, reportedly in charge of the cache of weapons deposited at the Criterion and headed for a rendezvous with General Stirling Price's Confederate forces marching on Missouri, the small worm of anxiety reappeared. Suspicion grew to conviction when word came to Denver about a military build-up going on in Texas.

Down in San Antonio, General Henry H. Sibley, CSA, was mustering a brigade of Texans and volunteers from other states for a march westward. His orders stemmed from a secret, bold but feasible strategy worked out by his superiors in Richmond and calculated to take the West for the Confederacy. Among the sparse population of the territories of Oregon, New Mexico, Utah, and Colorado, and the state of California, rebel agents found enough feeling against the North to encourage such an ambitious venture. The potential reward was great: gold-rich California and Colorado, a line of ports opening the door to the Pacific trade, and an interior zone of great resources. Added to this, under a secret agreement with Benito Juarez, then president of Mexico, would be the annexation of the northern Mexican states of Chihuahua, Sonora, and Baja California. It was a prospect that bedazzled the geopolitical idealists of the South and placed on the shoulders of Sibley and Price an awesome responsibility.

No less awed, had he known the extent of the Confederate plan, would have been William Gilpin. As it was, Gilpin knew only that a Confederate army was massing in Texas and that logically one of its first goals would be the taking of Colorado. With the Confederate eye obviously on Colorado gold, Gilpin, as a military man and as custodian of a federal territory, plainly saw his duty. While his legislative program went smoothly forward, the chief executive turned to the subject of defense. The nearest US troops were at Fort Wise, two hundred miles south of the capital. And uncooperative as he was, Sedgwick was telling the truth. His garrison and that at Fort Garland farther to the south were both understrength owing to the demand for men on the eastern fronts, leaving the door to Colorado's riches dangerously ajar. Clearly, if Colorado was to be secured for the North, she would have to rely on her own men to accomplish the deed.

With highly plausible and persuasively presented arguments, Gilpin sought permission from Washington to raise a regiment of volunteers to protect the territory. Authorization came promptly. What was lacking were funds to go with it. At this result, a lesser man might have given up the project. The territorial treasury was starkly empty. No tax monies were yet on hand to finance the venture and no miner-millionaire leaped into the breach to underwrite the cost of raising an army. But the governor was undeterred. He commissioned a set of officers—Denver attorney John Slough became the regiment's colonel; editorial writer for the *Herald,* Sam Tappan, got the nod as lieutenant colonel; and the militant minister John Chivington was tapped for regimental major. In July and August these three ardent officers sent recruiting teams to comb all the major mining districts in the territory. The timing was perfect. In May, 1861, the transcontinental telegraph line was completed as far west as Fort Kearney. Now Colorado was brought two days closer to the outside world. It was but four days from Fort Kearney to Denver by stage and every arrival brought stacks of eastern papers. By August just enough war news had reached the outlying diggings to fire up the passions of the miners and for once the pull of patriotism overrode the urge for a quick fortune. Men threw down their tools and followed their consciences. Southerners, many of them Georgians, headed for their home states to serve in Lee's armies. Some of the Yankees left Colorado to make their way to the northern states to join the regular and volunteer units. The rest flocked to enlist in the First Colorado Volunteer Infantry.

Down out of the mountains they came in tens and twenties to the board barracks at Camp Weld, a hastily built training center situated in the willows of the bottomland two miles upstream from Denver on the

Platte near the Byers ranch. Here they traded picks for gleaming Sharps rifles and their well-worn flannel and buckskin for new brass-buttoned uniforms of Union blue. The small matter of who was to pay for the new wardrobe and the arms was solved by Gilpin with characteristic directness. He reasoned simply that since the volunteers were in the service of the US government, it was the duty of that government to sustain the cost of such service. Whereupon he issued drafts on the US Treasury to the tune of $375,000. Businessmen in Denver who supplied the new regiment readily accepted the "Gilpin drafts" as payment for goods and services rendered. And why not? Wasn't Gilpin a man of honor? And wasn't he a friend of Lincoln's? Certainly the president must have known what his distinguished appointee was doing.

But as Tom Wildman and his brother Augustus found out to their despair, Washington not only did not know what Gilpin was doing, the Treasury refused to honor the drafts when they were presented for payment. The Wildman brothers had found mining neither attractive nor profitable so they came out of Gregory Gulch to Denver to take up anything that seemed to them to be a reasonable and remunerative employment. While trying their hand as clerks in Dick Whitsett's recorder's office they saw the accelerating need for local cartage as Gilpin's war preparations became a part of the capital scene. Consequently, they bought a couple of wagons and teams and hired themselves out as draymen. They supplied 100,000 board feet of lumber for the building of Camp Weld, enough uniforms to tote up a return of $2,000, and 70,000 pounds of corn to feed the embryo infantry. But the profit was all on paper. The Wildmans and others like them who held Gilpin's drafts suddenly found that instead of operating a going business they were close to bankruptcy.

Meanwhile, on the streets of Denver, the new recruits of the First Colorado did not act at all like an army that could not pay its bills. From the first day of muster it was clear that the cocky, freedom-fancying miners who made up the main body of the regiment would take orders when and from whom they chose. Their impudence was legend in a matter of weeks. When, for example, they found that guard detail was selected from the left of a line of men, they formed up in circles. When the whiskey ration was denied them after a drunk and disorderly spree in town, one company marched to the company commander's quarters and gave him three loud groans. When Company F was detailed to Fort Laramie to collect a shipment of arms, they plundered their way from one settlement to the next, lifting chickens, pigs, fresh vegetables, and stoneware jugs a-gurgle with tanglefoot whiskey. On leave in town they tippled and scrapped their way through the saloons and variety halls leaving a trail of broken glass, furniture, and teeth. When Gibson's *Herald* scolded the Firsters for their looting and lawlessness, a squad broke into the newsroom one night and made off with the entire stock of newsprint, turning the paper loose to the capricious winds on the mesa east of town. But as summer wore on, raw defiance of authority gave way to a restlessness born of pent-up energy and downright frustration. The men of the First had joined up to fight, but from sunup to sundown all they did was drill, drill, drill. Days of march and countermarch followed days of more of the same. The Firsters became so good at it that the nightly ritual of the dress parade saw cashmere-suited dandies from town squiring ladies chicly draped in sacques of fine calico and Mozambique cloth, and daubed with the latest fragrance, "Balm of a thousand bayonets," crowded around the edge of the parade ground to take in the show.

The zoolike quality of the soldiers' lives was enhanced by gangs of spectators far more irksome than the society folk. Not all the secesh boys left Colorado to join their Confederate brethren in ranks. A number stayed on, some to work actively if clandestinely for their cause, but most just to raise hell as they had always done as bummers. Soon after the formation of regimental companies, every time one of them marched into Denver to take up guard duty at the headquarters of the military district on Larimer Street, it was followed by bands of secesh who brazenly shouted taunts and abuse at the volunteers until scarlet color crept up troopers' necks and inevitably one or more broke ranks to attack their tormentors. But in the confusion that always followed, the rebels slipped away, only to turn up again the next time to bait the men in blue.

After two months of exhibition and harassment, some of the rawboned and thin-skinned Firsters decided that the enemy was a lot closer than Texas and should be dealt with without further delay.

In the tradition of soldiers from the dawn of war, the Firsters had often favored with their company the friendly parlor of Ada Lamont's house. They had done so as civilians and as soldiers they saw no reason to stop, even if there was a definite aura of gray instead of blue about the place. So it was here that B Company started its own war against the South. On the night of August 21, they strode into Ada's bordello, and when the first secesh boy cried, "Well, if it ain't one of Gilpin's Pet Lambs!", the fight was on. Shouts and oaths mixed with the high-pitched shrieks of Ada's boarders drowned out the tinkling piano, crashing glassware, and splintering furniture. The bartender and the piano player dove under the bar. Ada's girls fled into any corner they could find, screaming at the top of their lungs. In a matter of twenty minutes, the North had won a decisive victory. Leaving the rebels to lie where they fell, B Company straightened their tunics, dusted off their britches, and straggled over to the Planter House to toast their victory, and then returned triumphantly to camp.

After they came to, the southerners crawled up the street to the Criterion to report their defeat to Charley Harrison. The gambler checked the barrels of both of his pearl-handled pistols, gathered up a handful of henchmen, and set out on the vengeance trail. A lone sentry guarding one of the barracks attached to headquarters of the military district took the full brunt of their fury. He was found the next morning out on the prairie, alive but beaten within an inch of his Yankee life.

After the sentry's beating all Denver was alerted for trouble. Colonel Slough, hoping to forestall a showdown between his army and the people they were intended to protect, tried to persuade civic officials to disarm civilians. No one except the army liked that idea. Instead, an extra force of police was deputized to patrol the town's streets. They were, however, not in the right place at the right time when trouble came. The following Saturday night after the affair at Ada Lamont's, B Company headed for the Criterion to settle the southern question for a second and, they were determined it would be, a final time.

When B Company, fully armed, presented itself at the Criterion, its entry was quickly blocked by Harrison and his men. There were words. Someone threw a punch and the powder keg exploded into a flying-fisted free-for-all. The fight moved in and out of the saloon in waves of crashing fists, thudding bodies, rebel yells, and gutteral oaths. Suddenly, as one of the outward-bound waves of punching, rolling, wrestling men hit the dirt street, gunfire chattered from the upper story of the Criterion. Private George McCullough dropped with a bullet-shattered ankle, while another soldier grabbed his bleeding left ear. The rest of the men, rebel and Yank, retreated under the portal overhanging the porch. There was a shouted order and B Company, in a remarkable show of obedience, brought their rifles to the ready and surrounded the saloon while the southerners ran to take up positions at the windows inside. At another order, several of the Firsters broke away and went after the cannon that decorated the entrance to the headquarters on Larimer Street. When the marshal and his pack of mounted deputies arrived, having been summoned by a passerby, the gun was in place, loaded, and aimed at the front door of the Criterion.

Marshal Pollock quickly shoved his way through the line of Firsters and marched resolutely into the Criterion where he arrested Harrison and his lieutenants and took them into custody to await trial. The army packed up its cannon, sponged off its wounds, and returned to camp not unhappy at the outcome of the evening. The next day Harrison tried to free himself on a writ of *habeas corpus,* but it was denied on the grounds that his actions were evidence of complicity in the rebellion of the South against the US government.

Harrison's trial was a model of judicial propriety. The newly appointed Chief Justice Benjamin Hall read to the grand jury Harrison's rights under the Constitution and followed them up with dark allusions relating to the nature of treason and the forceful overthrow of government. After two weeks of testimony, the jury found the one-time bummer boss guilty of obstructing the functions of a duly-elected government. Rather than consign Harrison to the martyrdom of prison and perhaps thereby bring on Denver a full-scale uprising of local secesh, which was rumored to be a plan of revenge by the rebels, Hall fined the gambler $5,000 and gave him two days to sell out his holdings and leave town. Contained and suave to the end, Harrison did just that. From Leavenworth, where the stage from Denver deposited him, Harrison slipped south to join Colonel Emmett McDonald's Confederate Fourth Missouri Cavalry. Later, he and his old friend Major Park McClure rose together as leaders of a band of guerrillas roaming Kansas and Missouri. At Lightning Creek in southeast Kansas, on May 16, 1864, they were attacked by an Osage war party who found the encroachment by soldiers as abhorrent as that of settlers. McClure died with a tomahawk buried in his skull. Harrison, his handsome face blown away, fell to his knees still firing blindly as he was ridden down and scalped by the enraged red men.

Back at Camp Weld, Gilpin's volunteers groused through the early part of the Colorado winter more like caged lions than pet lambs, but their day was soon to come. In January of 1862, Denver heard that Sibley was on the move north to challenge Colonel E. R. S. Canby's Union forces in New Mexico, the important stepping stone to Colorado. Canby confronted the Confederates at Valverde in the Rio Grande Valley below Albuquerque on February 21. Sibley's rebels squashed Canby's forces and chased the survivors northward, capturing Albuquerque and Santa Fe in rapid succession. The southerners' next northward goal was Fort Union, beyond Glorieta Pass through the Sangre de Cristo Mountains. As soon as word of Canby's rout at Valverde reached Denver, the itchy-nitch Colorado First got its marching orders. For the moment, jackanapes stunts were forgotten as the Firsters hurriedly stuffed gear into packs, buckled on bayonets and canteens, loaded wagons and mules with stores and ammunition, and shouldered their rifles.

At the hour of departure, companies of men in dusty blue serge lined up by two's in the midday sun. Halfway down the column, twenty-three-year-old Mary Sanford stood holding the hand of her young husband, Lieutenant Byron Sanford, a miner from Boulder. Neither spoke. Their eyes told their feelings as the tears welled up and trickled down their cheeks. All along the line were similar scenes as wives and husbands realized the days of parades and pomp were over and that these could be their last goodbyes. Suddenly the shrill notes of a bugle cut into the silent colloquies and the First moved out, stepping briskly to the rolling notes of their marching song:

> Way out upon the Platte, near Pike's Peak we were told
> There by a little digging we could get a pile of gold,
> So we bundled up our duds, resolved at least to try
> And tempt old Madame Fortune, root hog, or die!

Under orders to make a forced march, Slough led his men a breakneck pace. Thirty to forty miles a day they tramped, up Cherry Creek over the divide, down Fountain Creek to old Fort Pueblo, east along the Arkansas to Fort Wise, and then south again over tortuous Raton Pass to Fort Union. Here the Firsters revictualed and rested for ten days, whooping it up in their usual fashion, stealing whiskey and delicacies from the officers' mess, ducking off post to consort with the Mexican women at a small settlement five or six miles from camp, and generally staying drunk and defiant for the entire time. Somewhat refreshed, the Colorado troops set out again toward Glorieta Pass on March 21 to find Sibley's Texans.

As the troops under Slough approached the Pass, Major Chivington and a force of four hundred men were sent ahead as an advance party. Chivington's explicit orders were to reconnoiter to locate the enemy's position but to avoid an engagement at all costs. Chivington however took orders no better than the rank and file. The day after his sortie began, the militant preacher ran into a force of Texans sent out to find the Union troops. Against orders, Chivington bellowed for a charge, and fortunately for him his miners sent the Texans pell-mell in retreat toward their main force after three hours of hard fighting.

On the twenty-seventh, Slough's Firsters joined Chivington's advance party at the eastern end of Glorieta Pass. From the prisoners Chivington had taken, the Colorado colonel knew that Sibley had sent Colonel W. R. Scurry out from Santa Fe with nearly twelve hundred men. With Slough's thirteen hundred, that meant the two forces were very nearly equal in number. But Slough reasoned that in narrow Glorieta Pass numbers would not count so much as strategy in a face-to-face encounter. After some deliberation, Slough again detailed Chivington and his four-hundred-man brigade to slip ahead of his main force, around the mountains, and to attack the Texans from the rear. While Chivington raced over the lower slopes of the mountains in a flanking movement, Scurry's men and a now much diminished force of Firsters met and fought toe to toe in the mouth of the pass. Gradually, Slough retreated in the face of superior numbers and heavy casualties. But his men gave as good as they got and gray uniformed bodies littered the piñon-studded hillsides along with the blue. Then abruptly at 5 PM, the Texans waved a white flag. Chivington had done his work well. When Scurry learned that the Union forces had skirted the seven-thousand-foot pass and had gotten behind him to burn his eighty-wagon supply train, bayonet his five hundred horses and pack mules, and capture his rear guard, he realized he had lost the day and was in danger of annihilation with federal troops harrying both his front and rear. As fast as they could travel through the sage- and pinon-covered terrain in the coming darkness, Scurry and his men scrambled the eleven miles back to Santa Fe. When Sibley heard Scurry's report of his losses, he caught the contagion of defeat and began a mass withdrawal from New Mexico back to the safety of the Texas border. Colorado was thus saved from the Confederacy. Chivington, the hero of Glorieta, was made a colonel, taking Slough's place as regimental commander after the zealous attorney resigned in a fit of pique because Canby refused to let him chase Sibley's army down the Rio Grande and finish the job.

❧❧❧

Denver went wild at the news of the Confederate defeat at the hands of the Firsters, and to celebrate there was an immediate wave of social activities. Balls, church festivals, private parties, and even taffy boilings filled the rest of the year. "All women were belles," recalled the wife of Governor Gilpin's surveyor general, but there weren't half enough of them. At a recherché ball held at the Tremont House there were so few women that "when quadrilles were danced they were mostly formed by two women and six men, two of the men with handkerchiefs tied on the left arm personating ladies in order that they might in 'the swing at the corner' and the 'grand right and left' at least touch the hands of women and, if possible, secure one as a partner for a later dance." Not so enthusiastic was the bride of Tom Wildman. Said she primly in a letter to her sister-in-law in Danbury, "I derive no pleasure whatever from attending these promiscuous assemblages, for such they *decidedly* are, as we find it an utter impossibility to attempt having a select gathering at the present time..."

Either Mollie Kehler Wildman's standards were too high or she was trying to keep from her husband's family the fact that, owing to Gilpin's freewheeling financing, they were too broke to entertain. In any case, there were at least thirty young matrons and maidens out of Denver's three thousand residents whose backgrounds were as impeccably circumspect as Mollie's. And every incoming coach brought more of them. Among Denver's social elite, each newly arrived wife, bride-to-be, sister, niece, and cousin was feted in a style aimed at if not quite the same as that found in the eastern homes they had left behind. The center of this whirl of social events was the large double cabin shared by the families of the Cass brothers, Joe and Oscar, who opened a bank in Denver shortly after Clark and Gruber opened theirs. Both well-to-do bankers had wives who quickly became renowned for the delightful parties they gave in their roughly built but finely furnished homes. Guests of the Cass families were seated on Sheraton chairs before a table laid with linen nappery and gleaming silver that the bankers had had freighted across the plains from the East.

Since, at first, household help was unattainable, the mesdames Cass, as did all Denver hostesses, called on their close friends to prepare cakes, salads, and confections for a scheduled affair. Later, the situation improved when coveys of young single girls braved the eight-day trip from Omaha to Denver in a rough riding stage ostensibly to benefit mankind by teaching school or doing mission work for their church, but really intending to latch onto a prosperous gold miner. Some, such as Miss Indiana Sopris, whose brother was one of the first prospectors on Cherry Creek and went on to become Denver's mayor after serving with distinction in the Colorado First, succeeded in both endeavors. She was the first woman to teach school in the little settlement, and she married handsome, and rich, Sam Cushman. Some others, who found their work as teachers and missionaries brought in scarcely enough money to keep them respectable, turned to jobs as domestics. But these were not the usual hired help and they insisted on all the privileges of the family. To the matrons of Denver, six hundred miles away from a ready supply of maids, cooks, and mother's helpers, hired help of any kind was worth the price they had to pay for the uncommon domestics. Competition for the services of these girls was keen. Each family tried to outdo the other in offering incentives, and it was explicit that privileges of the family included courting time. "Wanted," read a typical ad in the *Rocky Mountain News,* "a girl to do housework. She will be permitted to receive company every day in the week; a good substantial fence will be provided to lean against while courting and ample time will be accorded for that recreation, but no piano will be furnished."

The lack of a piano did not rule out musical entertainment. Down at Cibola Hall, the Christy Minstrels played to packed houses every night of the week. Before the performance, a brass band promenaded the streets blaring out ear-splitting renditions of the Anvil Chorus to drum up business and provide evening strollers with a sample of the night's entertainment. The Jack Langrishe-Mike Dougherty company held forth nightly in their elegantly remodeled Denver Theatre. The cast was often supplemented by a few dashing young amateurs like ex-Sheriff Ned Wynkoop, home on leave from duty with the Firsters. Offerings of "rich, racy, and refreshing songs, Dances, Gags, Local Allusions, and Pikes Peak Perpetrations," were intermixed with serious drama. There was no lack of patronage despite the occasional interruption in an evening's show. Once the male cast members, reacting in their own way to Denver's early shortage of females, went on strike demanding the import of actresses. On another occasion, a repeat performance of *Camille* had to be given because in the middle of the first act, the audience was suddenly called out to mount a posse to chase a murderer who had just killed a man across the street from the theater.

While the theaters and variety halls flourished during the war, business was generally dull. There were enough miners and enough gold coming into Denver to keep the pockets of the gamblers, saloon keepers, madams, and actors well filled, but commerce in real estate was at a standstill, no new businesses opened their doors, and those that were open stagnated. Part of the depression stemmed from the notorious Gilpin drafts. The refusal of the United States Treasury to honor them sent successive waves of anger, panic, and distrust through the business community of Denver. Overnight the popular Gilpin was labeled a cutthroat and traitor. Angry citizens got up a half dozen petitions demanding his recall.

In Washington, where he was called to explain his effrontery, Gilpin used his mesmerizing tongue to persuade the government that his action was totally justified in view of the imminent Confederate danger to Colorado. Behind the thick paneling of the hearing room doors, his listeners grudgingly consented that Gilpin's motives were pure, but they decreed that his behavior was ill-advised, and they reached a decision based on an old formula of political expedience: Gilpin's head for payment of the claims. Pat as it sounded, it was in fact not that simple. All those in Denver holding drafts were supposed to be duly paid, if they could submit fully itemized bills. However, many could present the bills but not the drafts because they had been passed from hand to hand at a discount and were counterendorsed so many times that those who originally held them had nothing on which to base their claims. To those who were left unpaid, there was some consolation in seeing Gilpin surrender the governor's chair to the eminent Dr. John Evans of Illinois.

Evans came to Colorado with a long list of credits. Friend of Lincoln, organizer of the Republican party in Illinois, founder of Northwestern University at Evanston (the name similarity was no coincidence—the town was named for him), he was at the same time a renowned physician who had led the way in the humane treatment of mental patients and in discovering the contagious nature of cholera. Last but not least, he was a man of wealth. If Gilpin had been the right man to secure Colorado for the Union from a military standpoint, Evans was the man to set the new territory on the road to civil stability. In his inaugural address to the territorial legislature in July, 1862, meeting this time not in Golden but down in Colorado City, a little log cabin town that sprouted at the foot of Pikes Peak, Evans made it plain that in his view Colorado's future depended on the development of three major areas—mining, agriculture, and transport.

It was in the latter area that his greatest interest lay, and among leading Denverites, he found many allies. Railroad-building in the sixties was a national pastime to rival town-booming. Railroading west was a part of the Manifest Destiny epidemic and a transcontinental track was the keystone to full realization of the potential of western expansion. In 1848, the United States had only six thousand miles of railroad, nearly all of it along the Atlantic seaboard. By 1858, the nation could boast twice that much track with ten railheads dotted along the length of the Mississippi. But there, on the eastern bank of the river, the network of rails ended. It stopped short not because of the barrier of the river, but because Congress and the supporters of what was adulatingly called the "Pacific Railroad" could not decide on a route across the rest of the country.

Colorado's interest in a transcontinental railroad was fostered less by national pride than by pragmatism. Mine-owners and speculators alike quickly saw the importance of a rail connection between the mines and the markets and supply centers of the East. And with the irresistible motivation of self-interest, Colorado people, with Evans in the lead, lobbied vociferously for congressional passage of the Pacific Railroad Act. In the meantime, to give some basis for their insistence, Colorado citizens commissioned Captain E. L. Berthoud and the old Indian scout, Jim Bridger, to find a pass through the Rockies that would be a compelling route and that would put Denver on the mainline of the transcontinental railroad. Berthoud and Bridger found a pass they thought would do nicely, and Evans went to Washington armed with charts and statistics on grades and weather conditions to step up the pressure on the slow-moving Congress. There was great rejoicing in Denver when the *News* triumphantly announced that Congress had at last passed the act authorizing construction of the long-vaunted railroad and subsidization of an organization to be called the Union Pacific Railroad Company to carry out the construction. The route however still lay in the balance, subject to more surveys.

Back in Colorado, William A. H. Loveland, miner and merchant of Central City, enlisted the help of Henry Teller, who had taken his law practice up to Central where the money was, in his plan to be ready with a vital link from the Clear Creek mines to the mainline, whenever it got to Denver. Loveland and Teller wangled a charter from the territorial legislature to build the Colorado Central and Pacific Railroad (any respectable railroader of the day put the word "Pacific" in the name of his road), from Golden, up Clear Creek to Central City, over to Boulder, and back down to Denver. A later amendment allowed construction westward to the Utah border. On the strength of the amendment, the two neophyte railroaders approached the directors of the newly formed Union Pacific Railroad Company and got backing to make surveys of ten mountain passes, including Berthoud's, through the Rockies to secure a route west to Utah. Not one of the routes made the grade with the eastern directors, so to speak, and the Union Pacific's interest in Loveland's project flagged. Its interest waned altogether when the Pacific Railroad Act was revised in 1864 allowing the Central Pacific Railroad (no relation to Loveland's project), a-building eastward from Sacramento, to extend its construction until it linked up with the westbound Union Pacific. Out of this amendment came a race of railroaders to see who could lay the most track in the shortest time. Expensive, time-consuming mountain construction in Colorado was ruled out—the first transcontinental track would follow the shorter, flatter route to the north, through Cheyenne, Wyoming, leaving Colorado, its multimillion dollar mines, and its dedicated believers in Manifest Destiny, high and dry.

ᖆᖱᖱᖱ

Denver had other attention-getting events to contend with closer at home than the westward inching of an iron ribbon of rails. Between 2 and 3 AM on the morning of April 19, 1863, flames burst through the roof of the Cherokee House on Blake Street in the middle of town. In the fitful gusts of the spring wind, the fire leaped from false front to false front, setting the sky alight with licking tongues of flame and billowing pine-pitch-smelling smoke. Men tumbled out of their beds to form ragged bucket brigades from the South Platte River up to the fire. But their efforts were too feeble to quench the searing flames and the plops of water hissed away as steam as the searing flames roared on from building to building and street to street. As they watched their town burn, the citizens cursed the city government for dragging its heels in buying fire equipment. Pumps, hoses, and carts had been authorized the year before but as yet there was still no sign of them. As structure after structure was consumed, buckets were abandoned as firefighters caught up crowbars and axes and harnessed teams to drag down buildings in the path of the fire. At dawn, the heart of Denver lay smouldering in ashes. Soot-blackened and woebegone, the residents stood on the edge of the ruins gazing dejectedly at the results of the holocaust. There were few human casualties, but many of the major business houses were destroyed. Property consumed in the three hours that the fire ate its way through town totaled $350,000, nearly half of the entire building investment in Denver.

With the resilience that marked their response to other vicissitudes of pioneer life, Denverites got to work to rebuild their town, little knowing they would have to go through the same process all over again the next year, one month and one day after the anniversary of the fire.

Back in 1860, when the first permanent buildings went up adjacent the placid little brook that was Cherry Creek, an old Arapaho, watching from his portable tepee village down on the Platte, commented ominously to William Byers that in his lifetime he had seen a wall of water taller than a man's head come down the creek and spread over the bottomland like a blanket. Neither the editor of the *News* nor anyone else took the old man seriously. To the settler from the East the phenomenon of the flash-flood was generally unknown. For several days in mid-May of 1864, far upstream on Cherry Creek, near the divide, a localized downpour deluged the area, swelling the soil to saturation around the creek's headwaters. A few showers fell down in Denver, but not enough to worry the townspeople, even though the *News* reported on May 17 that the creek appeared to be rising. Then shortly after the moonlit midnight of May 19, with "...the roaring of Niagara...the rumbling of an enraged Etna," a gigantic wave surged down Cherry Creek, bursting over its banks and carrying with it "broken buildings, tables, bedsteads, baggage, boulders, mammoth trees, leviathan logs and human beings..." The description is Professor Goldrick's, and for once he did not exaggerate. The flood swept away stores, stock, books, safes, and in cutting a new channel, the creek ate up what were the day before priceless pieces of real estate. D. C. Oakes saw his saw mill torn apart and carried away, the city hall and all of its records were swallowed up by the torrent, and scores of ranchers lost hundreds of head of sheep and cattle. The force of the flood carried away the three-thousand-pound steam press Byers used to print the *News* as well as the building housing it perched beside the creek. Goldrick's rhapsodic account of the event appeared in a special edition of the *News* published through the good offices of another Denver daily that survived the deluge in good condition.

Not only was the *News* building and press swept away, but its editor and his family were nearly lost in the flood. Asleep in their modest ranch home above Denver, they were awakened by the noise of the hurtling water and climbed onto the roof of the house just before the swirling current charged through the windows and doors of the house. After several hours of clinging to their precariously swaying roost, they were rescued by Byers' close friend Colonel Chivington, who arrived with two enlisted men and a rowboat to take the newsman and his family to the safety of higher ground.

Once more Denverites cleaned up the mess of a major catastrophe and began rebuilding their town. Despite their bad luck, residents kept their humor. The day after the flood there appeared in one of the daily papers the following notice: "Lost, on the night of the 19th four first class building lots...anyone who will overtake and return them will be liberally rewarded."

The ravages of fire and flood could be obliterated but in late November came an event the memory of which left a permanent scar.

When the first gold-seekers charged into Cherry Creek they cared little that they were squatting on lands the US government had given to the Arapaho and Cheyenne in the Treaty of Fort Laramie in 1851. Of the white men in Denver, only a few other than the mountain men and traders had any respect for the Indian. The rest of the whites looked upon the red men who camped downstream on the South Platte as sullen, lice- and disease-ridden beggars who loved horse-racing and whiskey above all else, and who would as soon murder a man as look at him. At the same time, the white man was desperately eager to stay in this new Eldorado to reap the profits of its mineral wealth, and he clearly saw the advantage of remaining friends with the Indians. To the Indians, at first, the settlers were a curiosity from whom they could beg, trade, or if need be, steal their favorite commodities—sugar, tobacco, and whiskey—and who periodically treated them to great feasts of antelope. But as the waves of gold-rush emigrants in 1859 and 1860 brought hundreds and thousands of white men over the plains to usurp the red man's grazing and buffalo land, the enormity of their betrayal slowly dawned on young and headstrong Indians and sage old chieftains alike. Some of the more astute chiefs saw the futility of fighting, saw that it would only lead to extermination of the tribes, and sought peace in their own way. Chief Left Hand of the Arapaho stood up at the intermission of a show given by the Denver Amateur Dramatic Association in 1861 and made an eloquent if halting plea for the cementing of friendship between the white and red man. The audience roared its approval, applauding loudly and stamping their hobnailed boots on the wooden floor, thinking his appearance was part of the play-acting onstage.

Not all chiefs were listened to or wanted to settle their differences with the white man peaceably. Some did not believe in negotiation, and to them the Civil War in which white man was pitted against white man provided the right backdrop for Indian retaliation. From his vantage point at Big Timbers on the Arkansas River, William Bent, one of the few white men considered a friend by the Cheyenne, watched in anguish as the Indians lashed out time after time in the only kind of defense they understood to pillage and kill among the settlements and wagon trains of the ever westward-moving white man. Bent begged the US government to define Indian rights for once and all. It did. The Treaty of Fort Wise in 1861 took away all of the land of Colorado east of the Rockies that was given to the Arapaho and Cheyenne in 1851, leaving them a small triangular-shaped reservation roughly one hundred miles across each leg of the triangle. The new reservation was bounded on the south by the Arkansas River, on the east and north by the meandering Sand Creek, and on the west by an arbitrary line drawn straight south from the creek to the Arkansas River, five miles east of the mouth of the Huerfano River. In the center of the southern boundary of the reservation on the Sante Fe Trail stood Fort Wise, which was renamed in 1861 to honor General Nathaniel Lyon, the first Union general killed in the Civil War. This fort, with its well-fortified outerworks, roomy officers' quarters, and sutler's store, was the logical administrative center for the reservation. Here were stocked the annual $30,000 in foodstuffs and other goods, including farm implements, that went along with the treaty. It was the hope of the US government that the impulsive, nomadic red man could be transformed overnight into a placid agrarian.

The Treaty of Fort Wise was ratified by Congress, and the Indian leaders signed it in August, 1861. But is implementation was slow. Bands of both tribes were roaming the plains, hunting and warring against their Indian enemies, the Ute, in the mountains of central Colorado. During 1862, in the process of their wandering and warring the Cheyenne and Arapaho occasionally raided settlers' farms for food and stock. Most of the trouble was in Wyoming, but the news of Indian raids even that close to Denver reawoke the latent fear in the townspeople. What really frightened them was the fact that their pride and joy protective force, the First Colorado, was still down in New Mexico, too far away to be of any help to Denver in an emergency. Their fears were somewhat allayed by the authorization of the War Department for the formation of a second regiment of Colorado volunteers. When it was at half strength, the First arrived home. Denver breathed a sigh of relief and turned its attention gratefully from Indian troubles to another drumming on the tom-tom, the project of statehood. The election at which the issue would be voted on was still two years away, but already the politicians were at work.

On the strength of the public acclaim he had received after Glorieta, blowhard Colonel Chivington settled his ambitious eye on both the job of territorial delegate to Congress and on the job of senator in that body should the motion of statehood carry. While he waited for the election to roll around, Chivington, as head of the Military District of Colorado, set out to make a name for himself on the Indian question. About this time, the mountain Ute began a series of raids in west-central Colorado Territory, and once more Denver became preoccupied with the danger presented by the red man. Chivington warmed to his work. He recast the Colorado First in the role of cavalry troops and split its forces, sending five companies under the newly promoted Major Wynkoop to chastise the Ute and the rest he sent eastward under the mild-mannered but firm-handed Lieutenant Colonel Sam Tappan to garrison Fort Lyon. The half-strength Second Colorado went even farther east to Fort Larned, Kansas, to try to protect the Santa Fe Trail against an increasing incidence of Cheyenne depredations. In Denver, as he manipulated his forces from the pinnacle of his office, Chivington made political hay out of the jitters of the people in his charge. A captivating speaker, the colonel made the rounds of lodge meetings, political rallies, church suppers, and ladies aidsociety gatherings, using his best evangelical tones to call for all Indians, "little and big," to be killed because "nits make lice." His stand was greeted enthusiastically, and before long nearly all Denver was convinced of his theory that one decisive defeat would bring the red man to his knees forever.

In early 1863, the Cheyenne raided a white rancher's spread up on the Cache le Poudre Creek. No blood was shed, but the red men got away with several head of cattle and successfully eluded the twenty Colorado First troopers ordered out to chase them. Another rancher, Isaac Van Wormer, who lived thirty-five miles east of Denver had some of his cattle run off by Arapaho. Late in March, at the instigation of the agent for the Cheyenne and Arapaho tribes at Fort Lyon, these two tribes and the Comanche and Kiowa sent a delegation of chiefs to Washington to meet with Lincoln in an attempt to find some road to peace. It was an amicable but fruitless meeting. The chieftains came back to their tribes to show medals and handsome canes given them by their hosts but without any concrete solutions. To add to the impasse, during the absence of the chiefs in Washington an event occurred that would have tragic ramifications. A Cheyenne brave, fired up with whiskey given him by a soldier at Fort Larned in exchange for sleeping privileges with a squaw, approached the trooper who was on guard duty outside the gate and demanded more firewater. The soldier refused. Enraged, the brave got on his horse and tried to run down the soldier at a full gallop. As the brave thundered toward him, the soldier raised his gun and put a minié ball through the Indian's heart, killing him instantly. There was no immediate reaction among the Cheyenne, but to play safe, Lieutenant Colonel Tappan, commanding Fort Lyon, sent some reinforcements to Fort Larned. For this, Chivington, who did not like subordinates to issue orders without his approval, removed Tappan to Fort Garland, well out of the way of the action on the Arkansas, and made Major Ned Wynkoop commandant at Fort Lyon.

Throughout the plains the atmosphere was taut with strain. Rumors of a coming uprising of the combined tribes skipped from settlement to settlement like prairie fire. Already bathed in the fear of a full-scale Indian attack on Denver, when word reached the governor of Colorado Territory in May, 1863, that the Cheyenne and Arapaho were planning just such a raid on the capital in concert with the Sioux, he was quick to call for a peace council of red and white men. Evans' informants were two squaw men, one of them William McGaa and the other a man called North. They told him that the Indians had made a pact with each other to go to war against the whites as soon as they could gather up enough ammunition to do it. Evans set a time and a place for the parley and traveled out into the plains to meet with the chiefs of the Cheyenne and Arapaho, but the red men failed to appear. Later, they told their agent at Fort Lyon that the buffalo hunt was going too well to break it off to talk just then. With no more assurances of the safety of his people than before, Evans returned to Denver, and the populace spent the rest of the year with guns at the ready.

Meanwhile, Evans had somewhat better luck parleying with the Ute. When Evans proposed a council to the Ute chief, Ouray, Ouray obliged and got his people to sign a treaty in which they agreed to relinquish the agriculturally rich San Luis Valley in return for $20,000 in annuities and some cattle, sheep, and horses. Now at least Denver and the mining camps to the west would not be struck in the back by the Ute if the settlers had to face an onslaught by the plains Indians in the spring.

When spring came, so did war with the Arapaho and Cheyenne. But it did not come quite as McGaa and North warned it would.

In April and May, in response to pleas for help by settlers in remote ranches and settlements, various companies of the Colorado First fought skirmishes with raiding parties of Cheyenne at Fremont's Orchard and at a ranch, about seventy miles northeast of Denver on the South Platte River, and at several places on the Republican and Smoky Hill rivers. Word of these engagements sent new waves of jitters through the people of Denver, but these waves were ripples compared with the abject terror that seized the town at the sight of four bodies trundled in on a buck-board on June 18. Holding the reins of the wagon's team was Isaac Van Wormer, to whom Indian raids were getting to be an old and grim story. In the wagon were the bodies of the tenant manager of his ranch, Nathan Hungate, Hungate's wife, Ellen, and their two daughters, Florence, six, and Laura, three. On the day before, Hungate and Miller, a hired man, had set out from the ranch to look for stray cattle. When they reached a rise east of the ranch house, they looked back to see smoke billowing from the building. Without asking the question, they knew it was an Indian raid. Miller spurred his horse for town to get help, while Hungate turned back to the flaming house to find his family. In town, Miller tracked down Van Wormer and together they rode out to the ranch. About one hundred yards from the house they found Mrs. Hungate's body and those of the two girls. The mother had been stabbed several times, scalped, and raped. The girls' throats were cut, nearly severing their heads from their bodies. A short time later, Van Wormer and Miller found Hungate's horribly mutilated body with the scalp ripped from his head.

The corpses of the Hungates were placed on exhibit in a shed on Blake Street and all day long solemn, scared townspeople filed past the remains of the dead settlers. Afterward small groups of people gathered on street corners, in stores and saloons, and in darkened parlors to talk of the outrage. Chivington ordered his troops into the field to look for the marauders, but the trail was cold and the soldiers turned up no trace of the Indians. The shock of the Hungate affair had not worn off when a few days later a man rode hellbent into town just as the sun set, shouting, "The Indians are coming!" Reacting on impulse, men gathered up women and children and swept them onto the second stories of the two safest buildings in town, Clark and Gruber's mint and the Army commissary. The local militia, formed after the first Indian scare found Denver without troops, gathered up its arms and patrolled the town beside the regular soldiers as wagonload after wagonload of white-faced settlers rumbled in from the outskirts. By midnight, some four hours after the first alarm, no Indians had appeared and scouts were sent to reconnoiter outside the town. They found no signs of the red men. Gradually, the panic receded, and families returned to their homes, tired and unnerved. Later, Denver learned that the rout was started by a jittery rancher who mistook a group of freighters moving around their campsite for Indians.

A false alarm was one thing, but there was a stark reality about the Hungate murders that convinced Evans that the threat to Denver was more serious than ever before. To meet the situation, he promptly issued a proclamation urging all friendly Cheyenne and Arapaho to put themselves under the protection of the Army by going to Fort Lyon, the Kiowas and Comanches to go to Fort Larned, and told all citizens of Colorado Territory to "organize for the defense of your homes and families against the merciless savages." Martial law was clamped on Denver, including a 6:30 PM curfew on all businesses, and Evans ordered all able-bodied men to drill daily with arms. Then he strapped on a pistol and wired Washington for permission to raise a third regiment, this one to be of one-hundred-day volunteers, expressly for the purpose of fighting the Indians. After cannonading Washington with letters and telegrams, each one more explicit on the impending danger from Indians than the last one, and all carrying the same general message—"unless the authority is given we will be destroyed. It is impossible to exaggerate our danger"—Evans got the permission he sought.

But the governor was exaggerating. The picture he drew of an ever closing ring of savagery around Denver was unfounded. There were the two raids by the Cheyenne up on the South Platte and the murder of the Hungates, but as it was later revealed there was no concerted plan by the plains tribes to sack Denver and slaughter its inhabitants.

To the east, however, there was no doubt but that the Indians of the plains had begun large-scale retaliation against the encroachment of the white man. From May through August of 1864, settlers' homes and wagon trains throughout southwestern Nebraska and northwestern Kansas were subjected to one bloody raid after another by Indians, chiefly the Cheyenne, who could not forget the killing of one of their braves by the soldier at Fort Larned the year before. In fourteen separate attacks, the Indians killed fifty white men, captured eight women and children, stole hundreds of horses, and set the torch to scores of cabins and wagons. Heading up the search and destroy mission in the field was Major General Samuel Curtis, commander of the Kansas military district. Under his command was part of the Second Colorado regiment, including his son Major Sam Curtis, one of the first of the Pikes Peakers to reach Cherry Creek after Green Russell's discovery of gold. Despite the eagerness of the troops and the ability of their commanders, the illusive, quick-striking Indians successfully evaded the pursuing cavalry.

Although Denver itself was not in danger, the town suffered the effects of the depredations the Indians carried out against the trade and communication routes that linked her with the rest of the country. The telegraph line to Julesburg in northeastern Colorado was cut with the regularity of a threshing machine. Freight trains and stagecoaches were driven off the trails by the angry red men. Travel to Denver came to a halt. Supplies grew short and prices sky-rocketed. And every day new reports of raids filtered into town. As these reports came to Chivington, he instructed Ned Wynkoop, the new commander of Fort Lyon, to watch for Cheyenne and "If any of them are caught in your vicinity kill them..." For Wynkoop, not yet twenty-six, but already a veteran of command and combat at Glorieta, the first part of those instructions was needless. He was watching diligently for the Cheyenne. He had read Evans' proclamation, and he was holding Fort Lyon open to any friendly Indians who sought out the fort as a refuge in accord with Evans' urging.

Suddenly in September, Black Kettle, one of the primary Cheyenne chiefs, sent word to Wynkoop that his tribe and the Arapaho, Comanche, Kiowa, and Sioux wanted a parley to exchange prisoners and talk about peace. Wynkoop was skeptical. The summer buffalo hunt was over, and the Indians would be looking to the long winter, perhaps without enough food to see them through it. Maybe, reasoned Wynkoop, they were only interested in making peace in order to get the annuities due to arrive from Washington, and then in the spring they would go on the rampage again when they could live off the land. Again, Black Kettle's invitation could be no more than a trap to lure Wynkoop and his men out of the fort and into annihilation. The young commander, however, listened carefully to Black Kettle's envoy as the man vowed he would rather die than lie, and that if he had been lied to by his own people he would not want to live, anyway. When the Indian then eloquently offered himself as a hostage, Wynkoop was suddenly humbled, and after a quick consultation with his staff agreed to Black Kettle's proposal to meet.

With 135 men plus his Indian hostage Wynkoop marched north-eastward from Fort Lyon to the Smoky Hill River to find the Cheyenne. On the fourth day, Wynkoop's men suddenly came upon the Indians, some five hundred of them, all in war dress. There was a touchy moment when the Indians formed themselves for an attack, but Wynkoop detached his hostage and sent him to call out Black Kettle, and what might have been a massacre was not. Black Kettle, his fellow Cheyenne chiefs, and Chief Left Hand of the Arapaho all came to Wynkoop's camp and the parley was conducted in an atmosphere of reason and fairness. The upshot, to the credit of both sides, was that the Cheyenne and Arapaho gave up their white prisoners, all of them women, in exchange for a safe conduct to and from a meeting with Governor Evans to discuss peace terms.

The chieftains rode into Denver on September 26 in an open wagon accompanied by a flock of citizens who, with the prospect of peace in the air and a brand-new regiment of cavalry close at hand to protect them, had eagerly ridden out to meet the visitors. The conference was held at Camp Weld, and regrettably accomplished little. Black Kettle, the aged White Antelope, and a vigorous young brave named Bull Bear spoke for the Cheyenne, while four close relatives represented the Arapaho chiefs, Little Raven and Left Hand, who had stayed behind. Categorically, the Indians vowed never to have made a war alliance against the whites with the Sioux or any other plains tribe, and testified that they wanted to ally themselves with the white man for the good he could do their people. Their people, the chiefs said, were destitute and hungry and they desperately needed to remain near the buffalo and to receive the annuities of food and goods that would come with a new treaty. Instead of grasping the opportunity to enlist the aid of the friendly Cheyenne and Arapaho to fight on the side of the Army to bring the other war-making plains tribes to the peace table, a venture which the chiefs indicated by their testimony they would be willing to undertake, Evans harped on the atrocities committed by the Arapaho and Cheyenne and on the fact that he had no authority to conclude a peace treaty. He was followed on the floor by Chivington who hammered away at the importance of their laying down their arms and submitting to military authority and coldly reminded the Indians that when they were ready for peace all they had to do was to put themselves under the protection of the Army at Fort Lyon. The truth of the matter was that Evans at this juncture was loath to do much about solving the Indian question. If he had, his frantic entreaties to Washington but a short time before based on his appraisal of Denver's danger and about the need for new troops would have made him look grossly uninformed if not foolish. As for Chivington, in his mind there was only one road to peace, not through negotiation and compromise, but through a decisive meeting on the battlefield.

After the meeting, the chiefs, rather than being discouraged, seemed to feel that the air had been cleared and they cheerfully went back to their tribesmen. In a show of good faith, they promptly moved their encampment onto the Big Bend of Sand Creek, about fifty miles northeast of Fort Lyon. Wynkoop now became the mentor of the Indians, who had voluntarily put themselves under his jurisdiction. When he found to his chagrin that Washington had failed to supply the post with the annuities specified in the Treaty of Fort Wise, he passed out prisoner allowances to the hungry tribesmen and let them have free run of the post, allowing them to trade buffalo robes and other articles for whatever they could get for them from passing freighters and travelers. As a result, it was not long until sorehead Indian-haters started stories about Wynkoop's generosity and slackness toward the red men. When it was rumored snidely in Denver that Black Kettle, not Wynkoop, was the real commander of Fort Lyon, Wynkoop was summarily removed from his command and sent to a staff job at Fort

Riley, Kansas. Major Scott J. Anthony, Wynkoop's successor, publicly announced he would follow Wynkoop's pacification policy, while privately he wrote that he meant to do so only "until troops can be sent out to take the field against all tribes." He had not long to wait.

Back in Denver, Chivington divided his time between shaping up the Colorado Third and campaigning for office. The statehood issue was a heated one. In July, the Republican Congress had passed an enabling act allowing Colorado to petition for statehood; any new states would add to Republican weight in Congress. A large group of Coloradans, led by the town's three leading Methodists, Evans, Byers, and Chivington, immediately got behind the movement. The clique was confident that the populace would insist on statehood, and proceeded to railroad through nominations for the new "state" officers. Evans put himself on the ballot as a candidate for senator; Chivington was nominated as a representative. He was also running as a delegate to Congress from Colorado Territory in case statehood failed to carry. Everything was going just the way the triumvirate planned it until the disgruntled men who failed to be nominated started an active opposition on the complaint that statehood would mean grossly higher taxes. To the daily wage earner, the sluice-box miner, and the one-horse farmer, the argument hit home at the time when the cost of living owing to the interruption of trade by the Indian war was already sky-high. When election day came in September, Chivington's caisson to Congress lost a wheel. He and statehood were defeated by a ratio of 4 to 1.

But the messianic Chivington had another blow to strike for Colorado. On November 14, he issued marching orders to the three companies of the Colorado First stationed in Denver and to the entire Colorado Third. The enlistment of the one-hundred-day men would soon run out, and to Chivington it would have been a waste of the volunteers if they had not had a taste of battle action before being mustered out.

It was a heterogeneous force to say the least that set out from Denver—companies of eager but green one-hundred-day men mounted on everything from plow horses to gaited geldings and armed for the most part with old-fashioned Austrian muzzle loaders, led by such men as Lieutenant Harry Richmond, the darling of Jack Langrishe's dramatic troupe, and Major Hal Sayre, who ran a busy assay and abstract office in Central City; a great contrast to the grim, hardbitten Firsters such as Chivington's aide, Captain Silas Soule, a Glorieta veteran who knew what war was all about and who now wondered what his meglomaniac colonel was up to. Their destination Chivington kept closely to himself. The march began southward, and at Pueblo the column turned eastward to follow the Arkansas River. At each settlement they came to, including Bent's Fort, the colonel ordered it put under guard, to prevent anyone from leaving and revealing the troops' whereabouts.

When Chivington's force arrived at Fort Lyon on November 28 it was a welcome surprise for Major Anthony. Even more welcome was the colonel's announcement that the First and Third regiments, supplemented by Anthony's garrison, would that very night march northward to attack Black Kettle's encampment on Sand Creek and deliver once and for all the decisive blow against the predatory red man. Appalled at what Chivington was saying, Soule, who had been with Wynkoop at the meeting with Black Kettle out on the plains the previous September, broke in to protest vehemently. He reminded his superior that the Cheyenne had put themselves, as they were told to do, under the protection of the US Army by coming to Sand Creek. For his effrontery, Chivington threatened to court-martial his aide. Soule said no more. Before darkness fell the tired, footsore Colorado troops, now numbering nearly seven hundred men, gulped down their sowbelly and hardtack, stowed away in their packs three days' rations, and fell in.

It was shortly after dawn when Chivington's army reached the sandy bluffs bordering Sand Creek on the south. About a mile to the west, on the north side of the bend of the creek there shone in the first rays of the sun the 115 bleached hide lodges of Black Kettle's village and the eight lodges of Arapaho next to them. With bugles blaring, the three companies of Firsters that had come from Denver with Chivington galloped headlong toward the village, cutting off the Indians' access to their tethered horses. Meanwhile, Anthony's companies dismounted and started firing into the village at will. Chivington had ordered no prisoners be taken, and the soldiers aimed at any figure they saw, be it a man, woman, or child. In the village there was panic. The Cheyenne poured from their tepees, scattering in all directions, some falling under the assault of bullets, some who escaped the slashing volleys dashing first in one direction and then in another. Suddenly Black Kettle himself swung into view holding a long pole with an American flag on it. Beneath the American banner fluttered a smaller white flag. Standing in front of the first row of lodges, the Cheyenne chief braced one end of the pole on his thigh and waved the two flags back and forth in a long sweeping motion. He shouted to his people not to be frightened, that it was a mistake, that the soldiers would stop shooting when they saw the Indians were friendly. But the soldiers did not stop, and all around him his people fell, moving quickly with life one minute, grotesquely still the next, swatches of crimson patching their bodies. Old Chief White Antelope, moving with the early morning stiffess of rheumatism, walked toward the troops holding his hands above his head and calling for them not to shoot. A fusillade from half a dozen Starr carbines dropped him in the muddy creek bed. A group of braves, herding women and children before them, made a dash upstream to take cover in some shallow depressions in the sand. On their heels were the soldiers of the Colorado Third, and as the

Indians turned to fight they were cut down by hundreds of yelping, frenzied white troops who overran the sand holes, finishing off with bayonets those they missed with bullets.

The fighting was over by three o'clock in the afternoon with enough daylight left for the blooddrunk First and Third to perform their own atrocities. Captain Soule, who refused to let his men fire a shot during the massacre, watched it all in horror—the murder of the wounded, the scalping and mutilation of bodies. Men in blue took turns shooting down Indian toddlers who wandered aimlessly amid the carnage searching for parents. Soldiers ran from body to body, hacking off fingers to get at the silver rings worn by the red men. From White Antelope's body Harry Richmond and Hal Sayre triumphantly cut off the scalp and testicles. From the bodies of squaws, soldiers sliced off breasts and wore them rakishly atop their hats or stretched over saddlebows.

Estimates by eyewitnesses of the number of Indians in camp when Chivington struck varied from four hundred to two hundred, but out of however many there were, only six prisoners were taken— two women and four children. Some few of the Cheyenne and Arapaho managed to escape, all of them on foot. One of these was Black Kettle, whose continued faith in the possibility of living in peace with the white man was destined to lead him to his death at the hands of Custer's Seventh Cavalry four years later.

At sundown, Chivington ordered the torch put to the village and then the victorious Colorado volunteers retired to Fort Lyon where the exminister sent off a triumphant dispatch to General Curtis.

[A]t daylight this morning attacked Cheyenne villages...killed between 400 and 500...Indians and captured as many ponies and mules. Our loss...9 killed, 38 wounded. All did nobly.

On December 22 in Denver, crowds of people enthusiastically turned out to greet the returning heroes of what the *Rocky Mountain News* crowed was "Among the brilliant feats of arms in Indian warfare..." On December 28, Harry Richmond appeared to a standing-room-only audience in the Langrishe staging of what was billed as a "great Indian drama," wearing the "splendid war bonnet of Chief White Antelope."

Not all those in Colorado felt the way Denver's majority did toward Chivington's action. One of these was Ned Wynkoop at Fort Riley whose rage and anguish at hearing the news of Sand Creek reached all the way to Washington. His letters and those from "high officials" in the territory, including Chief Justice Benjamin Hall, brought on three separate investigations. The Congress of the United States, the Army Department, and the commandant of the Military District of Colorado, the latter in charge of Chivington's replacement after he was removed pending the outcome of the investigations, all took testimony on what was soon called the Sand Creek Massacre. In each inquiry, dozens of witnesses were called from every level: Governor Evans, Colonel Chivington, Major Anthony, Ned Wynkoop, and on down to the lowliest private who had been on the field that November day at Sand Creek. During the hearings in Denver, there was an open animosity toward those who accused Chivington and his troops of barbarity, but no one knew how deep the feeling ran until Silas Soule, after his outspoken testimony damning many of the Colorado First and Third regiments by name and action, was assassinated on Lawrence Street in broad daylight by one of the regiments' soldiers.

Evans, Chivington, and Anthony were all censured, with the heaviest blame laid on Chivington as the chief architect of the massacre. But inexplicably, no official action was taken against him. When the hearings were over, the bellicose colonel formally resigned his commission and faded away, his prestige and power gone. Years afterward, however, he was still defending his action, claiming that his troops had found an Indian blanket hung with freshly taken white womens' scalps in one of the Cheyenne lodges after the shooting which proved Sand Creek was a hostile encampment. Evans resigned the governor's chair the following year to concentrate on his mining and railroading interests and angle for a senatorship in the still viable project of statehood. Wynkoop, mustered out of the service, became the agent for the Cheyenne and Arapaho tribes, while the news of the massacre so grieved Sam Tappan that he adopted one of the Indian children orphaned at Sand Creek.

Now the schism between white man and red man was wider and deeper than ever, when it might have been closed or at least narrowed if men like Sam Tappan and Ned Wynkoop had been given the authority Chivington enjoyed. The Cheyenne, Arapaho, Sioux, Kiowa, and Comanche all waged a harder war, and Colorado was not spared. The settlements along its eastern border were under constant siege, climaxing with the bloody sacking of the stage and telegraph station at Julesburg. Once more martial law came to Denver, and its citizens realized that Chivington's "one decisive blow" had only worsened the outlook for the infant territory. For twenty more years, Colorado would have Indian troubles.

After Glorieta and the rescue of Colorado Territory from the acquisitive tentacles of the troops of the Confederacy, the Civil War grew remote in the minds of Colorado residents. They had all the violence they needed to think about close at home in the form of Indian raids. So when the war ended on April 9, 1865, there was little demonstration in Denver. But when the news of Lincoln's assassination made its way over the two-year-old telegraph line from Julesburg, Denver was stunned. The governor decreed memorial services be held in the theaters since there were no churches large enough to hold the throngs of mourners. To Jack Langrishe's elegant Denver Theatre on April 18 wound a solemn and thoughtful procession past the stores, saloons, and dance halls, for once silent, their doors closed and draped in black. The men filed into rows of garnet plush seats in the orchestra and balcony, while the women sat erect on gilt chairs in the dress circle that were reserved for them, to hear the dead president eulogized as the man who brought Colorado into the Union.

For some of the mourners the eulogy served to remind them once more of the question of statehood for the new Eldorado. The issue was brought before the people once again and this time it passed, with ex-Governor Gilpin nominated as chief of state (apparently the business community forgave him his monetary machinations); John Evans won the nomination for the senate along with Jerome Chaffee, the stamp mill Midas of Central City. But in Washington, President Johnson ignored the new "state" of Colorado and the issue went to Congress. In 1867, Congress got around to voting statehood for the Territory of Colorado, but the bill was promptly vetoed by Johnson on the basis that the new Eldorado still suffered from a want of population.

And so it did. The official census taken in 1861 showed the territory contained 26,000 people. Four years later the figure was nearly the same. To easterners it did not look as if Colorado had a promising future, and conditions in Denver bore out the prediction. After a brief surge upward immediately after the war, business slumped alarmingly. A board of trade was established in 1867, but there was little for it to do. Indian raids on the cross-country roads kept Denver nearly isolated from her eastern sources of supply. Prices in town rose accordingly as stocks of goods on hand were depleted and replacements failed to arrive. In addition, the backbone of Colorado's economy—gold-mining—was in serious trouble. With the easily worked surface deposits of blossom rock giving way to refractory sulphide ores, countless mines were abandoned, their owners answering the call of placer fields discovered on the Salmon River and in Montana. To make matters worse, the price of gold was fluctuating wildly as the battle raged between the government and the public over whether or not the inflated, war-issue greenbacks should be redeemed in gold. With less gold being produced and its price level subject to daily uncertainty, Colorado's money-metal income plummeted from the 1861 figure of $7 million to $2 million in 1868. In Denver prosperity dwindled at the same rate and she wallowed into a full-scale depression. Even Uncle Dick Wootton gave up on his adopted town. He, who had not meant to stay when he pulled into the little tent-cabin settlement in 1859, now left Denver convinced that the town would never recover from its decline. He sold out his Western Saloon, loaded up his wagon with a few possessions, and moved down to the foot of Raton Pass on the New Mexico-Colorado border. Here the cagey old trader, with ever an eye toward profit, built a twenty-seven-mile toll road over the pass and held forth in his tollhouse tavern like the lord of a fiefdom. Countless others left as well, and a traveler passing through Denver in 1869 remarked, "The old mining excitement has ceased. The old Overland stage has stopped and its business rushes past on a railroad one hundred miles away. Business is dull, the town is quiet…plenty of places 'To Let.'"

7

The Little Kingdom of Gilpin

ALTHOUGH DENVER BY THE END OF 1861 REFLECTED THE DWINDLING PRODUCTIVITY of Colorado's gold mines, the mournful evidence of the decline was everywhere to be seen in the once booming Clear Creek diggings. In Gilpin County, among the gulches stripped naked of trees and freckled with abandoned mine shafts, sluices, and cockeyed raw-pine cabins, the mineral future looked as dismal as the political future of its namesake. As the well-weathered and easily amalgamated surface ores were stripped away, miners dug shafts to follow the gold-bearing veins below the surface. Often, however, they dug down only a few feet when they discovered that the ores suddenly became poor in gold and, what was worse, the gold was locked in a sulphide matrix. These vexing "sulphurets" refused to give up the precious metal either by water-washing in sluices or by pulverizing in the rudimentary stone crushers.

Not only were these "rebellious" ores a problem on Clear Creek, but some veins under pick and shovel suddenly pinched out into widths too narrow to mine for profit. To compound the setbacks, too many claims had been allowed on one lode and before a man had realized any but a bare daily wage he might find that his allotted claim was wholly depleted. Droves of miners gave up in disgust and left the gulches, some to join the Army to take out their frustration on the enemy in the Civil War, others to follow rumors of new gold strikes in Idaho and Montana.

As the coveys of prospectors left, another faction took their place. The combination of the war with its boom in metals and war-product manufacturing and a money market flooded with greenbacks of fluctuating value set off a wave of speculation in gold mines and the solid value they represented. Gold stocks skyrocketed as investors picked up gulch claims right and left. Scores of mining men on Clear Creek jumped at the chance to sell in the face of receding profits. Others who had borrowed heavily to continue operating had to sell to get out from under their liabilities or were foreclosed. In the midst of the chaos of speculation it was hard to tell if the stock offered for sale was bona fide or not, and many a broker made a fortune unloading shares in nonexistent claims. So frantic was the exchange of property at the height of the boom in 1864 that the Central City recorder's office stayed open around the clock to keep up with the changes in title.

Toward the end of the Civil War, the population of the gulches on Clear Creek swelled to forty thousand. New owners of mines in Gregory and the adjacent gulches set about putting their properties in order, consolidating claims, and attending to the matter of improving gold yield from the refractory ores. To most of the newcomers, who were practical businessmen and not mining engineers, the obvious answer to the loss of gold in the extraction process was to crush the resistant sulphide ore as fine as possible. Consequently, thousands of dollars went into the purchase, transport, and erection of ponderous stamp mills up and down Clear Creek from Black Hawk at the bottom to Nevada City at the top. This eruption of stamp mills did not happen overnight. There were long and costly waits for equipment because midwestern and eastern foundries were turning out high priority machines of war, and the "Indian troubles" on the plains caused delays in shipments and an exorbitant rise in freight rates.

But eventually, at the end of the war, the mills were established and operating with the hopes of the whole county resting on the new processing in which the ores were subjected to periods of crushing and exposure to mercury far longer than before. This reduced the amount of ore a mill could handle per day, but it also gradually increased to 75 percent the amount of gold realized from the sulphide ores. Miners and millmen were encouraged, but still the return was not enough to bring back the boom times of the diggings' first two years, and the gulches continued to decline. After a few months of operation mines began closing and mills shut down. Some mills, like the monstrous one-hundred-stamp operation put together under the cavalier superintendency of Colonel Fitzjohn, never opened, and had to be sold to defray the cost of hauling the giant boilers and massive stamps up Clear Creek canyon to Black Hawk. In June, 1866, Bayard Taylor, while on a lecture tour of Colorado, noted in passing through Central City and her satellite camps that "The deserted mills, idle wheels, and empty shafts and drifts along this and adjoining ravines—the general decrease of population everywhere in the mountains —indicate a period of doubt and transition...."

A period of doubt and transition it surely was. But one man kept faith with the promise he knew lay below the surface and stuck out the interval of uncertainty to reap a rich reward. George Randolph was twenty-four when he came to Central City in 1864. He came to manage the mines on Pat Casey's old lode for the Ophir Mining Company of New York and the Boston Milling Company, an Ophir holding. Randolph typified the breed of bright young men in the second generation of mining whose duty it was to keep production high and operating costs low. His mild manner belied the fact that he was a hero of Gettysburg as captain of a battery of US artillery that held off the rebels under the most suppressing of odds. Descendant of the Virginia Randolphs, young George found his father's influential voice as a major stockholder in the Ophir Company to be precisely the send-off a newlywed son could ask for. When the managership was offered to him he was delighted. The West held no fears for him. He was used to roughing it in the Army. But for his bride, Harriet, being thrust into the raw and makeshift life of Gregory Gulch after the genteel surroundings of her paternal home in Providence was little less than insult. She got her first taste of the rigors of the West on the trip out when the first stage they took from Fort Kearney was attacked by a band of shrieking redskins who came so close to the coach that Harriet could see the color of the hair on the scalps they wore around their waists. The stage turned back to Fort Kearney and for a time it looked as if Harriet might have to give up her golden tresses as the Indian ponies crept closer in chase, but in the nick of time the stage thundered into the midst of a well manned and armed wagon train and the Indians, whooping and *kiyi*-ing, wheeled and dashed away. When Harriet saw the mud-chinked cabin just north of Central City in a cluster of similar dwellings calling itself Nevada City and found that was to be her honeymoon cottage there was little of the idyllic romanticism left in her outlook on the prospects in store for the Randolphs.

Harriet loved parties and musicals and fun but at first there was little society in the graceless outpost to which her benedict had brought her. There were few enough women in Nevada City or anywhere in the gulches to create a social sphere, and the struggle of homemaking where a kitchen range with all of its obtuse features was a luxury, where the streets ran with mud in the winter, between freezes, and were dustpits in summer, and where the only water available was that sold door-to-door by the lard can left little time for the amenities of social intercourse. Besides, the incentive for gaiety was nonexistent in the doldrum days of the mid-sixties.

George Randolph's efforts to get the Casey mines working again were stalemated for lack of funds received from New York and even his normally reliable optimism flagged when there appeared on the scene one E. Humphrey, a young man who bore papers claiming he was to take over the management of the Ophir properties. Abashed, George relinquished his files and cast about the gulch for a job. The Randolphs of Gregory Gulch were nearly penniless. When Harriet pleaded with George to return to the East she was crestfallen to find that they did not even have the price of the fare as far as Cheyenne. Then there came a benefactor and the fortunes of George and Harriet took a turn for the better. Jerome Chaffee, who had gotten rich by buying one of the first stamp mills in the gulch, had taken himself and his money to Denver to enter the political arena and he hired George Randolph to run his Clear Creek mines.

About this same time, the mine-owners, local and absentee, took a hard look at what had been christened the "richest square mile on earth" and redoubled their efforts to return Gilpin County to its former production and prosperity. Out of the California and Nevada gold rushes had come a bagful of ore-reduction processes involving smelting. In this relatively new and experimental field of gold ore production was one process that promised great reward. Based on the simple principle that sulphur will burn readily, the Keith process called for roasting of the finely crushed ore until all of the sulphur was burned off, leaving melted gold to amalgamate with mercury. Several mill-owners in the gulch coughed up $5,000 for the right to use the process and quickly introduced it into their mills. Although the principle on which the Keith process was based was sound, in practice the process was grossly inefficient and was written off as a "miserable and expensive failure" by mill men and miners alike.

Another scheme, this one for an ore concentrator, was put into practice in the gulch by James E. Lyon and his partner and it also failed to give the results hoped for by mine-owners up and down the crooked cleft that was Clear Creek.

After these and other innovations had been introduced, all with the same lackluster result, it became painfully clear that the rebellious ores of Clear Creek were as resistant as ever to reduction.

But unknown to the discouraged residents of the gulches of Clear Creek, the solution was underway. As early as 1864, the very same year that the Randolphs came to seek their fortunes in Gregory Gulch, a group of eastern capitalists who had invested in Clear Creek mines at the height of the speculation boom turned from an empirical approach, one of trial and error, to research for the answer to the problem of refractory ores. To attack the problem they chose round-faced, mustachioed Nathaniel P. Hill, a professor of chemistry and metallurgy at Brown University. Despite the hazards of plains travel in those troublesome days, Hill fixed plain gold studs to his cuffs, hung his watch chain on his vest, donned his chip hat, and came out to the gulch to collect specimens of the problem ore to take back to his laboratory for analysis. He made the trip safely and two more besides until he had established without a doubt the composition of the ore. He then took a steamer to Southampton and a train to Swansea, Wales, to pick the brains of the experts at that world-renowned mining center. He carried with him the lading list for seventy-two tons of Gilpin County ores shipped ahead so he could watch what happened when the ores were melted in the smelter used at Swansea. Later, he crossed the Channel and went down the Rhine to study at Germany's crack metallurgy school at Freiburg. By 1867, Hill's careful and costly studies bore fruit. With $275,000 backing by industrialists in Providence and Boston, Hill incorporated the Boston and Colorado Smelter Company and built a plant at Black Hawk using a smelting process patterned after that used at Swansea.

The gold in the gulches was found in the presence of copper and a small amount of silver. In Hill's process, the ore was crushed, roasted, and fused in a furnace, leaving the gold and silver concentrated in a copper matte. The matte was then shipped across the plains and over the Atlantic Ocean to Wales where the gold was extracted from its copper bed. Out of ten tons of ore, about one ton of matte was produced.

Hill was no ivory-tower type; he ran the smelter on a strictly profit-making basis, charging $60 a ton to smelt ore that assayed at $120 a ton. For gulch mine-owners the price of Hill's smelting services was high enough so that only the highest-grade ores could profitably be smelted. A compensating feature, however, was that the price of copper was sufficiently high that it alone paid for the cost of transporting the matte to Wales, and the gold extracted in the copper refining process amounted to pure profit.

Hill's smelter was an unqualified success. To help him keep abreast of the latest European ore-reduction techniques, Hill brought Richard Pearce from Swansea and Herman Baeger from Freiburg. With those two metallurgists on hand, Hill was able to further improve his process. The productive results and professional tone of Hill's operation infused the gulch with new life. All around signs of emerging prosperity greeted visitors. Peter McFarlane put up a foundry in Central City and the Hendrie brothers opened one in Black Hawk to build crushing, hoisting, milling, and smelting machinery to supply the needs of gulch mine-owners. Gold stock prices and dividends moved steadily upward as "The Little Kingdom of Gilpin" climbed out of its depression.

With the yield of gulch ores improved by smelting, mine-owners now turned their attention to increasing the production of ore. Once more the experience of Welsh mining was tapped, and from the mines of Cornwall to Clear Creek, some by way of the Lake Superior iron mines, came troop after troop of hard-rock Cornish miners. The Cornish miner was the world-wide avowed champion of miners. By the time he was twenty the average Cornishman had spent twelve years in the dank, murky tunnels of Cornwall's cavernous tin mines. Industrious and highly competent, the Cornish were a happy-natured breed who laughed off the dangers of their work, who nonchalantly sang the rollicking songs of their country in their characteristically rich natural voices as they daily loaded themselves down with enough paraphernalia preparatory to going into the mines to pass for walking bombs. They tucked sticks of dynamite into their boots, stuffed detonating caps into tunic pockets, hung coils of fuse around their necks, picked up their lunch boxes in one hand, and grabbed a lighted candle with the other, all to the lyrics of "I could not find my Baby-O!" or the lighthearted "Three Jolly Hunters."

It was the Cornish who introduced leasing into Gregory Gulch mining. Under this system, one or two miners leased for an unspecified time a section of a mine where the ore was too low in gold content to be mined at a profit by a large operation. For their work, the "tributers" working the "tribute pitch" got a percentage of the take, ranging from 10 to 50 percent. This innovation, along with the solution to handling the sulphide ores, sent the gold output of Gilpin County soaring, and suddenly the Cornish were in great demand.

Nearly every one of the Cornish in the gulches had a "cousin John" ready and willing to come to Colorado, and before long the cheerful newcomers were universally called Cousin Jacks, and their wives, Cousin Jennies. As the population of Cousin Jacks increased, there crept into the daily life of Central City residents the ear-tickling Celtic dialect of the Cornish with a steady outpouring of tales involving "tommyknockers," the elfin creatures every true Cousin Jack believed inhabited the underground workings. And the wry wit, beguiling candor, and irrepressible good spirits of the Cornish soon gave the gulch an unending fund of anecdotes to be told over convivial glasses of beer at the Granite House bar. When portly Ben Eaton decided to run for governor, his brother-in-law, Doctor Paul, invited him up to Central to meet his future constituents. Paul proudly walked Eaton down Main Street, presenting him to passersby. When he came to a Cousin Jack, Paul stopped him. "Here, my good man, this is my brother-in-law, Ben Eaton," said Paul. The Cornishman eyed the beaming, rotund candidate and broke into a grin.

"Been eatin', 'as 'e? Looks to me 'as 'e's been drinkin'."

Cornish customs became gulch customs. On May Day there were flowered baskets, choral singing, and group dancing, the rapping of clogs on boardwalks mingling with the continuous thudding of ore stamps. On Christmas Eve in the clear mountain night there was always a procession of carolers among the shafthouses carrying hoops of evergreen lit by a candle in the center, and afterward a bite of pastie and saffron cake and a cup of steaming wassail in the simply furnished cabins scrubbed to gleaming cleanliness by the fastidious Cousin Jennies.

Compared with its rollicking market place down on Cherry Creek, the now booming Central City was the epitome of respectability. There were scattered up and down its streets all the typical trappings of a boisterous, wide open mining camp—saloons and variety halls, madams and their fancy houses, gamblers and their casinos—but the raw violence that marked other camps was nearly nonexistent in Central. Saturday nights saw drunks by the dozen hauled off to jail but in general there were so few imprisoned criminals that a popular story around town told of a lonesome jailbreaker who scratched on the wall of his cell before he escaped, "Can't stand nobody around, so I'm leaving."

One reason for the relative sobriety may have been the presence in Central of a substantial number of distinguished upholders of the law, led by Judge James B. Belford and the young advocate, Henry M. Teller. Belford came to Central City in 1870 as a Republican appointed to the territorial supreme court. A stern-visaged, outspoken man, he had little respect for the arcane convention of courtrooms. He characteristically heard his cases slouched down in his chair with his two large feet planted squarely atop the bench and delivered his judgments in stentorian tones that could be clearly heard by passersby on the boardwalk below the second-story courtroom. Soon after his arrival, the judge became a frequent figure on speakers' platforms. With his carrot-colored full beard bobbing up and down as he worked himself into an oratorical lather in support of such unpopular causes as woman's suffrage, Belford was a sight to behold. After watching a half dozen of his performances, one captivated newsman dubbed the judge the "Red Rooster of the Rockies," and the graphic, if demeaning, sobriquet followed the jurist all the way to Washington when he became Colorado's first congressman.

Henry Moore Teller, another member of the law fraternity who put down roots in Central City was thirty-one, temperate, and unflappable when he arrived in 1861. A New Yorker by birth, Teller was a graduate of Alfred University in his home state and had been a practicing attorney for three years before he came to Gregory Gulch, first in New York State and then in Illinois, where he got a taste of politics in the spirited election of 1860 and liked it. But Illinois was full of lawyers and the chance of another one getting to the White House in the near future seemed to young Teller to be reasonably remote. So he looked around for a place where lawyers were few and the opportunities for recognition were many. He chose well. After a brief sojourn in Denver, he took his legal talents straight up the rocky trail along Clear Creek to the well-spring of Colorado's future.

Scarcely had Teller opened the door of his office on Central's Eureka Street than he was besieged with clients. Before many weeks passed the tall, frockcoated attorney, his smooth-shaven face ringed with a chin-clinging beard, and his head crowned by a shock of stiff black hair that refused to lie flat, was a daily fixture in the courts of Gilpin County as he found that litigation over claim ownership and boundaries was as much a part of mining as was the extraction and processing of the ore. In time, Teller found himself part owner of a hodge-podge of claims and stampmills from shares taken in payment for some of his legal services. His business enterprises, like his law practice, returned him a comfortable income and Teller branched out to a tentative course of investment. First he erected a stamp mill. It thrived. Encouraged, he decided to diversify. When Teller announced that his second venture would be a theater there were a number of raised eyebrows among Central City's highbrows, despite the fact that he had taken on Judge Silas B. Hahn, one of the town's most conservative men, as a partner in the effort. When Teller and Hahn opened the Belvidere, however, the elite of Central breathed a collective sigh of relief—the young attorney had not violated the code. The opening production, a "Melo Dramatic Pantomime of Little Red Riding Hood," would hardly feature bare-legged, low-bodiced dancing girls. Opening night was a sellout and although it lacked the ta-rah-rah-boom-de-ahy spirit of Jack Langrishe's Montana, the Belvidere, like the rest of Teller's investments, flourished to become the evening entertainment center for Central's bon ton.

Among the new aristocracy of the prim gold camp were Harriet and George Randolph. After George took over the managership of Jerome Chaffee's Adrian and Clinton mines, he did well and was just about over his embarrassment at what he thought was his dismissal by the New York owners of the Ophir mines when it was discovered that Humphrey, the man who came as his replacement, was supposed to have assumed the managership of the Ophir Company's claims in Nevada City, *California,* not Nevada City, *Colorado.* The mistake was soon rectified and once more George assumed control of Pat Casey's old claims on the Burroughs lode, turning them into high-paying mines with the help of Nathaniel Hill's smelter down in Black Hawk. During the first five months of his resumed stewardship 2,437 tons of ore were taken from the Ophir with a gold yield totaling $45,567. By 1870, through the addition of skillful leasing to his already competent management, George brought the annual production of the Ophir group up to nearly $100,000.

The Randolphs' share of the Ophir profit was generous, and Harriet was ecstatic. They moved from Nevada City down to Central to a picturesque frame house on fashionable Casey Street, and George got himself elected mayor. His office called for round after round of entertaining and that suited the party-loving Harriet just fine. One of the Randolphs' first guests was Scottish poet James Thomson. When Harriet heard he was coming, her excitement manifested itself in scores of menus and dozens of guest lists, each one torn up in favor of another until at last the right combination struck her. When the appointed evening came, Thomson spellbound his dinner partners with his mellifluous voice and cultured mannerisms. As she listened to the poet's repartee, Harriet hoped to find the nerve to ask him to give an evening of poetic readings for Central's newly formed literary club. Thomson, however, had not come as a lecturing author but as the agent for the Champion Gold and Silver Mines, Ltd., an English company with claims in Gregory Gulch. The poet's job was to send back to London weekly reports on the prospects and the progress of the company's mines. Thomson arrived in May, 1872, and stayed for seven months, dutifully sending home reports and writing to his friends in poetic prose his impressions of the Little Kingdom of

Gilpin. To fellow romanticist William Rossetti he wrote that "The hills surrounding us have been flayed of their glass and scalped of their timber; and they are scarred and gashed and ulcerated all over from past mining operations; so ferociously does little man scratch at the breasts of his great calm mother when he thinks that jewels are there hidden." But it was to his young friend Alice Bradlaugh in London that he confided some observations on Central City's womenfolk that might well have turned Harriet Randolph's enchantment with her guest to gall. Some of Central's women, Thomson wrote, "...would be very graceful if the absurd boots did not make them hobble broken-backed with their Dolly Varden bunches behind. They incline to leanness, which one fears must grow somewhat scraggy with age..."

Central's upper crust, as well as the rank and file Cousin Jennies and Cousin Jacks, were never at a loss for entertainment. In winter there was a gala round of skating parties down on Black Hawk's ice rink, and every Saturday the town turned out to watch the bobsled races that saw teams from all the major mines in the district career their lightning-fast sleds down Gregory Gulch all the way from Nevada City to Black Hawk, stampeding mule teams and sending unwary pedestrians diving into the nearest snowbank for safety. In summer, there were moonlight picnics up on Gunnell Hill and in the groves of aspen on the high road to Idaho Springs. A stream in Lake Gulch was dammed for those who found canoeing to their liking, and for those whose evenings were not complete without a go at a teeth-rattling clog dance or a turn or two of a waltz, there was Lemkuhl's beer garden. Brewer Lemkuhl's newspaper ad reading "Hotels, families, and bars supplied with beer on short notice" kept him happily solvent enough to provide a band of musicians in his beer garden until midnight in the summer months.

But for all of the residents of the gulch and for every other mining camp, the big day of the year was the Fourth of July. The preparations started months ahead. Up and down the gulch, volunteer fire companies polished the ornamental brass on their carts till it shone as brightly as the bars of bullion from the Burroughs lode. Firemen practiced their drills until every move could be counted in seconds. The townswomen sewed miles of gold braid on uniforms of the members of the various miners' brass bands, and on June evenings the bleating notes of flutes and cornets issued forth from open windows of the two-room schoolhouse where the bands rehearsed. As the day came closer, town officials put together grandiloquent speeches likening Colorado's desire for statehood to the spirit of '76. Smoke poured endlessly from kitchen chimneys as fat hams baked cozily in wood-stoked ranges to a savory doneness, while the aroma of apple and berry pies bubbling in ovens tickled the salivary glands of passersby. The sounds of hammering echoed up and down the gulch as carpenters erected a reviewing stand at the junction of Eureka and

Main streets. Teamsters rushed extra barrels of beer and whiskey to every saloon, gamblers broke in new decks of cards, and pickpockets and sneak-thieves made ready to ply their trade among the crowds that lined the street for the Independence Day parade.

The festivities got underway early in the morning of the Fourth. By 9:30 AM, the townspeople were turned out in their Sunday best, a dozen deep on the boardwalks, small boys perched atop fathers' shoulders, brightly hued parasols and sun bonnets adding even more color to the masses of red, white, and blue bunting draped over every protuberance in sight. On the soft mountain breeze off Casto Hill down to the excited throngs on the streets below floated the first intermittent notes of a march, and then standing on tiptoe, eyes straining to see up the gulch, the crowd broke into wild cheers as the high-stepping color guard leading the parade strode into view over the crest of the hill. The thrill of drums and horns and flapping flags and smartly marching mines' guardsmen brimmed eyes with tears of pride. After the parade, the watchers crowded to the reviewing stand to hear the undulating words of the day's speakers. Caught up in the homilies and stirring appeals, the listeners hip-hip-hoorayed until they were hoarse, and when the oratory was over they went away cleansed in spirit to indulge in the welcome restorative of a family picnic with all the trimmings.

But the day had just begun. No one would linger so long over a pot of clabbered cheese or a succulent chicken leg that he missed the main events of the afternoon—the firemen's foot race and the rock-drilling contest. Eight spans of strapping volunteer firemen, their muscles rippling under their union suits and their chests heaving against the harness, pulled each hose cart in a race against time that sent men, women, and children into shouting, cheering frenzy as the racers thundered down the hardpacked dirt street. After the races, the crowd moved on to the roped-off pavilion where brawny drillers warmed up for their test of skill and speed. Each camp in the gulch had a champion and hundreds of gold dollars backed every one of them. For two weeks before the Fourth of July, the champions drilled night and day, developing their lung power for the fifteen-minute contest. When the moment came, a sinewy miner spat on his hands, picked up an eight-pound hammer, and swung it over his head a few times, glancing now and then at the audience, perhaps winking at a blond-haired beauty whose scarlet-tinged cheeks could not be hidden even by a parasol. Meanwhile, the drill-turner selected hisfavorite drill. The drills by regulation were seven-eighths of an inch in diameter. No such uniformity was prescribed for their sharpening, and drill-turners rarely confided to others their particular techniques for putting an edge on the steel that would bite through the hard slab of granite. When the drill team decided that the anticipation of the crowd was at its peak, they got into position. The drill-turner kneeled in front of the slab of stone, his hands on the drill. The hammerman, his hammer raised, stood over the drill. At the nod of the timekeeper's head, the hammer swung down and the count started. Down, up, down, up, swung the hammer with ever-increasing frequency until it reached sixty-eight times a minute, then seventy times a minute, then seventy-two, and when the count reached seventy-five blows a minute the audience broke into cheers. Sweat ran down the hammerman's forehead and into his eyes. Granite dust eddied over the drill-turner's vibrating hands. When at last the hammerman's lungs seemed about to burst and his arm muscles screamed for relief, the timekeeper called "Time!" The drill-turner's hands, reflexively shaking, lifted the drill out of the hole that reached nearly thirty-six inches into the granite slab. The judge pronounced the hole evenly round and deep enough to qualify for the $500 prize money. While the audience applauded the hammerman leaned on his hammer, gasping for breath, deaf to all but the pounding of his heart and the ringing in his ears.

Professor Hill's research brought a new day to Gregory Gulch. Mines on the major lodes, the Gregory, the Bobtail, the Buell, and the Burroughs went into day and night production. So did the stamp mills and the smelters. Up and down the gulch the crooked streets of the camps were filled with the hubbub of production, and a continuous pall of yellow smoke from the smelters drifted over the rooftops of the gingerbread homes and bright new brick buildings. By 1871, the Little Kingdom of Gilpin was producing nearly $400,000 worth of pay dirt a year. With the upswing in production there came to Central City and its neighboring gold camps new personalities, new enterprises. A. G. Rhoads began the manufacture of soda crackers in his little two-oven bakery. In time, his Black Hawk Cracker Works supplied most of the West's demand for that homely commodity. Dave Henderson, whose talents leaned in the direction of inventing, thought up an automatic way to make screens of any required mesh size for use in stamp mills. But the most popular product to come out of his busy machine shop was a dinner pail made in the shape of a kidney to accommodate the meat pastie that was the mealtime mainstay of the Cornish miners. Rival banks, Thatcher Brothers' and Kountze Brothers', vied for Central's growing banking business, each submitting to the *Register* a daily tab of deposits for publication in the hope of attracting new customers. More and more consumer goods reached the gulch. George Randolph found that he did not have to travel all the way down to Denver to buy his suits. Abe Rachofsky's New York Store on Main Street stocked a full line, and George liked Rachofsky's personable young clerk, Jacob Sandelowsky.

Jacob Sandelowsky was only nineteen but he knew the men's clothing business as well as George, considerably his senior, knew mining. Jake had arrived in Gregory Gulch in 1870, armed with a desire to make good and a letter of introduction to Rachofsky. Jake's father had emigrated to America from Poland with Abe Rachofsky, but they parted company soon after landing. Abe came west, while the elder Sandelowsky set up a clothing store in Utica, New York. Young Jake began his training in his father's store and when he had learned all he could, he too headed west to make his fortune. Within a year the handsome, dark-eyed youth had earned enough commission to be well on his way to amassing enough capital to open his own shop. But he bided his time, getting all the experience he could in his benefactor's store.

When James E. Lyon hit Gregory Gulch, he took on another bright-eyed young man to be his partner. Their first enterprise was buying and selling gold dust. This led to a stint in banking until Lyon got the idea for an ore concentrator. When that scheme failed, the two men turned belatedly to mining itself, developing with considerable success part of John Gregory's original lode. Throughout the Civil War, Lyon's partner came up with a steady stream of money-making ideas, but neither man could find backers; money was being invested in items of intrinsic value such as gold mines and not in such visionary schemes as a luxuriously appointed hotel room mounted on railroad wheels. But after the war, Lyon's partner found the money he needed to pursue his dream, and by the mid 1870s, George Mortimer Pullman's name was synonymous with the most elegant form of travel accommodations yet seen in this country. His richly furnished palace cars with their plush-covered seats, thick carpeting, gleaming silver fixtures, disappearing berths with soft down mattresses, and damask-shaded windows were the epitome of travel luxury.

No such luxury, however, awaited the traveler to and from Central City and Denver in the late sixties. The big Concord stages could hold ten or more people and all of their luggage, but without the ballast of a full load, it was a perilous ride with passengers flying from side to side and onto the floor or into each other's laps as the huge compartment bounced and swung on its leather suspension straps. The steep, narrow, and twisting rock-studded road crossed the bed of Clear Creek fifty-eight times in eight miles. Atop the box the rough and ready driver grasped the reins of the three plunging teams in one hand and a long black snake of a whip in the other. In port, a stagedriver was notoriously taciturn; on the road he was as voluble as a jackdaw, bellowing at the world at large and at the frenzied animals as he cracked the whip over their haunches to keep up the steady ten-mile-an-hour pace. Lamented Englishwoman Isabella Bird after a harrowing ride up Clear Creek, "The driver never spoke without an oath, and though two ladies were passengers, cursed his splendid animals the whole time."

The complaints of a visitor counted for naught, but the dissatisfaction of a local resident, especially one with money and influence was a different story. Henry Teller, who had ridden the convulsively errant down-coach to Denver once too often—he had been badly cut

and bruised on one occasion when it ran off a bridge—was the first
to agree that a railroad to the gulch was long overdue. In addition,
Teller was by now president of the impertinent Colorado Central
Railroad and engaged cheek by jowl with William A. H. Loveland in
the track-laying shenanigans that were the vogue of the day. Loveland
and Teller had been out-maneuvered by their rivals down in Denver in an attempt to connect Golden with the Union Pacific mainline
at Cheyenne. But they were determined not to be done out of the
Gregory Gulch ore business. Financed by a $250,000 bond issue from
the people of the Little Kingdom of Gilpin, they laid down a narrow gauge track from Golden to Black Hawk in seventeen months, as
fast as the formidable terrain of Clear Creek canyon would allow. On
December 15, 1872, to an enthusiastic welcome by gulch residents,
the first train rolled into Black Hawk, hissing steam and spewing cinders as it came to a groaning stop in its new terminal. Compared with
depots of other bustling mining camps, Black Hawk's was, through a
fluke, incongruously elegant. When the Colorado Central's engineers
reported that the abandoned stamp mill known around town as "Fitzjohn's Folly" stood smack in the way of the oncoming track-laying
crews, the company merely bought the mill, ordered a hole cut in
each end of the building, and laid the tracks through the middle of it.

But a train to Black Hawk was not a train to Central City, and
the citizens of the latter clamored constantly for the last leg of the
promised route to be finished. There was only a mile to go, but there
was the tricky matter of a forty-foot difference in elevation that would
require three miles of track in a grand switchback and $65,000 to
overcome. Unfortunately, the time to begin building the Central City
extension and the nationwide depression of 1873 coincided and the
Colorado Central could not scrape together the required funds. It
could not in fact round up the $65,000 for another four years, and
it was not until 1878 that Central City got direct service to Denver.

In the meantime, the line to Black Hawk flourished in spite of Isabella Bird's complaint that never had she seen "such churlishness and
incivility as in the officials of that railroad…or met with such preposterous charges." The scenery on the two-hour trip through Clear
Creek canyon was spectacular. The little cars moved up from Golden
along a grade that rose two hundred feet to the mile, causing Sam
Cushman, a frequent passenger, to remark wryly that "one is never in
doubt whether he is coming…or going…"

The railroad added to the renewed prosperity of the gulch. Freight
costs dropped nearly 60 percent, low-grade ore was now profitable to
mine, and retailers no longer had to inflate the price of goods brought
up to the Little Kingdom. Prices and wages fell accordingly. Skilled
miners who had been paid $12 a day in 1865 now drew $4.50 a day.
But Harriet Randolph and the Cousin Jennies could buy a hundred
pounds of flour for $2.50 instead of $20.

With more and more visitors storming the mountain gold camp, the Central City *Register* ran a series of editorials agitating for a decent hostelry to be built to house the town's new crop of guests. To the rescue came good old Henry Teller, who offered to build a $60,000 edifice if the citizens of Central would put up $10,000 toward the purchase of the lot. Three days later Thatcher Brothers' Bank informed Teller that the subscription had been met to the penny. Teller bought a lot on Eureka Street, Central's main thoroughfare, and ordered work begun immediately. The foundation was laid in July, and in the following July, the Teller House had its grand opening ball. To provide the necessary notes of the light fantastic, Barnum's Cornet and String Band was hired to come all the way to Central City from Fish's Long Island Sound Steamship Line for the one-night stand. The interior of the hotel was sumptuously fitted out—its ninety sleeping rooms featured down mattresses, deep-pile rugs, and velour draperies; its parlors were paneled in walnut, and "bathrooms and water closets" were placed on the second and third floors, "the first for the gentlemen, the latter for the ladies." Whether the positioning of these accommodations was based on a scientific study of the frequency of their use or the relative capability of males and females in stair climbing is not known. Down in the basement of the hotel Jones and Urich operated three barber chairs, and advertised hot and cold showers at 50 cents each, day or night.

Despite the appeal of its interior, the architecture of Teller House failed to create any aesthetic raptures. Banker Frank Young, writing a history of his town a few years after the hotel was built, decided that it "might easily be taken for a New England factory and cut off a story or two and it might pass fairly well for a cavalry barracks." Nevertheless, its completion was in time for the arrival of the gulch's most distinguished visitor.

Three months after his election to a second term, President Ulysses S. Grant, on a bright April day, with the snow of winter still thawing along the roadside, rode the Colorado Central up Clear Creek canyon to Black Hawk. Functionaries met the *Presidential Special* and whisked the party to Central, to the door of the Teller House where manager Bill Bush extended Grant a cordial welcome. Crowds of curious bystanders pressed closer to catch the reaction of the president when he alighted from his gleaming carriage and found he would walk across a sidewalk laid with shining silver ingots representing nearly $13,000, smelted from the Caribou mine over the mountains to the east. The presence of silver on this auspicious day was prophetic, although the gold camps of Gregory Gulch had yet to realize it. Soon silver, not gold, would be the siren of a new rush to the Rockies.

In the meantime, gold continued to pour out of gulch mines, creating an aura of permanence in the minds of the residents. Central City, next to Denver, was the largest town in the territory. Its seven thousand inhabitants were served by schools, two banks, six churches, and half a dozen fraternal lodges. No one heeded the warning of Amos Cummings, managing editor of the New York *Sun,* when, after a fact-finding trip to Central during which he noted the flimsy wooden structures and the meager firefighting equipment, he reported that "One of these days the city will be wiped out by a conflagration." On May 21, 1874, the town wished it had listened to Cumming's warning.

On the northwest side of the closely structured town Avas a collection of small frame shanties occupied by some Chinese who had come to the gulch to work in the placer mines and to lay track for the Colorado Central and who stayed on to eke out a living, some as hand launderers. About 10 AM on that May morning, a small wisp of smoke trailed skyward from one of the Chinese laundries. At first no one paid any attention to it, thinking it was smoke from a chimney. The first alarm came when a spire of flame shot into the light morning sky. In a matter of moments, the fire spread to the rest of the wooden shanties, and then tongues of flame lapped at the structures facing Main Street. Sizzling embers and acrid smoke filled the air as shopowners, gamblers, bankers, harlots, and everyone else in the path of the fire fled the buildings lying on both sides of Main Street. At the first alarm, bucket brigades and hoses came into play but suddenly the cisterns ran dry and the firefighters could only flee in front of the advancing flames. As the fire inched closer to Eureka Street, terrified horses whinnied and bolted from their traces, upsetting wagonloads of goods that merchants were frantically attempting to save from the conflagration. Chinese women and men whose homes were among the first to be consumed clutched a few belongings in one arm and a squalling child or two in the other to dart up Main Street, their eyes red-rimmed and bleary from the stinging smoke. Down to Golden went a frantic telegraphic plea for help. A special train hastily coupled together carried carloads of volunteers hellbent for Central, but on a sharp curve one of the firemen fell off the train, causing a hopeless delay. In the meantime, brawny miners, called out of the bowels of their mines to help fight the fire, swung axes to cut down frame buildings and dynamited sturdier structures in the path of the voracious flames. At the Teller House, the firefighters made a new stand, beating out the fingers of gold and crimson fire that flickered against the red brick building and manning hoses to play water taken from the hotel well on the walls of the hotel and adjacent buildings and on the street in front of it. Finally, some six hours after it began the fire was controlled, and slowly it died down to leave the heart of Central in smoking ruin.

A thousand or more luckless residents camped on Casto Hill east of town that night and for many nights afterward while Central rushed to rebuild itself. The Colorado Central put on extra trains to carry building supplies up Clear Creek. Brick kilns, stone quarries, and saw mills operated around the clock, and within nine days the first new brick and stone building was ready for occupancy. Broad stone sidewalks were laid on Eureka and Main streets replacing the wooden ones, and the streets were freshly graded. As a result of the fire, some old partnerships were dissolved and some new ones were formed among the merchants in town.

One of those who took advantage of the opportunity to make a fresh start was Jake Sandelowsky. With Abe Rachofsky's New York Store burned out, Jake was suddenly on his own, and he quickly made the most of it. With Sam Pelton, another young bachelor in town, he found space for rent on Black Hawk's Gregory Street and opened Sandelowsky, Pelton & Company, men's clothiers. The store did well in bouncy Black Hawk, and the twenty-five-year-old, fun-loving Jake soon found he had the time and the money to invest in his favorite diversions, gambling and dancing. He lacked but one thing, someone to help him pursue his pastimes with him. That impasse however was soon to be overcome.

Into reborn and busy Central City on a July day in 1877 came Mr. and Mrs. William Harvey Doe, Jr., a pair of newlyweds whose personal histories, now a mixture of fact and legend, would be forever told whenever Colorado's past was recounted.

The name of Doe was not new to Central. In 1860, William H. Doe, Sr., and his brother, Methodist preacher F. B. Doe, joined the rush to Gregory Gulch from their home in Wisconsin. William was going to prospect for gold, and the parson for sinners. While William combed the sides of the gulch for hopeful signs of blossom rock, F. B. called together a congregation on his first Sunday in camp and preached a stirring message, exhorting the crowd to follow in the paths of truth and righteousness, and inviting his hearers to attend a service to be held in the evening in his tent. A large number of prospectors turned out that night and made for F. B.'s tent. When they got there, they found that the good reverend would as leave play Black Jack as preach and the collected company spent a convivial evening over the flashing cards. In the process, the Doe brothers picked up a high-grade claim on the Gunnell lode. With this and a half dozen more rich deposits of blossom rock acquired with the profits of the Gunnell property, the brothers built a stamp mill and some boarding houses, and retired rich men in 1865, after incorporating their holdings as the Sierra Madre Investment Company. William Harvey, Sr., returned to Oshkosh to enter the timber business, supervise the Sierra Madre Company, and dabble in politics.

When the senior Doe's only son, Harvey, Jr., gave evidence of being on the brink of proposing to Elizabeth "Baby" McCourt, the pert and very pretty daughter of Oshkosh's prominent clothing merchant, the boy's father was more than pleased. His shy son could use a strong-willed woman behind him. For the boy's mother, the prospect was calamitous. Prominent or not, the McCourts were Irish Catholics, an anathema to the devoutly Protestant society leader. Nevertheless, the couple was married on June 27, 1877, in St. Peter's church in the poshest, most talked-of wedding ceremony of the Oshkosh season. To start the young people on their way to the good life, the senior Doe settled on Harvey a part interest in the Fourth of July, a promising claim he owned in Central City with Benoni Waterman, a man whose money-making twenty-four stamp mills up in Nevada City left him time and money to do a little speculating.

As the train bore the honeymooners west, their excitement grew with every turn of the iron wheels. In the breast pocket of his alpaca suit coat the slightly pudgy Harvey carried a sight draft for several hundred dollars his father had given him to use as working capital. In his mind, there danced the vision of immediate riches and little work. All he had to do was to get the verbal agreement his father had with Waterman drawn up and recorded, gather up a crew of miners, and start pulling up the profits. Under the agreement Harvey would lease Waterman's half of the mine for two years, agreeing to sink the shaft to a depth of two hundred feet, and when the mine began producing he had the option of buying Waterman's interest for $10,000 the first year or for $15,000 the second year. If, by some chance, the Fourth of July proved to be a bust, Waterman could sell his interest at any time he chose. To top off the benefice, Father Doe agreed to turn over the profits from his one-half interest in the mine to Harvey if his son made a success out of his first mining endeavor. For the golden-haired, sloe-eyed beauty at his side on the seat of the coach, the prospect of her future with Harvey was lit by the sparkle of jewels, shining silks, gleaming silver and china, and the splendid homes and carriages the gold of the Fourth of July would provide.

After a few days in Denver, where Harvey equipped himself with mining tools and textbooks, the Does took the Colorado Central up to Black Hawk and the stage to Central City to be met by Harvey's father who had preceded them west after the wedding. To Baby Doe, the barren gulches, treeless except for stumps dotting the landscape, the noise of falling stamps and the roar of ore wagons hurtling down the garlands of dusty roads that festooned the hillsides from mine shaft to mine shaft and the narrow winding streets of the gold camp were a far cry from the peaceful tree-shaded avenues of Oshkosh, but if to live in this environment meant that she would become a great and rich woman, Harvey's bride decided that she could meet the challenge.

After Papa Doe saw his young charges settled into rooms in one of his boarding houses, he made a point to introduce them to his friends in town, and then went back to Wisconsin, leaving Harvey to get on with the business of making money.

Unfortunately, Harvey was not a chip off the old block. As Baby Doe was too soon to learn, her twenty-four-year-old husband had no ambition, little ability, and less backbone. With the prop of his parent removed, Harvey shilly-shallied around talking to people about his mine and its great riches, impressing everyone with the fact that he was the son of Harvey Doe, the millionaire, and avoiding altogether anything that smacked of work. He made a half-inspired show of hiring a crew of miners, but he put off facing Waterman to get the agreement down on paper. After several weeks of watching his procrastination, Baby Doe in exasperation got out their two-seater buggy, hitched the team to it, and dragged the reluctant Harvey out to draw up and record the lease that would legally give them any profits from the Fourth of July. Once the deed was recorded, Harvey seemed to buck up, dutifully trotting out to the mine every morning to supervise the digging of the shaft. In the meantime, they rented a little cottage on Spring Street at the western edge of town and Baby amused herself decorating its rough board walls with family pictures and unpacking the boxes of wedding gifts they had brought out from Oshkosh, preparing to take her place in Central City society.

The summer stretched into a brisk fall but there was no sign that the mine was going to be the money-maker all had hoped it would be despite Harvey's apparent industry in keeping crews at work deepening and timbering the shaft of the Fourth of July. Assay after assay showed ore too low in gold content to mine at a profit. Soon Father Doe's grubstake was gone with nothing to show for it but a worthless rock pile. Harvey slipped into apathetic dejection. Disappointment had the opposite effect on the saucy Baby Doe. She tossed her naturally curly hair impatiently and became ever more determined that it was not the mine that failed them but Harvey's mishandling that caused their problem. After prodding and pushing her abject husband, she got him to borrow a thousand dollars from Thatcher Brothers' Bank, now renamed the First National, and urged him to continue digging for the elusive vein that would bear out the mine's earlier promise. To speed up the work, Baby cast propriety aside, climbed into a set of miner's coveralls, and joined Harvey at the mine. It was her idea to sink another shaft paralleling the one Harvey was working on. In that way, she reasoned, they would have a double chance of finding the vein. The moment she appeared at the mine she was a sensation. All of Gregory Gulch could read in the papers about her unlady-like behavior. "The young lady manages one half of the property while her liege lord manages the other...This is the first instance where a lady, and such she is, has managed a mining property. The mine is doing very well and produces some rich ore."

Baby Doe's status as a lady was in some question in the gulch but there was no question that her idea had been a good one. At last the Fourth of July was paying and the dream of rich indolence came to life again for the young couple. But in October, when Harvey's father, accompanied by his matriarchal mother, arrived to take up residence in Central, Baby Doe's sweet taste of success turned sour. Mrs. Doe, scandalized that her daughter-in-law would drag down the name of Doe by comporting herself as a common laborer, demanded that Baby retire to her hearthside where she belonged. In addition, the elder Doe was angry with Harvey for sinking two shafts instead of sticking by the agreement and deepening the one already started. The strained relations between the senior and junior Does were too much for Baby. She suggested to Harvey that they give up their cottage and leave Central's prim and stuffy atmosphere. When he agreed, Baby found a pair of second-story furnished rooms down in Black Hawk, four doors up from Jake Sandelowsky's clothing store.

In their new surroundings the newlyweds settled down to a routine and conventional domestic existence. Harvey daily labored at the Fourth of July under the watchful eye of his father, his hands growing calloused and the skin of his plain face toughening in the high dry air of the gulch, while Baby Doe mooned away the days in gloomy boredom. None of the likes of Harriet Randolph saw fit to call or to invite her to their teas and Harvey was so dragged out by day's end that he was hardly fit company. Her only pleasures were an occasional shopping tour, including a stop at Sandelowsky, Pelton & Company, and sometimes a sidesaddle canter on one of Harvey's span of horses, her elegant black worsted riding habit setting off her golden hair and creamy complexion in a cameo that drew the admiration of all the young bucks in the gulch. Perhaps no one was more attracted to the vivacious young woman than Jake Sandelowsky. As winter closed in about the gold camps, bringing swirling snows and freezing temperatures, Baby Doe found herself riding horseback less and dropping more and more often by the shop of the friendly clothier to pass the long hours of the day. At first their conversations revolved around the clothing business, with Baby Doe remembering all she had learned from her father's business. Later, as their familiarity grew, Jake discovered that Baby Doe shared his fondness for money, fine jewelry, and the exhilaration of a fast-stepping dance. Soon he was paying her gallant compliments and presenting her with small pieces of jewelry. When Baby Doe learned that Jake's birthday was the day after Christmas she invited the two clothiers to dinner. The evening was a great success. Even Harvey was drawn out of his usual morose shell by the laughter and the gaiety of the high-spirited young men. Jake taught the Does how to play poker, and the foursome played hand after hand far into the snowswept night.

The birthday party was the first of many evenings Jake, Sam, and the Does spent together. Occasionally, they went dancing at the Black Hawk Club. Harvey's talents, the few that he had, did not include dancing but Jake was as at home on the dance floor as he was over a poker hand and he was more than willing to whirl the lithe and laughing Baby around the dance floor until she was breathless. Sometimes after these outings Harvey treated for supper from his bank roll that was every day growing with the profits from the Fourth of July, and sometimes Jake and Sam picked up the tab. It mattered not to Baby Doe who paid. Blossoming under the attention of three devoted admirers, she was having the time of her life. Not even being snubbed by the bootstrap society matrons of the gulch bothered her any longer.

March came and with it the opening of Central City's new opera house. For months, Frank Young had talked up the need for a showhouse for the town. Central, he argued, was being left off the circuit by famous theatrical and musical troupes because of the lack of a decent hall. Since Teller's Belvidere and Jack Langrishe's Montana had been destroyed in the 1874 fire and were never rebuilt, Central lacked a first-class theater. Frank's listeners at the bar of the Elevator, Ed Lindsey's popular watering place on the ground floor of the Teller House, thought it was a fine idea. A public subscription toted up $12,000 and the project got underway without delay.

In designing the structure, the architect adapted a little Queen Anne, a little Jacobean, and a touch of the neoclassical to turn out a not unpleasing pile whose native stone walls were four feet thick and were periodically interrupted by Norman-arched windows and doors and were topped by three peaked roof tops. On opening night. Baby Doe and her three escorts marveled at the opulence of the interior—the patent opera chairs, frescoed walls, and a chandelier with one hundred gas jets illuminating the twin statues of Pegasus affixed one at each side of the stage. It was a gala evening and it marked the beginning of a long series of visits to Central City by such famous artists as Lotta Crabtree, Christine Nilsson, Helene Modjeska, and Edwin Booth. Unlike Denver, where grand opera was a flop, Central City thrived on it, and performances of *Romeo and Juliet, Norma,* and *Daughter of the Regiment* were certain sellouts in the Little Kingdom.

Within five years, however, Frank Young's stylish opera house was dark. It opened and closed sporadically thereafter for special occasions until 1933 when a group of Central City progeny and well-heeled Denverites got together to restore the now disheveled old theater and open it once again in a series of musical and dramatic productions that are a highlight of the Colorado summer season to this day.

What Central City lacked in lightheadedness its neighbor Black Hawk two miles down the gulch made up for in spades. Black Hawk became the milling and smelting center for the mines of the Little Kingdom, and its rough exterior was an exact representation of its inner character. For their pursuit of the cultural aspects of life, Black Hawk's mill men and smelter workers flocked to the popular Shoo Fly variety halls that cropped up in both Black Hawk and Central, to the utter mortification of the residents of the latter. Here were the scantily clad dancing girls, the vile whiskey, the raucous piano, the tinhorn gamblers, and jaded prostitutes. Here was a place a man could let himself go without facing the drab appearance and baleful looks of his wife that painfully reminded him that she could use an occasional respite from her daily drudgery over the wood stove and copper washtub.

Hearing Jake talk about the Shoo Fly piqued Baby Doe's curiosity but she resisted the temptation of asking to be taken to the variety hall and concentrated instead on the latest excitement rippling over the gulch— the completion of the Colorado Central track to Central City. For days, Baby Doe had watched the workmen hammering the last ties on the rail-bed at the Black Hawk end, and she eagerly grasped all the handbills announcing the day of the big event and brought them home for Harvey to see. When the day came on May 21, it was a day to remember. George Randolph, as mayor, led the deputation to meet the first five-car train that made its way into town, its steam whistle caroling its arrival. Another train of the same size followed the first, each one filled with dignitaries from all over Colorado. A parade of a dozen or more volunteer fire companies, their uniforms of bright colors indicating their home towns, and six brass bands snaked its way through Central's angled streets to the applause of hundreds standing on the sidewalks. After the hose companies staged their tournament for the benefit of the visitors and the speechmaking was finished, there was an oyster feast at the Granite House and in the evening a grand ball in Turner Hall. Here, under the benign eye of Harvey, Jake and Baby Doe danced to the lilting strains of the Centennial String Band until they were too tired to stand. For Baby Doe, it was the last evening of gaiety she would experience for some time to come.

At the end of May, the rich vein Harvey and his crew had been working in the Fourth of July abruptly pinched out. It took the young couple completely by surprise. The returns from the mine had been so ample and steady that they were lulled into believing the bonanza was endless, and neither Harvey nor his bride thought to put any money away as savings. And now there was no capital to once more dig deeper and cut drifts in search of the elusive vein. To risk Father Doe's wrath at his importunity by asking for another grubstake was something Harvey could not bring himself to do. Instead, he borrowed another $1,000 from the bank and tried to rediscover the source of their recent riches. He worked throughout the summer but his efforts came to nothing, and as the debts piled up—to the ore hauler, the grocer, and even to their landlord—Baby Doe for the first time began to worry about their future. Harvey turned bitter and petulant, scarcely acknowledging his pretty wife. Finding Harvey to be too caught up in his own worries to bother with hers, Baby Doe turned for solace to Jake Sandelowsky. After she confided in her friend the stringency of their circumstances, it was with great relief that she heard Jake offer her a sales job in the ladies' department of the new store he had opened in Central in June. When Harvey came in from the mine that evening, Baby Doe told him of Jake's offer and pleaded with him to let her take the job. Grudgingly, Harvey gave his permission, and to be near her work, they moved back up the gulch to Central City.

Baby Doe's wages were barely enough to keep them in food and shelter in spite of Jake's generosity with an occasional bonus. And rather than improving their lot as a team, Baby Doe's employment appeared to contribute to the further disintegration of their union. Harvey grew ever more withdrawn and spent long hours writing his woes to his mother and sister who now, with Father Doe, lived in Idaho Springs. When Harvey's father heard about the pinched-out Fourth of July through Harvey's letters to his mother, he wasted no time in coming over to Central to view the situation for himself. He arrived in the first snowfall of November, going directly to the mine where he found Harvey poking away desultorily at the wall of the shaft. The elder Doe took one look at the obviously unrewarding specimens of rock at Harvey's feet, squinted up at the snowflakes, and declared that the Fourth of July was closed for the winter. Then he negotiated a job for Harvey on the night shift at Benoni Waterman's stamp mill as an apprentice millhand. With his usual acquiescence, Harvey accepted both the shutdown of the mine and the job without a word.

Meanwhile, Baby Doe was doing well in Jake's store. She liked meeting people and showing the silk taffeta gowns and the frilly organza jabots cut in the latest New York styles. But what she liked most of all was the tender squeeze Jake gave her hand when he came up from Black Hawk to check the inventory and sales of the Central City store. With Harvey at the mill at night, Baby Doe dreaded to go home to their lonely rooms after store closing time, but whenever Jake was in town he saw to it that she did not have to go home alone. As the weeks went by, Jake found more and more reason to be in Central City. On Baby Doe's twenty-third birthday, Jake came up to Central and gave a small after-closing-hours party for his charming saleslady, presenting her with a nosegay of blue gentians during the festivities. After the party, if the coded message inked under the few dried flowers in one of eleven scrapbooks Baby Doe left has been accurately deciphered, Jake let his growing ardor take more positive form. "Jake gave me these September 25, 1878, the night of the festival in his store when we sat on the schoolhouse steps together," wrote Baby. "He kissed BT three times and oh! how he loved me and does now."

Before long Jake was seeing her home almost every night and, so it is told, there came a night when they stopped not at three kisses on the schoolhouse steps. To avoid as much as possible the watch of tattling gossips the lovers rendezvoused in a tiny cabin Jake rented in

Packard Gulch halfway between Black Hawk and Central City. Now every day Baby Doe waited anxiously for the delivery of the small envelope that bore Jake's fine lined script and that contained a message invariably the same, "Meet me, my darling, at ten."

In the few hours of the day that she saw her husband, Baby Doe found him to be his usual apathetic, morose self. At one time his mood would have angered her, but now she was almost glad to see it. It meant at least that he was still unaware of her affair with Jake.

In December, the beautiful Baby Doe found that she was pregnant. When Harvey heard of her condition he embraced not his wife but immediately sought out the comforting balm of the neighborhood saloon and the maudlin understanding of its blowsy shills where in time he heard what most of Central already knew despite the lovers' attempts to keep their trysts a secret. One night in a drunken fog, Harvey confronted Baby Doe with the accusation that the child she carried was not his but Jake Sandelowsky's. In fury, Baby Doe picked up a chunk of assay ore from the Fourth of July and heaved it at her leering, glassy-eyed mate, striking him in the neck. As the blood trickled down onto his collar, Harvey let loose a string of oaths and stormed out of the room. He beat it down to Sam Newell's feed store, borrowed $30, and caught the down train to Denver. Baby Doe, shaking with rage and humiliation, beat it into the readily enfolding arms of Jake Sandelowsky.

A day or so later, Baby Doe recovered her presence of mind and decided she could not stay in Gregory Gulch a moment longer, and despite Jake's entreaties she packed herself off home to Oshkosh and the welcoming bosom of Mother McCourt.

When the word of the breach between his son and his daughter-in-law reached the senior Doe over in Idaho Springs, he set about immediately arranging a reconciliation. Through the rest of the winter and spring, telegrams and letters flashed back and forth between Oshkosh and Denver and Idaho Springs. And finally in June, a month before her child was due, a calmer, sweeter Baby Doe stepped off the train at Central City into the arms of a penitent Harvey. He had a new job as a mucker in the Bobtail Tunnel and he had not touched a drop of liquor for weeks. With a slate clear of ill feeling, at least on the surface, the couple began their lives together again, but on July 13, their world sagged around them once more. Baby Doe gave birth to a stillborn boy, with large blue eyes and "dark, dark hair, very curly," not unlike Jake Sandelowsky's.

In the weeks that followed, the amnesty between Harvey and Baby Doe eroded away. The haunting suspicion that the dead child was not his still ate at Harvey's consciousness. He began drinking again. The couple's debts began to pile up again—not even the midwife had been paid—and then suddenly, Harvey quit his job at the Bobtail and left for Denver to find a buyer for three unworked claims he had bought in the heyday of the first earnings from the Fourth of July. While he looked, Baby Doe disconsolately disposed of their silver and furnishings and her jewelry, some of it gifts from Jake, to keep enough cash in the house to live on. For once, there was no Jake to turn to for help—he was on a trip to New York to buy for the new store he intended to open in Leadville in November. As the weeks went by, Baby Doe heard nothing from her errant husband, and with every piece of jewelry she pawned her gorge rose a notch higher in disgust at the name of Doe.

Jake returned to Central on October 4, and after stopping overnight in Black Hawk he went immediately to Leadville without taking the time to call on the Does and was therefore unaware of Harvey's latest disappearing act. When he got to the fast-growing silver camp, Jake wrote Baby Doe suggesting that she and Harvey make a trip over to Leadville to check on the opportunities that would allow Harvey to make a new start. He even offered to pay all their expenses. Alternating between despair and yearning, Baby Doe at last threw caution to the winds, brushed out her serge traveling suit, and caught the narrow gauge to Georgetown where she took a stage the rest of the way over the mountains to Leadville.

When Jake learned that Harvey had again run out on Baby Doe, he talked to her like a Dutch uncle, and the upshot was that on the day after New Years' Baby Doe went to Denver, sought out the sodden Harvey, and told him she wanted a divorce. He pleaded for more time to prove himself and promised her the deeds to the three claims if she would stick by him. His pleas were sufficiently pitiful that Baby's resolve to free herself from him melted and she agreed to make one more try. But after weeks of the same old drunken, maudlin scenes with Harvey, she cast about for a means of escape. On March 2 she had the good luck to catch sight of the tottering Harvey reeling into Lizzie Preston's parlor house on Denver's notorious Market Street. Quickly rounding up a policeman, Baby Doe stormed the door of the bordello, found Harvey with one of Lizzie's "boarders," and on March 19 was awarded a divorce by Judge Amos Steck on the grounds of adultery and nonsupport. From the judge's chambers, Baby Doe went directly to the mining exchange and sold the three unworked claims Harvey had deeded to her. With the money from the sale, she bought a few new clothes, and plunked down $17 in gold that bought her passage on the morning Denver and South Park train to Leadville.

By 1880, the year of Baby Doe's divorce from Harvey, the boom days of Central City and the rest of the gulch camps were beginning their decline, one that this time would be permanent. The output of the

Gilpin County mines was still $2 million a year and business was very good. In Central City, shops carried full lines of the latest available merchandise. Even undertaker Henry Thompson, not to be left out of his share of the commerce, advertised in the *Register,* "No more delays...I promise to keep a full line of coffins and caskets in stock...I have a new corpse preserver." But not even Henry Thompson's corpse preserver could forestall the inevitable. There were no more big mineral strikes to be made in the gulch. No influx of investment capital or surge of speculation lay on the horizon. It was only a question of the pedestrian development of mines already stripped of the easy-to-reduce, high-grade ore. As a result, the exodus of the younger Does and Jacob Sandelowsky from Gregory Gulch was not unique. With their millions made and the old glamor and excitement gone, the mining and milling magnates also moved on. Nathaniel P. Hill moved his smelter down to the environs of Denver. Following him to the capital were George and Harriet Randolph, the Thatchers, and the Youngs, who found they could all carry on their businesses in Central by making day trips to the gulch while living a relatively more genteel existence in Denver. Henry Teller, tapped to be Colorado's first United States senator, divided his time between Washington and Denver. And some, still seeking the vibrant days of discovery, took up the new cry that was sweeping Colorado—"Silver!"

William Green Russell, *courtesy of the Denver Public Library, Western History Collection, F-12030*

Placer-mining for gold in Russell Gulch on Clear Creek, 1877, *courtesy of the Denver Public Library, Western History Collection, X-61290*

On the trail to the new Eldorado, 1860, *courtesy of the Denver Public Library, Western History Collection, X-21803*

William N. Byers, *courtesy of the Denver Public Library, Western History Collection, H. Goehner, Z-2345*

Central City, 1860, *courtesy of the Denver Public Library, Western History Collection, Henry Faul, X-2606*

Central City, 1881, *courtesy of the Denver Public Library, Western History Collection, X-11627*

A miner's cabin, *courtesy of the Denver Public Library, Western History Collection, X-61350*

Visitors to Alma in South Park, 1883 (woman with gunbelt is purportedly Annie Oakley), *courtesy of the Denver Public Library, Western History Collection, Otto Westerman, X-6484*

Mail carriers in winter, Breckenridge, Colorado, 1898, *courtesy of the Denver Public Library, Western History Collection, X-964*

Arrival of a supply train from Kansas on Market Street, Denver, 1886, *courtesy of the Denver Public Library, Western History Collection, W. G. Chamberlain, Z-224*

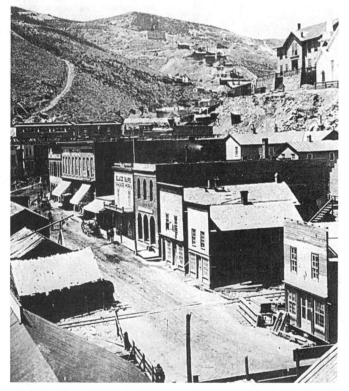

Black Hawk, 1880 (Colorado Central train on trestle enroute to Central City), *courtesy of the Denver Public Library, Western History Collection, Charles Weitfle, X-2004*

Smelting experts Richard A. Pearce (far left) and Nathaniel P. Hill (in doorway) exhibiting the silver ingots that were laid in the sidewalk in front of Teller House for President Ulysses S. Grant's visit to Central City, 1868, *courtesy of the Denver Public Library, Western History Collection, X-60031*

Central City Opera House shortly after completion, *courtesy of the Denver Public Library, Western History Collection, Z-3590*

Removal of the body of a miner killed in an avalanche, Silver Plume, Colorado, 1899, *courtesy of the Denver Public Library, Western History Collection, Louis Keck, F-20221*

Otto Mears's toll road between Silverton and Ouray, 1888. The Million Dollar Highway was built over this roadbed in the 1920s. *Courtesy of the Denver Public Library, Western History Collection, Charles Goodman, Z-7867*

Silver Plume (foreground) and Georgetown (background), 1900. Switchback trails lead to mines on Sherman and Republican mountains, *courtesy of the Denver Public Library, Western History Collection, Louis C. McClure, MCC-1168*

The Maxwell house, Georgetown, *courtesy of the Denver Public Library, Western History Collection, Elmer Moss, X-1294*

Infamous State Street, Leadville, 1940, *courtesy of the Denver Public Library,*
Western History Collection, X-6527

A bonanza baron's private militia,
the Silver Queen Guards, Georgetown, *courtesy of the Denver Public*
Library, Western History Collection,
X-1470

Thomas F. Walsh, *courtesy of the Denver Public Library, Western History Collection, Rose & Hopkins, H-99*

David H. Moffat, 1860s, *courtesy of the Denver Public Library, Western History Collection, Z-2879*

H. A. W. Tabor, *courtesy of the Denver Public Library, Western History Collection, F-7188*

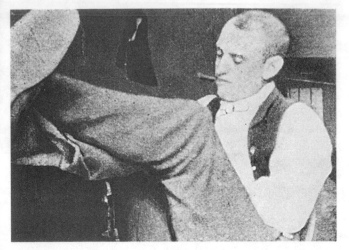

Eugene Field at his desk on the *Rocky Mountain News, courtesy of the Denver Public Library, Western History Collection, F-10394*

The Tabor residence, Denver, *courtesy of the Denver Public Library, Western History Collection, William Henry Jackson, WHJ-816*

Baby Doe Tabor, *courtesy of the Denver Public Library, Western History Collection, F-18591*

Grand Duke Alexis and his entourage, Denver, 1872 (second row: third from left, General Philip H. Sheridan; next to him, Alexis; and on the end, in cape, General George A. Custer), *courtesy of the Denver Public Library, Western History Collection, J. Lee Knight, F-9106*

General William H. Larimer, 1852, *courtesy of History Colorado (F-48)*

Jefferson Randolph (Soapy) Smith (hat in hand), *courtesy of the Denver Public Library, Western History Collection, Theodore E. Peiser, Z-8905*

Leadville's Ice Palace, 1895, *courtesy of History Colorado* (F-2437)

The City Hall War, Denver, March 15, 1894, (note Soapy Smith's henchmen at windows and on the roof), *courtesy of the Denver Public Library, Western History Collection, X-22120*

Some of the Socialites, Cripple Creek, 1894 (second from left, Spencer Penrose; at the end, Charles L. Tutt, Sr.), *courtesy of the Denver Public Library, Western History Collection, X-602*

Cripple Creek consumed by fire, April, 1896, *courtesy of History Colorado (Scan #20004517)*

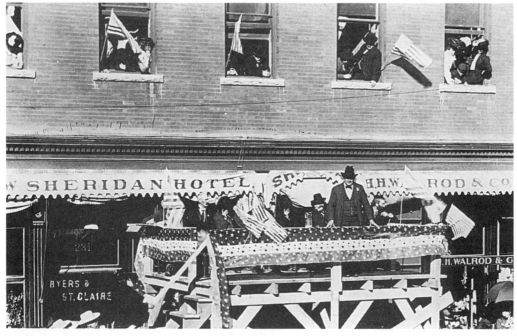

William Jennings Bryan rallying citizens of Telluride behind the free coinage of silver, 1902, *courtesy of the Denver Public Library, Western History Collection, Homer Reid, F-20023*

Margaret Tobin Brown (Unsinkable Molly), *courtesy of the Denver Public Library, Western History Collection, Cooley Studio, X-21699*

Frederick Gilmer Bonfils, *courtesy of History Colorado (Acc. #89.451.5497)*

PART II

SILVER, 1870–1890

White Metal

WHEN THE FIFTY-NINERS SWARMED OVER THE ROCKIES, THEY FOUND GOLD IN nearly every gulch and stream they came to. More often than not, along with the yellow money metal, they also found silver, but with the price of gold far above that of silver, few prospectors bothered to follow up traces of white metal they located. For ten years, from the time of the Gregory and Jackson gold strikes in 1859, silver took a back seat. But when the surface deposits of gold began to peter out and the refractory gold sulphides yielded but a paltry return, miners remembered the bits of silver float and the black, sluice-clogging sands that they had impatiently thrown aside in their rush to garner the glittering yellow ore.

The first notable silver discoveries of near bonanza proportions were made to the northeast and southwest of the Gregory diggings on Clear Creek. One fall day in 1860, Sam Conger, Indian scout and pal of Buffalo Bill Cody, got a yen for some fresh meat. He threw a pack over his broad shoulders, caught up his shotgun, and directed his moccasined feet westward from Boulder into the steep-sided and heavily forested foothills, looking for elk. Climbing steadily upward toward the continental divide, Conger took his time, following a branch of Boulder Creek, with his eyes and ears alert for a sign of his quarry. After a couple of days of fruitless tracking, Conger came to a meadow at the top of a small rise, forked by the stream he followed. At the edge of the meadow in a glade of pines, the scout made camp, determined to rest a day and content himself for the moment with a dinner of freshly caught trout instead of the broiled elk steak he craved. By the time he finished his meal and went to the stream to wash the frypan that served him both as a cookpot and a plate, the sun was low in the sky, its rays sliding through the trees to lie like fingers on the clear water, illuminating the bed of the creek. Suddenly, as Conger scrubbed his pan, his eye caught sight of a silvery flicker among the stones in the streambed. Plunging his hand into the icy water, he drew out a nugget of silver splotched with a brownish matrix. Peering into the water, he could see no other chunks of ore, so he pocketed his silver specimen as a curiosity and the next day took up his search for the elusive elk once again.

That is one version of how Conger stumbled onto the silver lode that ushered in Colorado's second money metal boom. A more romantic story has it that the Indian scout and a beauteous daughter of an Arapaho chief found themselves passionately in love and that the girl in the ardor of her devotion betrayed to her lover the location of the tribe's secret mine of silver.

However he came by the nugget, Conger apparently soon forgot about it. Then in 1869, as he passed through Laramie, he happened to walk by a carload of ore on a rail siding. Noting the appearance of the ore, he was suddenly struck by its similarity to the nugget of silver lying forgotten in his pocket. Quickly, he strode over to the ore car to pick out a chunk to compare it with his own specimen. He had barely taken a good look at both pieces of ore when he was rudely shoved aside by a scowling trainman.

"Leave that stuff alone, mister!" bellowed the brakeman. "That's high-grade silver from the Comstock lode in Nevada."

"Yes, sir," said Conger, backing off, his mind flashing back to that sylvan camp beside Boulder Creek. Hightailing it back to Colorado, Conger got together five of his friends and together they packed into the rugged wilderness looking for the site of his find. After days of climbing, on August 31, they reached the rise Conger was searching for, snuggled in the mountains at an altitude of nearly ten thousand feet. Conger and his party took a few hours off to acclimate themselves to the altitude and then they set to work prospecting. Carefully working the steam bed and bank up and down from the spot Conger recalled was where he had found the silver nugget, first one and then another man shouted out that he had found a pocket of promising ore. When the assays showed the ore to be rich in silver, the Conger party quickly recorded their claims, dubbed the district Caribou after the recent British-American discoveries in British Columbia's Cariboo region, and went to work mining. The first load of ore the scout and his friends trundled over the mountains to Hill's smelter at Black Hawk produced $16,000 worth of silver to the ton.

Despite attempts by Conger and his partners to keep their discovery quiet, a full-scale rush to the region was underway by June of 1870. A town was platted, a concentrate mill set up five miles east of town, and the area burgeoned into the now familiar boom camp of hard-working, hard-living miners, merchants, gamblers, blacksmiths, bullwhackers, innkeepers, and hurdy-gurdy girls.

After a few months, Conger traded his interest in the Caiibou to his partners for their shares in a neighboring mine and in turn they sold the Caribou to Abel Breed, a speculator from Cincinnati, for $125,000. Breed took $70,000 a season out of his new holding for three years and then sold it to Dutch interests in 1873 for $3 million. Huge chunks of rich Caribou ore were displayed to an appropriately awed crowd at the Philadelphia Centennial in 1876. The shining bricks of solid silver laid in the sidewalk of Central City in front of the Teller House for visiting President Grant to walk on were smelted out of Caribou ore. But spectacular as the Caribou strike was, poor management and legal complications in time forced foreclosure. And the once famous producer finally fell into the hands of Jerome Chaffee at a sheriff's sale for a fraction of its worth.

Eventually, after surviving a catastrophic fire and a diphtheria epidemic, the camp of Caribou lapsed at last into oblivion in 1893 when Congress wiped out the silver price support. One by one the mines on the Caribou lode, the Idaho that had produced $6,000 in the first twenty feet of shaft, the No-Name, the Spencer, the Seven-Thirty, all closed down, their once priceless hordes of silver as worthless as grains of sand.

⁂

The second of the early silver camps was destined to have a far longer and more notable life than Caribou Hill.

In June, 1859, George and Dave Griffith, two brothers fresh from their farm home in Kentucky, got to the Gregory diggings in time to find all of the choice real estate taken. Disappointed but not discouraged, the pair decided that their long trip halfway across the nation was worth one more try to locate a gold mine. With more instinct than instruction they started up Clear Creek in the direction of Idaho Springs, and from that bustling camp they moved even farther up the tumbling creek to a long hollow of a canyon flanked by soaring peaks, dotted with pines and aspen to timberline, and dusted with a thin covering of snow above it. Gophering diligently along the eastern edge of the canyon for several weeks, they one day came across what they were looking for—a pocket of pay dirt. The Griffiths quickly staked out their mine and a part of the meadow below it for a homestead.

As soon as the word of the Kentuckians' luck reached the camps farther down Clear Creek, hundreds of miners scurried toward the narrow canyon where George and Dave Griffith were fast becoming rich men. In a few weeks their homestead was known as Georgetown. Dozens of houses went up, mercantile stores and stables sprouted overnight, stamp mills straddled the creek, and Georgetown, like Central City, Breckenridge, and Caribou was transformed from a wilderness into a lusty, headlong boom town.

Unlike many of her sister camps that were inglorious collections of thrown-together wooden lean-to's and mud-chinked log cabins,

Georgetown from the beginning was a town of sturdy buildings, soundly built and appealing to the eye. This gave the aura of permanence to the town in spite of the fact that within a year or two her mines gave unmistakable signs of playing out. It was the same story as that on Gregory Gulch. The blossom rock on the surface gave way to the hard-to-reduce sulphides, mines were abandoned, and the familiar exodus began.

Then in September, 1864, a prospector came across the bonanza Belmont silver lode five miles northwest of town. For three years Georgetown lived on the edge of promise until 1867 when a hard-rock miner from the newly staked Anglo-Saxon mine clomped into Frank Dibdin's assay office and dumped a sack full of silver ore onto the counter. When the result of the assay was announced, $23,000 to the ton, all Georgetown knew that the future of the handsome little hamlet lay in the white metal.

Now the town really boomed. Every week the brand new newspaper carried announcements by the dozen of brand new silver strikes. Chamberlain and Dillingham opened a mill to add to those reopening after the fizzle of the gold rush. A smelter was built to treat the silver ore on the spot, and Professor Robert Orchard Old established the Colorado British Mining Bureau with headquarters in London and Georgetown, an organization that would have far-reaching consequences for Colorado's mining future. Old saw to it that impressive chunks of Georgetown silver ore were on display at the Paris Exposition of 1867, drawing foreign capital to invest in Colorado's infant silver industry. By the turn of the century it was estimated that the British alone had investments in Colorado mines to the tune of ten million pounds sterling. Nor were investors in the United States blind to the potential fortunes to be made in and around Georgetown. Jay Gould, Marshall Field, P. E. Studebaker, Felix Stoiber, and Charles Rothschild all cocked watchful eyes and open purses in the direction of the silver-rich Georgetown mines.

A stage a day (Chicago to Georgetown, one way, $196.50) brought countless newcomers: investors, miners, politicians, and tourists, including the celebrated lady of patent medicine, Lydia Pinkham.

Another woman traveler, Lady Isabella Bird, who in an age when the socially accepted posture of a gentlewoman was to be as helpless as possible, to dare nothing, and to swoon at the slightest provocation, had the remarkable spunk and fortitude to tour the Rocky Mountains by herself in the winter. She left a graphic description of Georgetown in 1873, at the beginning of its heyday:

> The road pursued the canyon to Idaho Springs...where we took a superb team of six horses, with which we attained a height of 10,000 feet and then a descent of 1,000 took us into Georgetown, crowded into as remarkable a gorge as was ever selected for the site of the

town...The area on which it is possible to build is so circumscribed and steep, and the unpainted gable-ended houses are so perched here and there, and the water rushes so impetuously among them, that it reminded me slightly of a Swiss town. All the smaller houses are shored up with young pines on one side, to prevent them from being blown away by the fierce gusts which sweep the canyon...I arrived at three but [Georgetown's] sun had set, and it lay deep in shadow...We drove through the narrow, piled-up irregular street, crowded with miners standing in groups, or drinking and gaming under the verandas, to a good hotel declivitously situated...

The "good hotel declivitously situated" was probably the spanking new Barton House where oysters on the half shell, richly flavored cigars, and vintage champagne were as easily come by as in New York's finest restaurants.

Some of the newly arrived filled their hours with entertainment at the Gayoso Saloon, but most beat a respectable path to McClellan's Opera House to see Helen Potter's Pleiades, a fabulous troupe of impersonators, or to listen in rapt delight to the tunes of the nimble-fingered Swiss Bell Ringers, whose pure notes seemed to hang suspended on the crystalline air in the high canyon of Clear Creek. Up to Georgetown from Denver came P. T. Barnum on a lecture tour with none other than the keen-minded Professor O. J. Goldrick as his Colorado manager. For each lecture the Prince of Showmen was to receive $75. The shrewd Goldrick knew what he was doing when he booked Barnum into Georgetown. To a large and avid audience who already knew how to achieve the last of his three topics, Barnum lectured on "How to Be Healthy, Happy, and Rich," and after paying the showman his fee, Goldrick pocketed a neat $247 profit.

As financial backing poured into the canyon to develop the mines, and in turn tons of high-grade silver ore were shipped down-creek to concentrators, stability came to the old Griffith diggings. Schools, churches, fraternal lodges, and banks were established. A water works opened in 1874 and four companies of crack volunteer firemen were formed soon afterward. To the owners of Georgetown's impressive array of frame-structured architectural wonders, it was comforting to know that the Alpine Hose Company, trained to perfection by debonair gambler Cort Thomson, won the first state tournament by streaking over the seven-hundred-foot course pulling a fully wound hose cart in 29:75 seconds.

A grocer named Potter grew rich on the appetites of Georgetown's newly rich, and with his wealth he created what is today considered to be one of the ten best examples of Victorian architecture in the United States. Unfortunately, Potter was able to live in his stylish mansard-roofed home only a few years before the silver crash forced him to sell it. When that occurred, he sold his elegant little house to

Frank Maxwell, a mining engineer who made his mark in the world not in engineering mines but in designing what became the region's chief tourist attraction in the 1880s—the Georgetown Loop.

When Loveland's Colorado Central nosed its first Baldwin locomotive into the new brick depot of lower Georgetown in August of 1877, the camp went wild. The celebration was hardly over before Georgetown looked around to see where else track could be laid to cinch its place on the roster of railroad metropoli. It rankled the busy little mining town to be the end of the line. Over Loveland Pass to the west lay Breckenridge, another booming camp, and beyond that was still another silver camp on the threshold of its great boom, Leadville, located just north of the old gold site of Oro City. In a burst of vigor, localites incorporated, capitalized, and put under construction the Georgetown, Breckenridge & Leadville Railroad. In spite of the enthusiasm of its backers, however, the GB&L went nowhere fast. It took five years to lay track from Georgetown to Silver Plume, a sister camp two miles up the canyon. But in those two miles grew a railroad engineering marvel of the world, thanks to Frank Maxwell's applied technology. Although Silver Plume was only two miles away from Georgetown, it was a thousand feet higher. To make the climb feasible for the little narrow gauge locomotives, Maxwell designed a series of loops using rails elevated on trestles, each trestle higher than the last. When it was finished, the curving track of the Georgetown Loop rose a giddy 143 feet to the mile for four and one-half miles, with the highest trestle a breath-catching three hundred feet above the bed of Clear Creek. To those who had held cherished hopes of a direct, money-making, over-the-mountain line to Leadville, the GB&L was a white elephant. Its total trackage finally reached eight and one-half miles out of a proposed 110. But for Jay Gould and the Union Pacific, whose appetite for Colorado short lines was gluttonous, it became a first-class tourist bonanza. At the height of the summer season seven and eight trains a day, crowded with sightseers, pulled out of Denver's Union Station on excursions to the famed wonder of the Georgetown Loop.

Georgetown's neighboring camp, Silver Plume, soon boasted a railway that matched the Loop in scenic wonder and engineering skill. Edward John Wilcox, at home as a mountain goat in the high country, found himself a silver mine near the crest of fourteen-thousand-foot Mt. McClellan, the sheer-sided peak that rose immediately to the south of Silver Plume. The district that grew out of his find became the Argentine, and his camp took the fancy name of Waldorf. Wilcox, impatient at the slow transport of ore by burro down the sixteen miles of precipitous trail to the canyon floor, decided that a narrow gauge railroad was what was needed to serve his and the other high-lying mines. With some money of his own and a generous amount contributed by others he dropped $300,000 into the project and the Argentine Central was born, holding the dubious record of being the highest railroad constructed in North America. The road rose to its eagle's nest terminous in a series of hairpin switchbacks on which Shay locomotives, fitted with gears to get them up the grades, valiantly inched their way up and down the side of the mountain, strings of ore cars in tow. But Wilcox's line was plagued with troubles, mechanical and financial, and within three years it was sold to the Colorado and Southern Railroad for a mere $44,000. With the mines it was meant to accommodate played out, the mountain line became another tourist attraction. A giddy ride over the Georgetown Loop and a breath-holding trip up and down the side of McClellan Mountain on the Argentine Central were enough to make a visit to Georgetown and Silver Plume unforgettable for even the most fearless of thrill-seekers.

On the heterogeneous multitude that swarmed into Georgetown in its halcyon days, one of its residents looked with a particularly selective eye. Only the cultured, the articulate, and the discerning were welcomed to Louis Dupuy's renowned Hotel de Paris. The moody, gray-eyed Frenchman arrived in Georgetown in late 1869 after a brief sojourn in Denver as a reporter on William Byers' *Rocky Mountain News*. Twenty-five, muscular, and ready to work, Dupuy was hired on the night shift at the Cold Stream mine high up on the east side of Sherman Mountain. It was his job as a mucker to go in and shovel out the ore and debris after the charges set in the mine wall by his partner were detonated. On the night of March 6, 1873, the taciturn Dupuy listened for the explosions and then trudged into the tunnel to do his work. When he reached the working area, he suddenly saw that one of the charges had failed to fire. Just as he opened his mouth to warn the other miners, the charge exploded, hurling the luckless Frenchman backward, breaking his collarbone, one rib, and badly damaging one eye.

During his long convalescence, Dupuy decided that his mining days were over, and he turned to a vocation that came more naturally. He bought Delmonico's bakery on Alpine Street in the center of Georgetown and converted it into the region's handsomest and most exclusive inn. Patterned after a provincial hotel of his native country, the Hotel de Paris boasted walls of masonry three feet thick, neoclassic pedimented windows, and an iron cheval-de-frize along the roof, which was topped at one end by a cast-iron statue of Justice. A vine-hung courtyard flanked the building on each side, one guarded by a reclining lion and the other by a stag of soldered zinc. Dupuy's guests slept in solid walnut beds, bathed railroad cinders from their eyes in marble-topped basins fitted with faucets producing both hot and cold running water, and took comfort from the fact that the hotel's "noiseless toilets," discreetly placed throughout the building, meant there was no need to brave the icy night air of the high mountain canyon for a trip to the outhouse.

But of all his luxurious appointments, Dupuy's dining room was the most favored. Painstakingly, the Frenchman laid the floor in a pattern of alternate strips of walnut and maple with the pattern continued to waist height along the wall as wainscoting. Bronze figures of nymphs and engravings of the western countryside adorned the walls. And in the center of the room surrounded by tables set with sparkling white linen and Haviland china was a pool continuously fed by a bubbling fountain. Here, Louis, whose sense of artistry spilled over into his culinary endeavors, kept small brook trout to catch and cook on short order for his guests. Or, if a patron preferred hardier fare, Dupuy's kitchen could turn out filet of beef garnished with mushrooms, truffles, and anchovies. Local quail, elk, and venison were always on the menu, their natural flavors sharpened with the Gallic touch as they passed through Dupuy's skilled hands. To accompany the delectable meals he served, the Frenchman offered the choice of a complete cellar. As the diner savored a vintage Saint fimilion, Medoc, Riesling, or zinfandel, Georgetown was transformed from a rough mining camp to a gracious and cosmopolitan spa.

After dinner, if Dupuy took a fancy to one of his guests, he might ask him, or her, to join him in the richly furnished library for a glass of cognac and an evening of conversation. One of his favorite guests was Madame Janauschek, the popular tragedienne. Whenever she was in town, she and Louis sat up all night conversing in French before the cheery fire in the tile-grated fireplace. Dupuy was a master of the subtle bon mot, yet he could tell a robust yarn with equal effect. But it was on philosophy that he shone. The books lining the library shelves betrayed the hotelier's materialistic leanings—Shaw's *Fabian Essays In Socialism,* Hull's *Irrepressible Conflict,* Tolstoy's *Work While Ye Have the Light,* Mackay's *The Anarchist,* Croille's *King Capital.* For Dr. James E. Russell, dean of Columbia's Teachers' College, an evening of such discourse with Dupuy sparked an entirely new addition to the college curriculum.

"Man is a machine for storing energy and giving it off in work," declared the Frenchman. "Man should work if he would keep the machine from rusting out but he needs fuel to keep it going. He has learned from experience what fuel is best—long before you damned professors cooked up theories about it—and there are a few specialists who know the secret. I am one of the greatest in this country."

Russell was captivated by Dupuy's theories on the importance to man, the machine, of the selection and preparation of food. All the way back to Columbia he mulled over the Frenchman's words, and when he once more sat in his dean's chair he formulated a syllabus and methodology for instruction in a totally new subject, domestic science.

For all of his public exposure, much mystery surrounded Dupuy. He hated crowds and was imperious to those whose manner or appearance displeased him. He refused to pay taxes and vowed to shoot anyone who came to collect them. His friends were few. And they were only his friends because he wanted them to be. Yet as eccentric and trenchant as he was, Dupuy could also be tender and thoughtful. A small child who fell and skinned his knee outside the Hotel de Paris was sure to find Louis quickly at hand to bathe the wound and send him on his way comforted with soft-spoken words and a handful of mints. At Christmas, Dupuy called upon his few friends bearing a bottle of rare wine and rich little cakes of almond paste he had baked especially for them.

At the death of a French miner in Georgetown, Dupuy took it upon himself to befriend the dead man's family. Over the years he supplied them with food, clothing, and funds. As the eldest daughter of the dead man grew to young womanhood, Dupuy took in hand her education, loaning her books from his library and letting her exercise her growing intelligence in long discussions of philosophy and science. Their close friendship continued after the girl graduated from normal school and began to teach in the Georgetown school. And then one day Dupuy realized that in the years that he had nurtured the developing mind of his ward his interest had grown from paternal to sentimental. He was in fact far-gone in love for the attractive, keen-minded girl. His passion erupted in a series of letters telling her of his admiration and devotion. Gone was his usual reticence under the freely flowing pen as he caressed in words his beloved, "whose beauty of face and candor of eye portrays purity of heart and nobility with understanding and charity of mind." But when Dupuy proposed, the young woman declined his offer. With the very candor he so much admired she told him that she could never love an avowed agnostic. At first, Louis continued to plead his case in lilting prose but when his love married another man and left Georgetown, the Frenchman put away his pen and withdrew even further behind a wall of reticence, a wall now daubed with ruefulness.

The only other who could claim a close friendship with Dupuy was Sophie Gaily, a small, round Frenchwoman, years older than Louis, who could neither read nor write and whose role in the hotelier's life kept the gossips perennially busy. How Sophie came to live in the Hotel de Paris is not clear; some say she was the widow of the cabinetmaker who helped Dupuy build his establishment. Louis always referred to her as a "guest" but he paid her an allowance of $20 a month and she spent more than twenty years in the hotel, purportedly as housekeeper and scullery worker.

In 1898, the fifty-six-year-old Dupuy journeyed down the canyon and across the plains to New York where he sailed to his native France, for the first time in thirty years, to claim a small legacy. Shortly after his return to Georgetown, he fell ill with pneumonia and died very suddenly in October, 1900. To Sophie Gaily he left the Hotel de Paris, his ranch, and all else he had.

It was only after his death that Jesse Randall, editor of the George-town *Courier* and one of Dupuy's few confidants, revealed the story of the Frenchman's early life. Born Adolphe Francis Gerard in 1844 in Alencon, the lace center of France, young Louis early fled the semi-nary school and the studies that destined him for the priesthood as dictated by his well-to-do parents, and ran away to the bright life of Paris. Here he washed dishes in a restaurant to earn his keep and was lucky enough to catch the eye of the chef who taught him the tradi-tions of French cooking. But at this stage in his life, a career in the kitchen was too far away from the action of the world, so he turned to journalism. His efforts were not noteworthy and when a $50,000 inheritance came along, Louis gave up working altogether. When he had squandered his money, he returned to the pen, this time in Lon-don where in learning to write English he copied the writings of oth-ers, sold them as his own, and was accused of plagiarism. To escape prosecution, Louis left England for America where he cloaked himself in the anonymity of the US Army. After a year of duty on the plains with the US Second Cavalry, he deserted in Cheyenne in April, 1869, dropped the name of Gerard in favor of Dupuy, and made his way to Denver and thence to Georgetown where he finally found his niche.

Old "Tante" Sophie lived but four and a half months after the death of her mentor. She was buried in the little Georgetown cem-etery next to the body of Louis. A common tombstone lay above their graves simply inscribed with the words *Deux Bons Amis.*

With its two picturesque proprietors gone, the elegant little hotel continued to thrive under new ownership until 1955 when it was converted into a museum. Today the bubbling fountain in the dining room is dry and the great brick ovens that turned out croissants and quiche by the dozen are cold. In the library, black and silent is the once crackling hearth that danced with light far into the cold clear night as French Louis discoursed over brandy on a favorite theme: "In our crowded societies, what we call pleasures are often pains in disguise."

The silver camp two miles to the west of Georgetown on the other end of the Georgetown Loop came to life in 1870 when avid prospec-tors uncovered vein after vein of Colorado's new money metal high on the sides of the Republican and Sherman mountains. The first miners to strike it rich marveled at the form of the ore—leaf silver in the form of a perfect feather—and named their camp Silver Plume.

Typically, the first habitations were tents and lean-to's strung along the side of the mountains, well above the swampy canyon floor overlaid with shrubs and undergrowth. But as the prospects of Silver Plume's future grew brighter, the miners drained the marshy bottom-land and built themselves a real town with the usual ratio of saloons to churches: 9 to 2. Groceries, barbershops, fraternal lodges, and a newspaper, the *Jack Rabbit,* soon made their appearance in false-fronted buildings to serve the two thousand residents. A teamster named Watkins, with a stable of seventy-five mules, and his rival Poirson, with forty-five mules, carried on a lively traffic hauling ore down the tortuous corkscrew trails that reached dizzyingly upward to the portals of the famed Mendota, Terrible, Seven-Thirty, and Payrock mines. Seven concentrating mills dotted the banks of upper Clear Creek and hard by one of them was the inevitable opera house, where on a Saturday night the populace could listen to the oom-pah-pah of the Payrock mine brass band. The band was renowned throughout the district and whether a man could mine or not he could get a job at the Payrock if he could tootle a horn.

Atop a rise on the south side of the narrow canyon at a spot where the rays of the sun lingered longest in the day, the people of Silver Plume laid out a small cemetery among the slender aspen to receive the remains of miners blown to bits by aged dynamite, crushed by collapsing shoring, smashed in the wreckage of runaway ore wagons, or murdered in the vendettas of disputed ownership that plagued the miner's world.

Few mines had a history of bloodshed to match the story of the infamous Pelican-Dives feud. In 1868, Owen Feenan struck out with his pick and pack of grub, hiking up steep Cherokee Gulch between Sherman and Republican mountains, hacking away at quartz outcrops as he went. After a few days he came across what looked like a promising sign of ore. Marking the spot on the crude map he kept of his wanderings, Feenan moved on up the gulch. But each day the going got rougher and Feenan found his strength ebbing. With his weakness came a pounding headache and a rising fever. After a few days it was all he could do to muster enough strength to retrace his steps down the gulch and crawl into his bunk at his boardinghouse. As fever racked his weakened body, sending him into fits of delirium, Owen's mind fastened on the image of the promising ore he had picked up on Cherokee Gulch. Mounds of shining silver leaped before his eyes, and he reached out for the sparkling nuggets only to have them vanish. In his lucid moments, memories of his dreams haunted him, and convinced as he was that he was going to die, Feenan grew determined to have his claim registered before he died so that his funeral, if not his life, could be enriched with the profits from his strike. Calling for two of his closest friends in a voice hollow from weakness, Owen gave his friends each a share in the claim if they would go to the recorder's office, register it, and work the mine. The two miners quickly agreed. But to his astonishment, and that of his two cronies, Feenan recovered. However, when he found that his erstwhile friends, who were also convinced that he was going to die, had failed to record his name as co-owner of the now high-paying claim, Feenan very nearly had a fatal relapse. As it turned out, he could be glad to be dealt out of the Pelican.

As soon as the word spread that the Pelican had tapped a vein of almost pure silver, scores of miners headed for the area. Adjacent the

Pelican a group of men staked out a claim calling it the Dives, and sunk a shaft to tap the same vein on which the Pelican was located. So close was the Dives shaft to the underground workings of the Pelican that tempers flared and a full-scale feud developed over the right of possession of the vein. Pelican people hotly charged the Dives miners with being unmitigated thieves. The Dives owners accused the Pelican men of trespassing, a cardinal sin. After a few hand-to-hand skirmishes among the miners of each side, the owners sent their men down into the tunnels armed with pistols. Meanwhile, on the surface, the owners hired a battery of mining engineers and geologists at $100 a day each to settle the question by resurveying the property lines. The results of the survey did nothing but fan the flames of the feud. The high-priced consultants reported that the original vein was forked by a mass of nonmetallic rock so that one spur of the vein lay in the Pelican property and one in the Dives property. At this the contenders showered the courts with suits to see where the apex of the vein lay. As the furor grew, miners from other mines, merchants, and scores of others to whom the outcome could make little difference threw themselves into the fight. Judge James Belford's court sessions were attended by so many high-strung observers that the judge was forced to keep a brace of pistols loaded and ready at his side.

As the lawyers chalked up hefty fees in the endless litigation over the right of possession of the vein, the mines were supposedly closed by injunction. In reality production soared as owners used every means they could devise to get the ore out in secret. From Denver to the Pelican came six specially constructed coffins, ostensibly to house the bodies of six caretaker-miners caught in the collapse of part of the tunnel. The coffins were loaded instead with thousands of dollars worth of high-grade ore that was safely and profitably deposited at the smelter. Over in the Dives, they evaded confiscation of the ore by hauling it away on Sundays, when the courthouse was closed and no writ of attachment could be issued. On one Sunday alone $65,000 worth of ore was brought down the gulch.

The bitterness between the two groups escalated to a point beyond reason and it was no surprise to the townsfolk when one afternoon an owner of the Dives was brained to death with a section of hitching post by one of the owners of the Pelican when he happened to meet him on a street in Georgetown. Later, Jacob Snider, one of the Pelican shareholders caught sight of Jack Bishop, one of the Dives armed guards. Snider got on the first horse he could lay his hands on and whipped it to a gallop heading down the canyon. But Bishop, on a faster mount, gave chase and caught up with Snider and peremptorily put a bullet in his head. A few months later, J. H. McMurdy, superintendent of the Dives, collapsed and died from unexplained causes. And so it went, from one year to the next—beatings, knifings, shootings, suits, countersuits, decisions, and appeals—and it was not until the vein first discovered by Owen Feenan started to pinch out that the hostilities ceased.

The only man to realize a fortune from the Dives without jeopardizing his life was district strongman William A. Hamill, who picked up the property at a sheriff's sale for a paltry $50,000.

Hamill was no hard-rock miner who struck it rich overnight. He was one of a type of men who began to make their appearance in Colorado in the seventies and eighties as resident managers of mines. As claims were sold to syndicates and ownership was consolidated, more often than not in eastern or foreign financial circles, there came the need for on-the-site management. Hamill was a natural to superintend the Terrible, which is what brought him to Georgetown. The Terrible was a high-grade producer in Silver Plume, owned by a group of capitalists in far-off London. Born in England, Hamill had lived for a number of years in Philadelphia where he kept active his ties with his English friends. Not only did he run the Terrible with highly profitable efficiency for his absentee bosses, but he found time to dabble in claim-buying in partnership with Jerome Chaffee and soon he was a millionaire in his own right.

Arrogant and imposing, Hamill laced his career in mining ventures with a swing into politics, organizing Clear Creek County for the Republicans. Again, his efficiency paid off. Republicans led the field of the candidates at every election. "I carry the whole county in my vest pocket," boasted Hamill unabashedly. When the northern Ute staged a rebellion in the mid-seventies, Governor Pitkin, in recognition of Hamill's leadership ability, dubbed the mining man a general of the militia, and sent him over the mountains with a force of irregulars to help quell the disturbance. In addition to his mining, politicking, and soldiering feats, Hamill devoted himself to learning the intricacies of mining law and became one of the leading authorities on the subject in the state.

Like grocer Potter, Frank Maxwell, and hotelier Dupuy, Hamill added to the architectural wonders of Georgetown. Hamill House was far ahead of its time as a residence. With a penchant for comfort, the general spared none of his ample bankroll to equip his bay-windowed, dormered home with the last word in turn-of-the-century convenience. A cast-iron furnace sent warm air swirling through ducts to all of the first-floor rooms, a carbide gas plant on the property gave Hamill House gas light before the rest of Georgetown knew what it was, and a tank on the roof held enough water to fill the full-length zinc-plated bathtub every day of the week. The fact that in summer the sun warmed the water in the tank so that hot bathwater filled the tub and in winter the cold days meant an icy bath was an eventuality on which the general's mind did not dwell. The furnace, the lights, and the bathtub were practical features, but Hamill did not forget the aesthetic aspects in his home. From the fully paneled dining room there opened a cheerful solarium with a terrazo floor and tiled fountain and row upon row of exotic plants. In the parlor were gold-dipped chandeliers hung with pendants of French crystal, and all the window catches in the house were plated with sparkling silver.

But of all the original appointments with which the general saw fit to adorn his home, the one that combined artistry and utility to a degree seldom seen in one structure was the outhouse. Square, meticulously clapboarded, its pitched roof edged by an ornately scalloped cornice and topped by a saucy cupola, this architect's caprice sat immediately behind the main house prominently calling attention to itself. The two doors, one facing the house, and one on the opposite side of the building, were each fitted with a highly decorated Queen Anne canopy supported by elaborately carved brackets. When nature called the family, they sat on three walnut-covered seats in three sizes within the door facing the house. When the spirit moved the servants, they repaired to the opposite side where three pine-covered seats, also in three sizes, awaited them.

As his wealth and importance grew so did Hamill's demagoguery, until his high-handed politicking became the scandal of the county. After several years of domination by Hamill, a few voters got up the courage to challenge his steely grip on their franchise. When county Republicans convened in Georgetown in 1882 to nominate a ticket for the coming elections, the Idaho Springs delegation came resolutely forth to unseat the general. Their resolution did them little good. Informed of their coming by a well-placed spy Hamill simply barricaded the town hall doors and ran his own convention with hand-picked delegates. Not satisfied with being able to select a slate to suit himself, Hamill further pronounced that no Democratic candidates would be allowed to file. It was the habit of office-seekers of the day not to file until the last minute, and Hamill, well aware of this, made sure of his oracular announcement by ordering the county clerk to close his office and make himself unavailable until the deadline for filing was past. A good party man, the clerk did as he was bidden. But Hamill's plan was short-circuited. He failed to remember that the sheriff was a Democrat and a feisty one at that, who had no compunctions about breaking into the clerk's office and placing the list of Democratic nominees on file.

When Jesse Randall, the fire-eating editor of the Georgetown *Courier,* came onto the story he threw caution out of the window and fearlessly printed all of the incriminating details on the front page of what Hamill had come to assume was the Republican party organ of the county. Enraged, the general set about to run Randall out of town, attempting to intimidate the editor with bursts from the only piece of artillery immediately at his disposal—the decorative little one-pounder sitting outside the doorway of the Alpine Fire Hose Company directly across the street from the *Courier's* offices.

But Randall was unbowed. Amid broken glass and flying splinters the scrappy editor got out his daily broadsides lambasting Hamill and his tactics right up to election day. When the votes had all been counted, the electorate, who had been in the habit of giving the Republicans a more than comfortable 700-vote majority, defiantly dropped a Democrat into every county office, thus unhorsing the fuming general. The people won, but the war was lost for Randall. Hamill lay siege to the *Courier* in the form of a boycott of the paper, and with most of his heavy advertisers in Hamill's camp Randall slipped into bankruptcy, sold the paper, and left Georgetown. The general's political machine recovered from its defeat and rolled along for a few more doctrinaire years and then Hamill, too, tasted defeat when the silver crash of 1893 toppled him and his regime ignominiously from prominence.

No demagogue but no more rewarded by his wealth than Hamill was Clifford Griffin. An Englishman by birth as was Hamill, Griffin was in his early thirties when he arrived in sequestered Silver Plume in the early seventies. From the beginning, he aroused the curiosity of his fellow miners with his gentle manner, his reluctance to talk about himself, and his strict preoccupation with prospecting. In the saloons and over back fences men and women speculated hour after hour about their politely aloof neighbor. There was no envy, however, among his few acquaintances when the pensive young Englishman discovered the silver-threaded Seven-Thirty lode halfway up the side of Brown Gulch on the western face of Sherman Mountain. Nor was there enmity as Griffin's mine, ore sack by ore sack, made him the richest man in town. Unlike Hamill, whose wealth made him expansive, Griffin's good fortune turned him more and more into a recluse. For companionship he relied on the bottle and the sweetly strung violin that hung on the wall of his log hut. Every summer evening as the shadows crept along the steep canyon walls long before the sun set, the townspeople paused to listen to the haunting melodies that carried clearly down the mountainside on the still air as Griffin gave vent in music to the feelings he kept so carefully hidden from his fellow men.

On the evening of June 10, 1887, as the violin notes wafted over the small mining camp they seemed to some to be more poignant than usual, as if the solitary Englishman had some special message to convey. Griffin played melody after melody and then abruptly the music stopped in mid-phrase. For a second or two silence filled the void. Then came the echoing crack of a single pistol shot. A party of miners scrambled up the narrow trail to Griffin's hut and found him lying face down in a rude tomb hacked out of the rock, a bullet hole in his heart. Beside him lay his violin and a note held down with a stone. "Clifford Griffin," read the note, "Son of Alfred Griffin, Esq. of Brand Hall, Shropshire, England. Born July 2, 1847. Died June 10, 1887, and in consideration of his own request buried here."

In deference to his wishes, his friends buried Griffin in his rocky grave and a monument was erected over the site bearing the words written in his note.

With Griffin's passing there blossomed scores of stories about his life and death. One was that he sought the remoteness of Silver Plume in an effort to obliterate the memory of the night before his planned wedding when his fiancee was found dead in his rooms. Another was that his younger brother had come from England, tracked him down to the faraway mountain camp, put the bullet through his heart, returned to England to claim their father's baronetcy, and later anonymously sent funds for the erection of the monument to ease his troubled conscience. So far the stories about the mysterious Griffin remain stories with no proven veracity. But where the truth lies is cause for a moment's reflection among those who, on a summer evening, stroll along Silver Plume's now nearly deserted streets and pause, thinking they hear the faint but unmistakable strains of a violin floating down the side of Sherman Mountain.

9

The Silvery San Juans

IT WAS VERY SOON CLEAR TO THE THOUSANDS OF PROSPECTORS WHO ROAMED THE Rockies that the mineral riches of Colorado lay in a roughly diagonal swath that ran from the northeast to the southwest across the mountain bastions of the territory. After the first strikes were made north of Denver near Central City and Boulder, waves of money-metal hunters clambered over the mountains to the west and south until they reached the desolate, Ute-ridden wilderness of the San Juan Mountains in the southwest corner of the territory. Here the Spanish friars Escalante and Dominguez prospected back in 1776. And here Charles Baker in 1860, following up a lead given him by an Indian, rooted around for pay dirt with no success, went off to fight in the Civil War, returned in 1868 to take up his search again, only to be killed in a Ute ambush before he located much more than a few flakes of color. The next year, on the strength of the stories of Baker's meager discoveries spread by trappers who had met the explorer before his death, two more parties braved the dangers of the region: the violent winds and formidable snows, the terrifying chasms and spires of the towering peaks, and the wrath of the Ute who brooked no threat by whites or other Indians to their age-old homeland. Still more came after these men, driven on by an unerring instinct that told them that somewhere in the granite majesty of the San Juans were fortunes in minerals waiting to be discovered. Down the valley of the Gunnison and up the narrow canyon of the Uncompahgre they came. They were joined by others crowding over Stony Pass from the east and surging up the Animas River from the south. Sooner or later the finds were made.

Miles T. Johnson dropped down into the Animas River Valley via a trail that descended a dizzy 2,300 feet in two miles to make the first big strike in the region, a vein assaying at nine hundred ounces of silver to the ton. Hunter Enos Hotchkiss, clawing his way over the hogback peaks at Slumgullion Pass east of Lake San Cristobal, uncovered a high-grade claim he called the Golden Fleece.

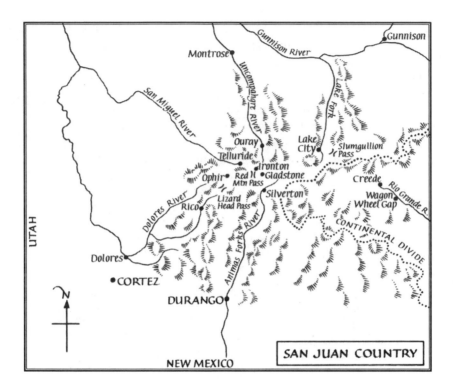

Prospectors and townmakers flocked to the region, and Lake City developed a population of five thousand silver-seekers in a matter of weeks. For Hotchkiss, it was a short-lived strike. The Golden Fleece soon pinched out and its owner reverted to a surer pursuit, becoming the champion mountain lion killer of the district. In one year he dispatched thirty-six of the predators, earning enough bounty and thanks of the besieged community to defray his disappointment at the early exhaustion of his mine.

That other pockets of high-paying ore were shot through the San Juans, Hotchkiss never for a moment doubted. But like his friend Otto Mears, who farmed wheat in Saguache east of Lake City, Hotchkiss saw that there was money to be made in providing services for other prospectors. So together, Meats and Hotchkiss put up the $5 registration fee required of those who intended to build a toll road and went to work. By hand they hacked out a trail from Saguache to Lake City, slapped a toll on pedestrians, wagons, and teams, and sat back to reap a tidy profit as the waves of emigrants rolled into San Juan country.

There was no requirement then to show the profile of the road in order to obtain a charter from the territorial legislature. All a man had to do was to indicate the settlements he intended to connect with his road. The fact that the route might be very nearly impassable because of its perpendicular ascents and descents was, in the eyes of the territorial lawmakers, the user's problem and not the government's. However, no such pitfalls greeted users of the roads built by Otto Mears. When he began his first road, connecting Saguache on the south fork of the Arkansas with the Denver–California road through South Park, he got some rewarding advice. Ex-Governor Gilpin, now a landowner in the Sangre de Cristo Grant in the lush grazing and farming valley of San Luis, came to the wiry Mears and told him he would be well advised to make any grade gradual enough so that railroad tracks could be laid over it. Mears savvied, putting Gilpin's counsel into practice and at the same time salting away in the back of his mind the intriguing notion that someday he might build a railroad himself.

Mears was a man of many parts, all of them entrepreneurial. He had learned early to make his own way. Born of a Russian mother and an English father, young Otto was orphaned at four, abandoned by his guardian uncle at eleven, and on his own, bumming around the California gold camps, throughout his teens. Somewhere in his travels the bantam-sized Mears picked up the art of tinsmithing and he went from camp to camp beating out appliances for the miners until the Civil War arrived to give him another outlet for his restless energies. He served in the New Mexico campaigns with the First California Volunteers, and when the war was over the young tinsmith turned to homesteading. As a soldier he had seen the southern part of Colorado Territory, especially the rich San Luis Valley, and it was here that he staked out his farm, on Saguache Creek. For Mears, subsistence alone was not enough. Recalling the needs of the miners in the California gold camps, he planted wheat, built a small flour mill, and carted his coarsely ground flour over the rutted path of freight wagons to sell it to the eager miners up in Oro City at the height of the gold boom in California Gulch. After a few of these jolting trips down steep-sided arroyos and over boulder-strewn slopes of the foothills, the idea of the toll road suggested itself to Mears. By 1886 he had constructed twelve roads, nearly all by his own hand, connecting the major mining camps in the San Juans. There was no obstacle that stymied Mears for long. Plunging into the unfriendly ravines and forests of the mountains, Mears even made friends with the hostile Ute, learning their difficult language which he spoke as he did English, with a Russian accent.

While Mears and Hotchkiss were building their toll road from Saguache to Lake City, Jim Kendall and a group of others scaled the mountainsides that surrounded Baker's Park, a broad alpine meadow on the Animas River named for the ill-fortuned early explorer. High above the park at altitudes of nearly thirteen thousand feet Kendall and the others located a series of claims, forming the Las Animas mining district in June of 1871. Miners' shanties, cafes, cribs, saloons, and stores sprang up at the base of the four mountains encircling the park. When Jim Kendall told a newcomer, "Hell, we've got silver by the ton!" the cluster of nondescript buildings had a name—Silverton.

The contagion of silver fever infected men with the same recklessness that accompanied their search for gold. Caution went by the board. In the San Juans, where travel in clement weather was risky and where the winters could be as devastating to the traveler as those in the Arctic or in Siberia, a sensible man did his prospecting in the late spring and summer when at least he could see that the next step might plunge him over a cliff and when starvation might be warded off by the snaring of game. Even in the relative safety of a mining camp the specter of starvation hung over the inhabitants as winter snows closed the rudimentary toll roads, cutting the miners off from sources of supply. As the stock of foodstuffs disappeared, one by one trail oxen went under the ax, and when those were eaten, soggy bullrushes torn from the frozen banks of streams became the daily fare, laced with an occasional rabbit snared when its curiosity brought it too close to camp. It was common to hear that travelers caught by winter's fury on the trail gnawed on buffalo robes, leather pants, and saddle straps to keep alive. But among all the stories of trail hardship and disaster, the one that has captured the fancy of generation after generation is the legend of Alfred E. Packer.

Packer's large head, characterized by a forehead that sloped back sharply from his heavy brows, sat on broad, hunched shoulders. Unlike the scruffy, full-beared miners of the San Juans, Packer's waxen face carried only a small goatee. To those who knew him, what remained in their memories was his peculiarly haunting voice, not high, not low, but hollow sounding. According to the legend, the man Packer made his memorable appearance in Colorado in February, 1874, as a guide for a party of eager silver-hunters from Utah. Some say Packer had been sprung from a Salt Lake jail to take on the job. In any case, Packer and five other men headed into Colorado from the west, picked up the Uncompahgre River where it joins the Gunnison and followed it south until they came to the Los Pinos Ute Indian Agency at the edge of the San Juans. Here, Ouray, the sagacious chief of the Uncompahgre Ute, cautioned Packer against going into the wilds of the formidable mountains until warmer weather came. But Packer scoffed at his advice, boasting to his friends that he knew the maze of gulches and waterways of the San Juans as well as the Ute. Packer's party were persuaded by his confidence, and within a day or two after buying supplies from the agency trader they set off into the snow-stilled wilderness.

Nine weeks later Packer strode into Los Pinos, carrying a handful of live coals in a coffeepot, to tell a harrowing tale of a sudden blinding snowstorm in which his five companions perished while he barely managed to stay alive by dint of his long experience in mountain country. Holding up the coffeepot of coals Packer claimed that had it not been for his alacrity in saving the coals from the fire at the first onrush of the blizzard winds he too would have died, for the storm turned all their matches to wet pulp and blew away their stock of supplies. Then in his sepulchral voice, he described how the others one by one had slipped into delirium from the lack of food and the saturating cold despite his attempts to keep a fire going with the rescued coals. At last, said Packer, each man died and he was left alone. When the storm finally broke, Packer related that he was able to recognize a landmark in the clearing air and once he got his directions straight he made his way back to the agency.

The Indian agent listened stonily to Packer's story, noting that for someone who had been through an ordeal as grim as the one Packer described the man was in curiously good condition. He showed no effects of starvation: no hollow cheeks, no skin shiny and taut over bones, and his clothes fit his large frame as snugly as they had when he first came to Los Pinos nine weeks before. But suspicious as he was, the agent could not crack Packer's glib story and so he let him leave the area, only making him promise to return in May after the snow had gone to help a search party look for the bodies of his dead companions.

In the meantime, another party of imprudent prospectors found the remains of Packer's friends. Four of the men had apparently been bludgeoned in their sleep, while there were indications that the fifth had put up a fight to save his skin. There was little decomposition owing to the icy temperatures and it was clear to the discoverers of the grisly tableau that pieces of flesh had been hacked from each body.

When the agent at Los Pinos heard the prospectors' tale, he sent some of his men to round up Packer and bring him in for questioning. When Packer arrived, the agent confronted him with the odious evidence and got a speedy confession. Packer admitted that in his abject fear of dying in the desolate storm-whipped mountains he had killed his companions and fed off their bodies to keep alive.

With the title of "man-eater" hung on him by the sensation-seeking papers of the period, Packer was manacled and taken to Saguache to jail from which he promptly escaped. For nine years the culprit evaded sheriff's posses and bounty hunters, and then one day someone spotted him in Wyoming and he was brought to Silverton to stand trial before Judge Melville B. Gerry, a rock-ribbed Democrat from Georgia whose notions of right and wrong revolved closely around the political alignments of his time. The trial was held in short order and the verdict was definitive: Packer drew the death sentence. He appealed the sentence, however, and got a new trial in which his sentence was reduced to forty years in prison for manslaughter. Packer served until 1901 when he came out on parole and lived quietly in a cottage in a Denver suburb for the rest of his pitiable life, a gentle old man remembered for his kindliness to neighborhood children.

What lives on irreverently and exquisitely in this otherwise grim story of human behavior in an environment of desperation are the words of the estimable Judge Gerry delivered to the only man to be convicted of cannibalism in the United States at the end of Packer's first trial.

"Damn you, Alfred Packer," purportedly bellowed the indignant judge, "you Republican son of a bitch! There were only seven Democrats in Hinsdale County and you had to go and eat five of 'em!"

Tribute to the universal and undying appeal of a good story is the petition by the students of the University of Colorado at Boulder to the school administration requesting that the name of the school restaurant be changed from the Roaring Fork to the Alfred E. Packer Memorial Grill. *Sic itur ad astra.*

Bad weather was a seasonal hazard in the San Juans. Ambush by the irate Ute was an all-year-round danger. From the first gold strikes on Cherry and Clear creeks, the miners had the Ute always at their backs. The soaring peaks, deep gulches, and grass-covered parks of the Rockies belonged to the Ute by the ancient right of conquest. And with chilling frequency they emphasized that right by conducting murderous raids on ranchers and prospecting parties. In 1864, to forestall a two-front war with the plains Indians to the east and the Ute to the west, Governor Evans got up a treaty with the mountain-dwelling

Indians in which they agreed to vacate the eastern slope of the Rockies. As the white man's search for mineral wealth expanded, Evans' guarantee was expanded in 1868 when Kit Carson and D. C. Oakes packed up a delegation of Ute and took them to Washington to get them to agree to claiming only that portion of their hereditary lands that lay below the fortieth parallel and west of the 107th meridian. At first the obstinate chiefs balked, but when Otto Mears was called in and spoke to them in their own tongue they were persuaded, for $60,000 in annuities, to retreat from their lands. The ink on the treaty was scarcely dry when the surge of white prospectors spilled over the continental divide into southwestern Colorado, once more trespassing on Ute treaty land. The Indians protested through their agent to Washington, and in February, 1872, the federal government issued a warning to white men to get out of Ute territory by June 1 or face arrest. Troops from Fort Garland and Fort Union were on their way to enforce the order when President Grant suddenly canceled the command after hearing that Chief Ouray was willing to sell part of his people's lands. In a few weeks, Ouray and his compatriots were called back to Washington for new treaty talks under the leadership of Felix Brunot, chief of the group of commissioners selected to bring peace to the San Juans. Once again. Otto Mears was asked to help in the negotiations. Otto Mears and Ouray had been friends from the days when the intrepid pathfinder first started constructing his toll roads. Both were fair-minded and farsighted men. And the realities of what the white man's presence meant to the Indian's way of life, painful as it was to contemplate, was not lost on either of them.

Ouray was no ordinary Indian. He possessed a grasp of social and political questions far beyond that of his fellow red men. He was born in Taos of a mixed union, an Apache mother and a Ute father. Very little is known about his early life except that he was quick to learn and could speak Spanish fluently and had mastered the rudiments of English. He emerged as the strong man of one of the three major bands of Ute in 1864 during the negotiations in which Washington pried the first of many chunks of land away from his people. After Ouray returned from the nation's capital to the Uncompahgre River he started a cattle ranch in an attempt to set an example for the other men in his tribe to follow in learning to lead the sedentary life which was the only kind of life under which Ouray knew that his people could survive.

Now, once again, the US government was asking the Ute to give up more of their land. Under the force of Mears's and Ouray's persuasion, the Indians grudgingly agreed by the Brunot Treaty of 1873 to relinquish approximately three million acres in a rough quadrangle encompassing the San Juan Mountains where hundreds of white men already operated in a frenzied chase for money metal. In payment for their lands, the Ute received $25,000 in annuities and Ouray, as chief, received an annual stipend of $1,000 to use as he saw fit in keeping his braves peaceable.

But it was an uneasy truce. In the main, the northern and southern Ute were still manifestly resentful of the white man's governing of their affairs and the white settlers in the San Juans were equally resentful of the presence of the dreaded red men in their vicinity.

The three bands of Ute were loosely federated, and although the white man considered Ouray to be the spokesman for all of the Ute he was in fact wholly accepted as a chief by only his own Uncompahgre unit, to a limited degree by the White River or northern Ute headed by chiefs Douglas, Captain Jack, and Colorow, and not at all by the southern Ute under Chief Ignacio. The chieftains did not share Ouray's vision of the fate of the red man if he refused to cooperate with the whites and they resisted any efforts to turn their people from the nomadic life to one at the plow. The white man however continued to insist on the transformation.

Particularly resentful of white efforts to change their ways were the northern Ute at White River whose agent, Nathan C. Meeker, was an ardent champion of the cause of bringing the Indian under the protection and guidance of the Christian paleface. Meeker, former agricultural editor of the New York *Tribune,* came to White River in May, 1878, after founding a model agrarian community in northern Colorado at the instigation of his boss, Horace Greeley. Only Meeker's zeal outran his misunderstanding of the Indian, and when shortly after his arrival at the White River agency he wrote to Henry Teller that he proposed to "...cut every Indian down to bare starvation point if he will not work..." Meeker stamped and sealed his own death warrant.

To carry out his policy of converting the Ute to model rural citizens. Meeker ordered work begun on an irrigation ditch and the clearing of land for planting. The Indians ignored his order. When Meeker threatened them with the loss of their annuities, some of the Ute set about desultorily digging the ditch. The rest hit the trail on an extended hunting trip. After the summer was over, the Indians straggled back onto the reservation and settled down for the winter, not as Meeker wanted them to do by getting behind the plow but by racing their fast ponies, gambling, and generally enjoying their leisure. The Indians either ignored their agent or treated him with studied insolence. The behavior of his charges preyed on Meeker's rigid mind, and one day in a fit of pique he had his four white assistants plow up the Ute's racecourse. In retaliation, one angry brave got out his gun and took a potshot at the agent. The bullet missed but it alarmed Meeker enough so that he decided to parley. He called the sullen leaders to his house and patiently explained the virtues of what he was trying to get them to do, but he made headway only when he promised the collected chiefs special favors if they would intercede with their people. In exchange, the chiefs gave Meeker their raceway to use for planting a crop. But only two days after the meeting, Meeker was badly beaten by one of the chiefs who felt that he had been slighted in the distribution of goods which comprised Meeker's promised "special favors." At this juncture the agent was thoroughly frightened and he wired the US Army for help. Down from Wyoming under forced march came Major T. T. Thornburgh with three companies of cavalry and one of infantry. When the Ute got wind of Meeker's action, all of their pent-up rebelliousness broke forth. One segment under Captain Jack ambushed Thornburgh and his men, killing the major and thirteen troopers. The rest of Thornburgh's command dug in and were pinned down by the menacing Ute for six days while another band of Indians under chiefs Douglas and Johnson attacked Meeker's compound, murdering the agent and the four other white men in his employ. After the slaughter they took captive Meeker's wife, his teenage daughter, and another white woman, and headed for the hinterland they knew so well.

When word of the uprising spread, fear gripped the settlers on the Western Slope. Panic leaped from settlement to settlement in a firestorm of rumor: the Uncompahgre and southern Ute were joining the rebellion, the Indians were systematically setting fire to the white men's camps, Chief Ouray was the author of a message warning the whites that his young men were straining to get into the fray and he feared that he could not control them. In the face of these rumors, miners begged the Colorado governor to issue arms and ammunition to civilians. Governor Pitkin obliged, advising the terror-stricken whites by telegraph that "Indians off their reservation, seeking to destroy your settlements by fire, are game to be hunted and destroyed like wild beasts...Gen. Hatch rushing in regulars to San Juan." Pitkin then created three military districts, one to cover each of the major bands of Ute, placing Georgetown's champion organizer William Hamill in charge of the trigger-itchy militia watching the White River Ute.

In the meantime, the scare caused by the rumors was gradually subsiding after it became known that Ignacio's southern Ute decided with some discretion to stay out of their northern brothers' feud, and that Ouray was in fact well on his way to persuading the rebellious northern chiefs to lay down their arms and surrender the captive women. He was helped by his sister, who was the wife of rebel Chief Johnson but who had a soft spot in her heart for the white man ever since a troop of cavalry rescued her from the clutches of an Arapaho war party some years before. Between the efforts of Ouray and his sister, the Meeker women were released to a rescue party, under regular Army General Wesley Merritt and former Uncompahgre Ute agent General Charles Adams, on October 15, sixteen days after the massacre. A government commission attempted to get the Ute to deliver up the rebel leaders for punishment but only Douglas was brought in to serve a few months in the prison stockade at Fort Leavenworth. None of the rest was ever brought to justice for the murder of Meeker and his men.

Although their panic subsided after the release of the Meeker women, the settlers were left with the adamant conviction that "the Utes must go." The clamor kept the wires to Washington humming. Finally, responsive once more to the settlers' mood, the federal government created a new commission to cart Ouray and a group of lesser chiefs to Washington early in 1880 to formulate a new treaty, the fourth one in fifteen years.

In the new agreement, the three bands of Ute were to be relocated: the southern group to go to the La Plata River area along the southwestern border of Colorado, the Uncompahgres to go to the mouth of the Gunnison between Colorado's western border and Utah, and the northern Ute to go to the Uintah reservation in the barren, windswept plateaus of Utah, as far away from Colorado as the commission could legally send them. In order to become effective, the new agreement had to be signed by three-fourths of the adult males of the tribe. All complied except Ouray's Uncompahgres who felt the wording was too indefinite and might be construed to mean that they could be settled in eastern Utah along with the northern Ute instead of in the verdant western portion of Colorado that had been their home. Once more, Otto Mears was called in to work his persuasive power with the Ute. His first act was to get Ouray to agree to sign, and then he paid $2 out of his own pocket to every adult male who would sign. The necessary signatures were quickly found. As a result of Mears's action, however, some of the commissioners accused him of bribery, but the charges against him were not pursued and the treaty went into effect.

Ouray, still clinging to his trust of the white man, returned to Colorado to supervise the relocation of his people. But shortly afterward, on a trip down to the southern Ute agency to confer with Ignacio, Ouray fell ill of Bright's disease and died. And by so doing he was spared the sight of seeing the worst fears of his people come to pass.

When Mears and the other commissioners took up the relocation problem, they decided that within the prescription of the treaty there was too little agricultural land in the area at the mouth of the Gunnison River for the Uncompahgres to subsist on and that they would therefore have to go to Utah. Colorow, who inherited Ouray's mantle as chief of the Uncompahgres, bitterly protested the decision but no one listened to his pleas.

The day of the Ute exodus was a warm, lazy Sunday in September, 1881. "Slowly and sullenly they filed along the Uncompahgre, down the Gunnison to the Colorado, and then westward on that sprawled-out river toward their new lands and home. Fourteen hundred fifty-eight homeless Indians including squaws, bucks, braves, and children, driving ahead of them over 10,000 sheep and goats, riding, leading, or herding 8,000 small ponies, made their way down the historic river, indifferently drinking in the beauties of the late summer sun playing on the mountains. Chief Colorow…was the last to leave the valley—a dull, prosaic dash of copper at the end of a long Indian sentence."

A few days after the Indians had gone, the Ouray *Times* echoed the sentiments of the white settlers: "How joyful it sounds and with what satisfaction one can say 'The Utes have gone!'" And the Army, on hand to conduct the Ute to their new home, reported that "The whites…were so eager and so unrestrained by common decency that it was absolutely necessary to use military force to keep them off the reservation until the Indians were fairly gone."

The behavior of the white man in the affair of the Ute was no different from the way it had always been. The throngs of prospectors who threw themselves over the continental divide into the gorges and onto the forested slopes of the mountains of southwestern Colorado did not consider themselves trespassers. To them it was as Levi Russell had said nearly twenty-five years before: they were "working to develop the resources of the great unknown West." And it was unthinkable that the white man with his superior intelligence and technology should be denied the opportunity of exploiting the mineral-rich Rockies because of some nebulous rights of a group of hunter-state savages unenlightened in the concepts born of the Industrial Revolution.

With each erosion of Ute land by a new treaty, a fresh tide of white fortune-hunters rolled into the bonanza lands of the San Juans. Within weeks after the signing of the Brunot Treaty of 1873, the recorder's office in Silverton alone was swamped by over three thousand mining claims. Big producers were the Sunnyside, Shenandoah, Silver Lakes, the Highland Mary, and the North Star, each of them so high up on the sides of the precipitous mountain peaks surrounding Baker's Park that aerial trams were necessary to get the ore down and men and equipment up to the mine portals. Typical of the tramways was the one to the silver-belching Silver Lakes group where portions of the grade were as high as 70 percent, nearly straight up. On this tram a miner rode to his work in a "long bucket," long enough to hold a man, swinging up and down as much as a hundred feet in a sickening rhythm above the sheer-sided walls of King Solomon Mountain.

Discoverers of the Silver Lakes chain of mines were the Stoiber brothers, Edward and Gustav, both of whom, like Richard Pearce up in Central City, had gotten their mining training at Freiburg. The two brothers were able to manage their mineral properties in perfect harmony. Managing their wives was something else altogether. Ed Stoiber's brassy frau Lena developed a deep aversion to Gus's wife and ultimately, to keep their own sanity, the two brothers severed their business relationship, with Gus taking a set of high-paying mines up a nearby gulch, and Ed left in sole charge of the Silver Lakes group. The split did nothing to stem the wealth that poured out of the Silver Lakes mines, and Frau Lena Stoiber, nicknamed Captain Jack by residents of Silverton for her temperamental similarity to the dreaded, fiery-minded chief of the northern Ute, found lots of places to spend it. One project was a monumental brick mansion erected near the mines that became the social center of Silverton. In a town where a tablecloth was considered the epitome of elegance, Captain Jack's "Waldheim," with its sixteen bedrooms, ballroom, game room, and theater complete with gaslights, raised stage, and handpainted drop curtain, became the town curio. But no matter to Captain Jack. She continued to issue fancy invitations to the people she liked, requesting their attendance at the frequent dances and dramatic productions, and she sealed off the vision of her taffeta-gowned and top-hatted guests from the ogling eyes of her neighbors with a fence two stories high.

The balls and parties at Waldheim were a far cry from the miners' whooplas that saw damsels in gingham arriving at the dance hall astride burros or on foot with skirts held to ankle height to keep them from dragging in the mud and manure of Silverton's main street. Music was provided by a fiddle and a banjo and the opening number was invariably the "San Juan Polka" which the *Silver World* vowed "resembled a Sioux war dance." While guests at Waldheim dined delicately on sage hen washed down with champagne, miners and their fancy girls fortified themselves for the all-night revelry with groundhog, bacon, doughnuts, and coffee. And no palate was so sensitive that it could not tolerate an occasional swig of fiery forty-rod rotgut from a bottle that passed from hand to hand while the dancers took a breath of fresh air out back.

Silver Lakes grossed $11 million under the Stoiber management, operating even through the silver panic of 1893. In 1901, the Guggenheims picked up the properties for $2,333,000. Shortly afterward, Ed Stoiber was killed in an accident in Paris, and Captain Jack in due time married Englishman Hugh Rood. In April, 1912, Rood booked passage on the brand new White Star liner *Titanic,* along with Ben Guggenheim and Leadville Johnny Brown's garrulous wife, Molly. When the sleek steam-turbine-powered liner hurled herself against a massive iceberg in the North Atlantic on an inky black midnight and sank, Guggenheim and Rood went down with the ship. Captain Jack and Unsinkable Molly, however, survived, with their fortunes tidily intact.

At the head of Cunningham Gulch adjoining Arastra Gulch, the site of Silver Lakes, lay the rich Highland Mary mine. No happenstance by the usual standards attended its discovery. In New York, a pair of affluent and compulsively acquisitive brothers named Ennis absorbed enough newspaper talk about the riches of the Pikes Peak region to catch an incurable case of gold fever. But they wanted more guarantee of success when they got out to the wilderness of the mountain West than the by-guess-and-by-gosh techniques of most prospectors. So they hunted up a spiritualist and paid her $50,000 in advance to look into her crystal ball and come up with a gold mine in the Rockies that would turn them into millionaires. The spiritualist cheerfully obliged the brothers by pinpointing a spot on the map on the side of King Solomon Mountain and embellished her hocus pocus with the pronouncement that there the brothers would find a lake of gold.

Following the medium's instructions, the Ennises immediately set out for the West and they could scarcely contain their eagerness as the stage rumbled over Otto Mears's toll road into Silverton on the last lap of their long journey from New York. The boys bought a prospecting outfit, hired a guide, and pointed out a series of alpine lakes on the map of the region as their destination. The medium had done her work well. Once on the designated spot, the Ennis brothers dug in with picks and had turned over only a few chunks of decomposed rock when they came up with promising color. Quickly, they staked out and recorded claims on the property they called the Highland Mary, and hired on a crew of miners. So pleased with their spiritualist's success were the brothers that they urged her to keep up the good work by gazing once again into her crystal ball and coming up with the direction the tunnels and shafts on the Highland Mary should take. Whether the Ennises communicated with their psychic friend by conventional means or through seances is not known, but what is known is that when it was communicated to the superstitious miners that the operation of the mine was being directed by a spiritualist in New York, they threw down their picks and skedaddled down to town where the most occult pitfall lying in wait for a man was an honest case of the d.t.'s.

By upping the wages they paid, the Ennis brothers induced a new crew of miners to work in the spooked Highland Mary, but that did not cure the brothers' troubles. The early signs of gold now gave way to other metals, some of it silver, which the Ennises in their determination to uncover a gold bonanza threw into the tailing ponds. In frenzied persistence the brothers poured close to $800,000, a fortune in itself, into developing the Highland Mary in a futile effort to find gold. Finally, financially bankrupt and physically spent, they sold out in 1885. With an irony that probably would have escaped the gold-blinded, single-minded brothers Ennis, the new owners turned the Highland Mary into the second largest silver producer in the Silverton area.

The silver fortunes that cascaded down the mountainsides around the flat that was Baker's Park left a trail of prosperity in the town. The pride of the populace was the Grand Imperial Hotel on Green Street, the straight, broad, main thoroughfare through Silverton. Built by the Thompson brothers, whose Sunnyside and Sunnyside Extension mines ten miles east of Silverton were big producers as early as 1871, the Grand featured fifty-four rooms, three baths, and a saloon that never closed. Here in the Hub, as it was called, a visitor sat at the bar one night morosely staring into the beveled French plate-glass mirror behind the mahogany bar when a wild-eyed woman burst through the swinging doors. "Is my man here?" shrieked the creature. "If I don't find him fast, there's goin' to be a hot time in this old town tonight!" At her words, the visitor's mood quickly changed to one of delight as he pulled out pencil and paper and began writing the ragtime tune that became an American classic.

Popular as the Hub was, the real concentration of gin mills and pleasure-seekers was one avenue away on infamous Blair Street. In two wide-open blocks there were no less than thirty-eight saloons, sporting houses, and dance halls. Obligingly, the Mikado, the National, the Sage Hen, and Diamond Kate's and Lola's, and a number of the rest stayed open around the clock to accommodate both day and night shift miners on the town. But if a miner was down on his luck and could not afford to frequent one of the more elegant watering houses, no matter how long they stayed open, there was no need for him to go thirsty. Benevolent brewer Charlie Fischer saw to that. On the front of the stone cellar he built into the base of Sultan Mountain, Charlie rigged a spigot and hung a tin cup and over it a sign advertising "FREE BEER, HELP YOURSELF."

As in any other boom camp the troublemakers came. There were daily shoot-outs and drunken brawls. So flagrant became the lewdness and lawlessness that respectable residents on the upper and lower ends of Blair Street renamed their blocks Empire Street to take away the taint of the bawdy on their addresses. When a pair of renegades murdered the night marshal on Blair Street, the citizenry formed a vigilance committee and sent for Bat Masterson, fresh from judicial triumphs in Dodge City. Under his rule, as a paper of the time put it, the "undesirable element dispersed."

With the imposition of law and order, Silverton's civic pride blossomed. Public subscription produced the Reese Hook and Ladder Company, the first in the San Juans, who thrilled onlookers with their prowess as they went through their drills decked out in their "neat and tasty uniforms." Culture came to the bonanza camp when someone heard that thrifty Andrew Carnegie was receptive to furnishing funds for the building of a library if a town could prove that it could support one. Silverton proved it by staging a benefit on Bobbie Burns Day that netted $100 for the purchase of books. In a few months a brick library graced Reese Avenue, and soon after a city hall, a county courthouse, and a county jail joined the Grand Imperial Hotel as brick and stone structures, imparting in the eyes of the town's inhabitants a solidity and a sense of purpose to their town.

Part of the spur to the building spree was the arrival in 1882 of the town's first railroad. Ever since the first cry of "Silver!" reached the outside world, the meandering Denver and Rio Grande Railroad set its tracks in the direction of the San Juans, reaching Alamosa in 1878, Durango in 1881, and accompanied by a wild welcome lasting two resounding days, it arrived in Silverton on July 18, 1882. The narrow gauge road was hacked out of the stubborn granite of the canyon of the Animas over one of the nation's most scenic routes. The train between Durango and Silverton remains in service today, holding the record for the longest continuously operated narrow gauge railroad in the country. Curiously enough, at a time when railroad officials poor-mouth the claim that the American public is sufficiently disinterested in rail travel that railroads are, by their statistics, forced to discontinue passenger service owing to increasing losses in revenue, the "Silverton" does the greatest passenger business per mile of any scheduled train in the United States. During its peak season in the summer, reservations are mandatory and four trains a day are necessary to carry the crowds of "disinterested" travelers over the forty-five miles of mountain grandeur between Durango and Silverton.

By the time the Denver and Rio Grande readied Silverton, Otto Mears, the wagon road-builder who never forgot William Gilpin's advice and kept the grades of his toll roads at an angle that readily allowed a railway car to be moved over them, was bitten by the railroad bug. And when he discovered that the D&RG directors threw up their hands in dismay at the surveyor's report on the feats of engineering and thereby the money that would be necessary to parallel Mears's toll road from Silverton north to tap the rich mines spewing forth silver on Red Mountain, the pathfinder decided to have a go at it himself. Mears's Silverton Railroad snaked over hairpin turns on its climb to eleven-thousand-foot Red Mountain Pass and slithered down the other side over a convoluted series of switchbacks and loops for sixteen miles, opening the door to the profitable transport of low-grade ores to the smelter. Four trains a day, each with armed guards riding on every ore car, were required to handle the burgeoning production of the Red Mountain mines.

To please the pack of illustrious investors that now saw fit to visit Red Mountain's silver-rich sides, Mears bought a fancy little combination sleeper-restaurant car, trimmed it in Eastlake décor, stocked its kitchen with choice beef and mountain game and fish, nested vintage wine and champagne in its liquor cabinet, named it Animas Forks, and hooked itonto the laboring line of gondola ore cars to wind its toonerville way in slightly tottering splendor over the three-foot-wide tracks on its two-hour journey. To most railroaders of the day it was the height of poor management to put a sleeper on such a short run, but they amended their judgment when the Silverton's passenger business grossed $20 a mile in contrast with the $1 per mile considered a profitable rate by larger roads. Other than having a tendency to overturn, Animas Forks was in all respects a first-class accommodation.

And only in winter was the failing generally discomforting, but even then its passengers loyally stood ready to leap into the snow to right it and were called upon to do so no less than six times in one day during one particularly rugged winter.

Mears, who was as indefatigable about railroading as he was about toll-roading, started the Silverton Northern eastward from Silverton toward Lake City with a little monetary help from Simon Guggenheim. But it never reached its destination, stopping at the camp of Animas Forks about twelve miles short of its goal. A declining ore market and snow-slides, the scourge of mountain railroaders, gave Mears one of his few entrepreneurial defeats.

Snowslides and mudslides on the precipitous slopes of the San Juans were devastating to man and machine alike. When one October day the D&RG into Silverton saw a section of its track ahead carried away by a mudslide after a heavy rain and the rest of the track threatened by the rising water of the Animas River, the telegrapher at Silverton was instructed to contact Mears, the nearest railroader, for help. Help he did. The wire found him at a formal dinner. Without a moment's hesitation, Mears clapped on his top hat, graciously took leave of his hosts, and with his white scarf flapping in the night he rushed men and equipment from his own line to aid the stricken D&RG. For nine weeks he directed the work, still in his formal attire, now mudcaked almost beyond recognition, commandeering coal from Silverton merchants and homes to fuel the locomotives hauling repair crews and supplies. As in all authentic stories of heroism and good works, the first violent blizzard thundered down on the canyon just as the last nail was driven into the last tie, and Silverton was saved from a winter's isolation.

One of the smelters to which the Silverton Railroad brought its brim-filled ore cars was improbably called the Martha Rose. Its operator was an Irish jack-of-all-trades from Clonmel, Tipperary, named Thomas F. Walsh. Nineteen when he left his famine-plagued homeland in the 1850s, the bright-eyed, quick-minded Irisher came to America and to the West where the exciting pace of life suited his restless temperament. Plying his trade as a woodworker and carpenter, he worked his way to Colorado and got a job building railway bridges for the expanding Colorado Central at Golden, keeping his eyes at all times open for a telltale sign of gold. But the days of discovery on Clear Creek were over when Tom Walsh arrived and building bridges soon palled as an occupation. So when rumors of gold riffled down from the Black Hills, he was one of the first to head for South Dakota. His luck was erratic. One of his claims brought him a tidy $75,000, and another took it away. After a few more ups and downs, Walsh was glad enough to harken to the new money-metal rush in Colorado. Packing up his tools, he headed for Leadville, where silver mines were popping into view with the frequency of a crop of mushrooms. Again his luck was uneven and about the only discovery of value he made in the Cloud City was a saucy, brown-haired beauty named Carrie Bell Reed. After they were married Walsh took his new responsibility with considerable seriousness and gave up prospecting for the less exciting but more certain activity of innkeeping. But Walsh was not a man to stick to a tame job. The lure of mining caught him once again, and he and Carrie loaded the wagon with their meager possessions to trundle them over the mountains to the Animas River where new silver strikes were being made each day. Between prospecting forays into the backcountry Walsh ran the Martha Rose at Animas Forks. The profits from treating the fifteen cars of crude ore Mears's Silverton Railroad daily dumped into the smelter furnaces provided the Walshes with a living and some left over for a grubstake or two. It was not a life of plenty, and on the whole Tom Walsh's prospects appeared decidedly limited. But in a few years that would all change and his connection with the Martha Rose would be a tie that Walsh would long remember with wonder and gratitude.

❧❦

Between 1882 and 1918, the mines of Silverton produced ore valued at more than $65 million. Despite the fact that some of the mines continued production into the twentieth century, the crash of 1893 delivered a blow to Silverton from which it never fully recovered. The camp did not die out all together, however. Vestiges of its early days live on—the daily steam and cinder arrival of the D&RG narrow gauge that disgorges hundreds of tourists who swarm into the reopened saloons of Blair Street and uptown to the quieter frontier atmosphere of the Grand Imperial Hotel. Around Silverton itself miners of the Dixieland Development Company in late 1968 began to open new veins of valuable metal in older, once abandoned properties. And as if to prove that the West of Bat Masterson's day is not altogether dead, in the late 1960s, Silverton had a classic visitation of outlawry that rolled back the clock a hundred years.

Because of Silverton's decline to a relatively sedate relic of its bonanza-day prosperity, the town was without a bank for a number of years before the boomlet of 1968 began. Consequently, every Thursday, the payday for the Idarado Company miners, the courthouse laid in a stock of cash amounting to thousands of dollars and for that one day served as a bank, cashing the miners' payroll checks. On a Thursday in September, 1967, a gang of men with red bandanas covering their faces in the style of all authentic western bad men entered the Silverton courthouse. Once inside they whipped out pistols, bound and gagged the clerks, and scooped up the stock of cash, safely making their getaway before anyone in town discovered the robbery. Four days later the outlaws were still at large, with the only clue being one red bandana found beside the road leading to Durango. The thieves left behind them disgruntled miners, a grim sheriff, and irate towns-people. But perhaps the hardest hit was a saloon down on Blair Street. One of the county clerks doubled as the saloon's piano player and business fell off alarmingly after the robbery because the outlaws had tied her wrists so tight her hands became too swollen to manipulate the keys of the piano.

≈≈≈≈

North, over Red Mountain Pass, and through the gorge of the river the Ute called Uncompahgre or "Red Lake," for the reddish water in the hot spring near its source, moseyed Gus Begole and Jack Eckles in July, 1875. Begole and Eckles were following their intuition that told them that somewhere in the steep canyons formed by the streams feeding the Uncompahgre they would find vein silver to match the float they found in the river. Some twenty hard-going miles from Silverton they suddenly came out into a small, bowl-shaped park, rimmed with elephantine peaks that were striated with layers of rock of subtle colors and dotted with patches of bright green alpine grass and ledges of dark green where scrub pines found a precarious foot-hold. Following the river as it wound its way along the west side of the verdant park, Eckles and Begole made camp in a box canyon and set out each day to prospect the area with systematic earnestness. In a week or so, about four hundred feet up the face of the canyon on the south they came across a lode shot through with large particles of native silver. The ore occurred in a series of parallel veins, like rows of potatoes in a field. Taking note of this characteristic, the excited pair of prospectors named their mine the Mineral Farm, and hurried back to Silverton to record their claim and restock their depleted food supply. On the way, they met A. J. Staley and Logan Whitlock packing into the wilderness bent on a catch of mountain trout and told them of their find. Staley and Whitlock promptly put away their fishing poles and scrambled up the box canyon looking for a mine of their own. They found not one but two rich claims and commemorated their luck by naming them the Trout and Fisherman lodes.

Between these four men, and two of their friends who happened along later, enough of the nomadic population of prospectors in the San Juans heard about the new strikes to start a stampede to the site. The normal course of development of a mining camp proceeded: the miners platted a town and called it Uncompahgre, after the river. The first building they put up housed the Star saloon, which on Sunday served as a church with kegs of beer as the altar and cases of whiskey as pews. By the spring of 1876, four hundred miners and merchants were firmly settled in. The little valley rang with construction sounds as four general stores, a saw mill, an ore sampling works, two hotels, and a post office rose on the neatly laid out, right-angled streets.

Like most of the rest of the San Juans, this beautiful little park in which the camp of Uncompahgre lay had been the property of the Ute and they were slow to relinquish it. Before their final removal, parties of Ute were regular visitors to the village mining camp, bringing horses to race against the white man's mounts and goods to trade. At first the relations between the whites and the Ute were cordial but soon there was an increasing incidence of drunken Indians raiding the homes of settlers, of unscrupulous white men buying Indian women for the night, of ambushed miners found slain on the mountain trails, and of Indians discovered in back alleys mysteriously dead of bullet wounds and knife thrusts. Tensions grew as the Ute became belligerent and the angry whites stood their ground. When the situation was close to fulmination, Chief Ouray rode into town, the gait of his horse held to a dignified walk, and his tawny visage classically impassive. He called for a parley. The town fathers were more than ready to oblige. The place chosen for the talks was the cabin that served as the recorder's office and town hall. After everyone was seated on benches and on the floor, a pungent-smelling pipe made its rounds from one man to another and Ouray laid out the solution to the problem in the simplest of terms. Let the white man stop selling the Ute braves whiskey and let him stop sleeping with Ute women and he would see to it that the settlers were not troubled further by his tribesmen. The offer was quickly accepted. The council of townsmen levied fines against offenders and both the red men and the white men promised to keep a tighter rein on their young men. In gratitude for the chief's efforts in creating a workable coexistence, the name of the bustling new camp was changed forthwith to Ouray.

Prospecting, despite the raging blizzards of a San Juan winter, kept up all through 1875-76, and by the following spring over a thousand people were holed up in the picturesque bowl that was Ouray. New silver strikes were announced daily. In October, a pair of snowshoeing mineral-hunters from Silverton spotted a likely outcrop on a windswept ledge at the twelve-thousand-foot level of fourteen-thousand-foot Mt. Sneffels. Their mine produced ore that assayed at 1,200 ounces of silver to the ton. After working the Wheel of Fortune, as they called their claim, for two years for a tidy profit they sold it for $160,000. The story of the Wheel of Fortune was typical: A lucky one- or two-man strike, a couple of years during which the owner, perhaps with a small crew of daily wage men to help him, worked the mine himself, then a sale to a large mining or smelting company. Even Eckles and Begole, whose Mineral Farm had rich ore so easy to extract that they "farmed it out like potatoes by quarrying or stripping veins at or near the surface," sold their high-grade producer after three years to the Norfolk and Ouray Reduction Company, the chief smelter in town. Begole then turned to the less romantic but less strenuous life of a grocer while Eckles disappeared from the scene.

Others, after the sale of one mine, went right back to prospecting to find another. One of these was George Jackson whose gold placer on Clear Creek in 1859 had started the cycle of mineral discoveries in Colorado. Jackson sold his mine on Chicago Bar and poked around the adjacent gulches looking for another lucky pocket when the Civil War distracted him from all but his southern heritage. He enlisted in the Confederate Army and when the war was over he once more came to Colorado, the addiction of mineral sleuthing sending him over the mountains to Ouray, the scene of the postwar bonanzas. The knack of discovery had not left him. On his first try he picked up the Saratoga lode and lived comfortably in the booming little town regaling small boys with his adventures as one of Quantrell's Raiders. To very close cronies he confided that he had secretly met his old saddlemate Jesse James on a remote trail above Ouray to give the outlaw a poke of money to speed him on his flight from capture.

Jackson, the old pro of prospecting, and hunting, succumbed in 1897 to the very hazard Luke Tierney had so carefully cautioned against in his guidebook to the gold fields circulated in the early days of the Pikes Peak rush. On a trip to a neighboring town in winter, Jackson saw some elk tracks in the snow beside the trail. He pulled up his sled and grabbed his shotgun by the barrel to draw it out of its case. As he did so, a single shot reverberated along the silent sides of the canyon sending snow showering down from laden pine branches. Jackson, a hole the size of an orange in his chest, tumbled backward into a snowdrift. When a farmer came upon the trail of his sled pony and found him, Jackson was dead.

George Jackson was sixty-five when he died, a ripe old age for a miner in the San Juans. Scores of others never reached middle age. The dangers of mining were routine. On a morning when the shift boss gave orders for a gang of miners to start slabbing a stope—shoring up an excavation with timbers so that the pillar of ore in the center that had given support to the rock above it could be removed—each man knew the hazard that lay before him. He could be buried alive if the slabbing failed, letting the earth collapse around him when the pillar of rock was drilled away. Then there was the ever-present danger of dynamite charges that failed to fire or that fired prematurely. Most of the miners, despite the preponderance of Irish and Cornish whose volatile emotions were renowned, greeted the moment of catastrophe with some stoicism. But not Big Paddy Burns. Paddy and two others, Robinson and Maloney, were loading dynamite in the Virginius tunnel when the charges suddenly went off. When Ouray's Doc Rowan was swung up the side of Mt. Sneffels on the tramway to the portal at 10,000 feet above the valley floor, he found Robinson decapitated, his chest laid open, showing the bright red matrix of his lungs, Maloney dead with his brains spilled out on the tunnel floor, and Big Paddy Burns sitting against the rubble with blood coursing down all sides of his head from surface cuts made by flying rock fragments. Unlike his two shift-mates, Paddy was very much alive and lamenting at the top of his fine tenor voice, "I'm dead. I'm dead! Oh, why did I not die at home with my father?" Paddy's clarion dirge kept up all the time the doctor bound his bleeding head, until at last Rowan, whose origins were as Irish as Paddy's, silenced the injured man with one exasperated outburst: "For the love of God, you Irish biddy, shut up! If it would get me to Heaven and keep you out, I'd pack you off to the old sod and see to it you drew your last breath in your father's house myself!"

A doctor's life in a mining camp was a busy and varied one. Whatever Doc Rowan's credentials, he was all Ouray had in the early days in the way of a medical man and he made the most of it, He manufactured his own pills, a different colored one for every disease he could think of, and he did a thriving business in his home-brewed patent medicines, often threatening his patients with a trip to the graveyard, which they euphemistically called Rowan's Ranch, if they did not partake of his elixirs.

Not the doctor in town but the undertaker was the one whose business boomed in the grim days of winter when the snowpack reached a depth deep enough to make it an avalanche hazard. Perched on the sides of the mountains, bunkhouses, mineshaft houses, and the vital trails that led up to them were subjected night and day to the threat of a death-dealing snowslide. The winter of 1878 was a lamentably memorable one—91 inches of snow was measured on the level with its depth in the bowls and gulches of the mountains deeper still. Every edition of the Ouray *Times* carried at least one notice of men having been swept away in the sudden thunder of an avalanche. On Christmas Eve, young Albert Morrison, far from his family and sweetheart back in New York, sat in his chair in the bunkhouse of a mine writing a letter home. "My Dear Mother," Morrison's letter began,

> I was obliged to take an option and embrace an opportunity seldom offered to a young man...Hence I resigned my position in the post office, put my little savings into the venture reserving only a sufficient amount to pay the next quarters interest on the farm mortgage, which I sent you by registered mail, and came out here to begin work of digging a fortune from the eternal hills, staking all upon the hazards. I came west, as you know, with the sole purpose of making enough money to raise that accursed mortgage, to put the place in good repair and once again provide you with the comforts you have not had since father died. That done, I felt that, with your blessing, I might claim Charlotte as my own, and settle down near you in the quiet enjoyment of a life of honest and fairly

requited toil. My Sainted Mother, share with me the confidence I feel...Sunshine will bring warmth and cheer to hearts that are doubting and desolate.

This letter is not one-half finished, but an old clock there, on our improvised cupboard is about to strike twelve; so, for tonight I will content myself with wishing you, dearest mother and Charlot...

And there the letter ended, cut short by an avalanche that snuffed out Morrison's life and the lives of his three companions. Two months later when rescue workers tunneled through fifty feet of snow into the cabin, they found Morrison still sitting upright in his chair, his fingers still holding his pen. One of his fellow unfortunates was in his bunk and the others sat at a table staring unseeing at the playing cards in their hands. The four men had been suffocated when the snow mass enveloped the building, creating a vacuum that sucked the air from their lungs within seconds. Their bodies had been perfectly preserved in their icy catacomb. Morrison, who would never pay off the mortgage on his mother's farm nor claim his beloved Charlotte, was laid to rest in Rowan's Ranch in a rough wooden coffin lined in cambric and fitted with a pillow made of aromatic pine shavings.

Some of those caught in the snowslides were lucky. Reported the *Times* in April:

On the 30th Mr. Barker of the Yellow Rose and Messers. Bell and Spencer of the Hoosier Girl were carried down the mountain several hundred feet while passing on the trail between the Millionaire and the US Deposit. Barker and Spencer escaped injury but Bell was buried several feet under the slide and almost smothered but was otherwise uninjured.

By the late 1870s, Ouray mines had given up over $7 million in silver, an incredible sum considering the difficulties of access to the ore and its transport to the larger smelters at Del Norte, Denver, and Pueblo. All of the ore was taken out on the backs of patient, personable burros, nicknamed "Rocky Mountain canaries" by the miners to commemorate the animals' constant braying. With the price of ore transport running as high as $80 a ton in the winter, and the value of the ore in the Ouray district varying between $100 and $2,000 a ton, it was as it was in other remote mining areas of Colorado—only the highest-grade rock could profitably be mined.

In 1881, the tireless Otto Mears built the most ambitious of all of his toll roads—the one connecting Ouray and Ironton, a camp north of Silverton that was the terminus of his Silverton Railroad. Carved out of nearly solid rock along the sheer-sided gorge high above the Uncompahgre River, it cost Mears as much as $40,000 a mile to build. When a concrete auto road was laid over his toll road in the 1920s, the price was still high enough so that today it goes by the name of the Million Dollar Highway. Mears's road was a great boon to Ouray. Ore transport costs dropped as wagons replaced burros, and owners of low-grade claims now found it paid them to open up their mines. With cheaper transport available, the demand for investment money and miners rose spectacularly and the town of Ouray took on airs. Streets were graded and wooden sidewalks were laid on the main street in front of McGaughey's Dry Goods Store and Mears's Hardware. A telegraph line linked Ouray with the outside world via Silverton and Durango, and Stoddard's stationery store printed invitations to the grand opening of the town's finest hotel.

The sumptuous Beaumont opened its mahogany-paneled doors to business in December, 1886. All of its furnishings were shipped west from Marshall Field's in Chicago, and for the gala inauguration the management of Chicago's Palmer House loaned cadres of its crack staff. Sleighs driven by mufflered drivers in mackinaws delivered guests wrapped in buffalo robes over formal attire to the foyer where they ascended the grand staircase to the rosewood dining room. Here a string orchestra supplied lively background music for the ringing of crystal goblets and the popping of champagne corks. Today, the Beaumont still dominates Main Street, its past a checkered one of openings and closings, complete with a resident ghost story. At midnight, so it is said, on the first Monday of each month, the ghost of a young woman murdered on that night in the 1890s by another woman because she won the love of the other's swain roams the creaking stairway and echoing hallways of the stately old hotel.

In its heyday, a room at the Beaumont cost $8. That was too much for the hard-rock miner who doffed his denim pants, work shirt, and rubber boots for his Sunday best serge and tripped down the trail on a Saturday night for a spree on the town. Tailored to his purse, if he slept at all that night, was the $4.50 charged by the popular Dixon House. After a refreshing bath in the hot mineral spring water piped into a bath house from Box Canyon, the miner stepped out, his feet invariably directed toward the rollicking variety houses down by the river. At the "220" and the Gold Belt he applauded the magic act and whistled and stamped his feet when the line of chorus girls waved their shapely red-gartered legs in his direction. From there he could make the rounds of thirty-five saloons, drop a dollar at Bob Butterfield's faro table at the White House, then stumblingly try a polka with one of the dance-hall girls for 25 cents, get his perspiring face slapped for stealing a kiss, and, in a moment of surrender, shell out a dollar for the consoling caresses of one of the one hundred girls operating "on the line" at such fancy houses as the Temple of Music, the Morning Star, and the Bird Cage.

Henry and William Ripley's staunchly Republican Ouray *Times* held the monopoly on news coverage for the first two years of the town's boom, and then in August, 1879, they saw fit to publish the following indignant notice:

> Ouray has had the noble red man, the six bit capitalist, the man with a new process for treating ore, the man who was going to put up reduction works but went back on us, Tommy Patterson, and now as though this was not enough is about to have a Democratic paper.

The Ripleys went on to sermonize that "there is not business here for two papers and it is a question of endurance [as to] which will hold the field."

Endurance, as it turned out, had nothing to do with it. The new paper, by the novelty of its name, the irreverence of its publisher, and the provocation of its contents, survived the Ripleys' pedestrian *Times* by six fat years.

David Frakes Day arrived in Ouray in 1878, a thirty-year-old firebrand journalist, temporarily, he was certain, on his uppers. He cajoled a job in double-quick time cutting cordwood while he sized up the town, pondering whether it was ready for his talents. Of Scotch-Irish descent, David Day was born near Cincinnati and left home and school at the first opportunity. At fifteen, he joined the Union Army in the second year of the Civil War, and for his adroitness at the front he was soon tapped for scout work. Somewhat of a handicap to the young soldier was his inability to read or write. But that shortcoming was taken care of when his performance sufficiently impressed General L. A. V. Rice so much that the general detailed one of his aides to school the unlettered Day. After being wounded several times in the course of his duties as a scout, captured by the Confederates three times, having escaped from three enemy prisons including the infamous Andersonville, and at one time blatantly traveling in Confederate uniform as a member of the Sixteenth South Carolina Infantry until he could get close enough to federal lines to rejoin his unit, Day was decorated for bravery and discharged at the age of nineteen.

After the war, Day followed his last commanding officer, General Frank Blair, to Missouri, where the general saw to it that the ex-scout put his newly acquired literacy to work as a cub reporter. It was in this job that Day developed the stiletto style that Colorado newspaper readers would clamor for. Fond as Day was of the reporting job, the monetary rewards were not enough to keep the wife and four children he had quickly acquired as a civilian. Someone talked him into going into the grocery business. But Day was not cut out to be a merchant, and the store did poorly. The crowning blow came when Day cosigned a loan with a supposed friend only to have his erstwhile friend disappear, leaving him with a hefty debt. With nowhere to go but up, the irrepressible Day thought again of a career in journalism, not as a reporter, but as a publisher, where the money was. With the nation agog at the new Eldorado in Colorado, the ex-scout decided that somewhere in those magic mountains lay his golden opportunity.

While the sawdust flew as he cut up the logs in the back lot of the fuel and feed store, Day observed that Ouray suffered from a superfluity of Republicanism, led by the Ripleys' *Times,* and this was precisely the opening he needed. Glib and persuasive, Day moved among local Democrats of means, convincing them of his loyalty and ability, and soon he had enough backing to give the Ripleys a run for their money. Casting about the mountain camps, Day found a little newssheet for sale in Lake City. He promptly bought it, hauled it over Mears's toll road in a wagon, gave to it the unconventional name of the *Solid Muldoon,* and launched it on its way to becoming one of the nation's most quoted papers of the period. Day chose the unlikely name from the currently popular song that Broadway's Ed Harrigan sang about the prize-ring hero, Bill Muldoon, which seemed to the journalist to sum up his editorial policy:

> For opposition or politician
> Take my word, I don't give a damn
> As I walk the street each friend I meet
> Says "There goes Muldoon—he's a Solid Man!"

For thirteen years Day regaled his readers with stinging sarcasm leveled at whomever and whatever he decided required his attention—fraudulent mining promoters, Republicans, self-righteous mining camps, Indian-baiting ranchers, and pompous congressmen.

One of Day's antipathies was the red-bearded Republican Judge James Belford, who opted for the seat of Colorado's congressman when the territory became a state. During one campaign swing around the mining camps, when the "Red Rooster of the Rockies" gave an address in Ouray's city park, Day appeared at the edge of the crowd and made a great show of lying down on the grass as the judge's words spun on and on in flights of preelection promises. When Belford spotted Day's action, he interrupted his speech to say pointedly, "Some of you seem to be tiring so I will bring my talk to a speedy end." Before he could continue, Day's voice cut lazily through the crowd of listeners, "Ah, don't mind me, Jim," he drawled. "I can lie down here as long as you can lie up there."

Weekly press day was eagerly awaited in Ouray. As soon as copies of the *Solid Muldoon* hit the stands crowds gathered on street corners to read aloud and guffaw at the definition of the Colorado state legislature: "...a conglomeration of Rural and Metropolitan Asses elevated by Misguided Suffrage to positions intended by the Constitution for Brains, Honor, and Manhood." Turning the page, they read that "the *Mining Register* of Lake City says that the jeweled garter craze has not yet reached Lake City. This information is entirely superfluous as the average Lake City woman's style of architecture requires no artificial stays or fastenings. They simply cut a hole in their stockings and button them over their knee caps."

To read Day's acid columns one would be led to believe that there was no other camp in the entire San Juans that could boast of ores as rich as Ouray's. Every new or old silver camp got the Day treatment. Once in a while, the victim fought back. When the *Solid Muldoon* called upon the new settlement of Ophir, lying to the west over Imogene Pass, to show that its mines were rich enough to warrant the camp carrying the name of the biblically famous mining area, the miners rose up in wrath. But without a newspaper Ophirites were stymied at finding a means of retribution, until one of their number struck upon an equally appropriate vehicle for vengeance. Rounding up a herd of burros, the miners of Ophir drove them over the pass and down onto Ouray's Main Street at midday on a busy Saturday. The name of a prominent citizen of Ouray was affixed to the rump of each of the Rocky Mountain canaries. After a deputation of those indignant and injured parties made a solemn call upon the impudent editor, Day called a truce with the surrounding camps.

At one time the scrappy publisher had forty-two libel suits pending against him, but no one won against him and he went merrily on his way building up a circulation in town and out that numbered in the thousands. Even Queen Victoria was a regular subscriber to the *Solid Muldoon* after Day journeyed to England to be presented at the Court of St. James's.

Day's serious campaigns as well as his frivolous ones were touched with his tongue-in-cheek magic.

The Mears toll road to Silverton had been a welcome development in lessening Ouray's feeling of isolation but in 1881 when the Denver and Rio Grande, moving westward along the Gunnison River, reached Montrose, some thirty-five miles north of Ouray, the silver boom town looked with lustful eyes on a branch line of its own. The railroad was glad to oblige if the citizens of Ouray would put up $40,000 to help defray the expense of construction. At this, the town balked. All save Dave Day. He bombarded his readers with editorial after editorial promoting the wisdom of anteing up. But all of his advice went adamantly unheeded until one day in the pages of the *Solid Muldoon* Ouray residents read that Day had purchased four large ranches on the Uncompahgre River down the valley to the north, and was advertising for settlers with the intention of subdividing the land and platting a new town. With the money from the sale of the lots Day promised to bring a spur of the D&RG to the new town, a maddeningly four miles short of Ouray. The reaction of the townspeople was immediate. Convinced by previous tests of his character that Day meant what he said, the people of Ouray quickly found the $40,000 required by the railroad. Day smugly deeded back the ranches to their original owners who were in on the hoax, and the first D&RG train chuffed into Ouray's hastily built depot in 1887.

The year 1895 found the restless Tom Walsh in Ouray. He had left his Martha Rose smelter in Silverton in the hands of a manager and had come over Mears's toll road to the new diggings hoping to cash in on the silver discoveries in the peaks surrounding the picturesque valley. But as usual, Lady Luck failed to look his way. And then the bottom dropped out of silver and the circumstances of the Walsh family, which by now included a daughter, Evalyn, and a son, Vinson, were modest at best. The Martha Rose was their chief support and its solvency was wavering. There was no silver coming in, and the high cost of transport dictated that only the highest-grade copper, zinc, and lead ores also treated by Walsh could be profitably smelted. As a consequence, shipments of these metals to the Martha Rose were few and the Irishman's income was considerably reduced compared with what it had been during the bonanza days of silver. To bolster his sagging enterprise, Tom Walsh cast about to find a way to improve the flux in use at his smelter so that lower-grade ores could be treated. His inquiry brought him word of some siliceous rock high up in the mountains along Canyon Creek that would produce first-rate fusion. Walsh quickly contacted Andy Richardson, a local prospector famed for his seat-of-the-pants ore savvy, and sent him up to find the rumored high-grade quartz. When he waved Richardson away little did Walsh dream that out of the prospector's sortie would come the biggest bonanza the San Juans would ever see, a mine that would become a legend in its own time and that would make the persevering fortune hunter from Clonmel a millionaire ten times over.

Richardson packed up his tin coffeepot and short pick and set out for Canyon Creek, climbing over fallen trees and man-sized boulders, scaling jagged rock walls to get around the tumbling, rushing stream, and nosing into every abandoned diggings he found. About three tortuous miles up the gorge he came to the site of two worked-out silver claims, the Gertrude and the Una, according to the map he carried from the recorder's office. The mines were originally prospected by Bill Weston and George Barber. Weston was in England in 1875 when a friend wrote him from Colorado telling him of the high-grade silver ores uncovered on Mt. Sneffels and jokingly suggesting that Weston take a cram course at London's School of Mines and hurry back to the States to get in on the rush. Weston did just that and came to Ouray where he met Barber. Together they located several claims, among them the Gertrude and the Una. After they extracted all of the outcropped ore, Weston and Barber hired a contractor to drive a tunnel into the side of the mountain to intersect the vein underground. When the tunnel was finished, it intersected the Gertrude and Una vein all right but it did so just at the point where it pinched out. Discouraged, Weston and Barber sold their rights to the claims in 1881 for $50,000, a sum they were more than glad to get, and left to the new owner the job of driving another tunnel to find the elusive vein. By the time the new owner got around to developing the mine further, the "silver question" was reaching its catastrophic climax and plans for the mine were speedily dropped.

When Richardson found the Gertrude, it was as it had been abandoned, with rusting machinery scattered among piles of tailings. Poking through the mine debris, the prospector examined the cast-off ore closely. The longer he stared at it, the better he liked the look of it. Bagging several canvas specimen sacks with ore, he lugged them down the canyon to Walsh, who stowed them under his bed. It was several weeks before Walsh took them to Silverton for assay. But when he did have the ore analyzed, he hurried back to Ouray, burst into the house, and caught up young Evalyn, whirling her around in an Irish jig. "Daughter," shouted Tom Walsh, "we've struck it rich!"

It was no blarney. Richardson's ore sense was true and at last Lady Luck looked Walsh square in the eye. The assay turned up $3,000 to the ton, not of silver, but of gold, in rare tellurium.

Quickly and quietly, Walsh bought all the old claims in the vicinity, shelling out $20,000 eked out of the Martha Rose, extracted from the sugar-bowl savings of thrifty Carrie Bell, and borrowed from Thatcher Brothers' Bank. He named his new properties the Camp Bird, the name the miners gave to the Canada jay that sought out the San Juan Mountains in the spring. By 1903, Walsh held 103 claims covering over 900 acres in his diamond cuff-linked arms. Between 1896 and 1902 alone the telluride ore of the Camp Bird converted to cash amounted to $4 million a year.

Never forgetting what it was like to be a hard-rock scrabbler, Walsh poured $5 million into surface improvements to the Camp Bird in five years. His four hundred miners lived in a three-storied boarding house that featured lights, steam heat, hot and cold running water, marble-topped basins, and even a reading room furnished with comfortable chairs. On the second floor was a company store where the men, marooned in their eleven-thousand-foot-high snow bowl for most of the winter, could buy woolen pants and shirts, gloves, helmets, carbide lamps, boots, mackinaws, and rubber suits to keep out the dankness of the underground workings. Here, too, a man could buy pencils and paper with which to communicate his yearnings to a belle of the town that lay three-thousand-feet down the narrow shelf of rock that twisted itself in sickening juts and angles along the side of the mountain to the valley below. When the dinner bell rang, miners took their places at the long, oilcloth-covered table laden with platters and bowls of solid, nourishing food. Like loggers, miners wasted no energy at table by talking. They concentrated in singleminded purpose on the business of stowing away the steaming victuals.

The same singleminded purpose was exhibited by the Walshes, who concentrated on spending their new wealth as fast as it came from the Camp Bird. As Lucius Beebe summed it up, the economic credo of the Walsh family was disarmingly simple: "They knew money existed to buy nice things and they had the money."

From Ouray, Colorado's newest millionaire and his family gravitated to Washington, DC. The nation's capital had come to be the natural habitat of the eclectic group of mineral-rich Rocky Mountain nabobs whose wealth led them not to philanthropy but to politics. Here, Tom Walsh flexed his new money muscles by investing heavily in Washington real estate, government bonds paying a safe 3 percent, and the Republican party; Carrie Bell triumphantly presided at afternoon receptions for diplomats, statesmen, and potentates; and young Evalyn proudly rode to school in her own private victoria driven by a dutiful coachman in well-brushed livery.

In 1899, at the bidding of President McKinley, Tom Walsh packed up his family and sailed for France as a commissioner to the Paris Exposition to reign royally, if not in title then in opulence. Camp Bird coin and invitations engraved with the great seal of the United States made it possible for the Walshes to enjoy the company of the world's hoity-toitiest society. Among their frequent guests was King Leopold II of Belgium with whom Tom soon found himself in rewarding conversation revolving quite naturally around what each man had the greatest admiration for— gold: Tom's pile in Colorado, and Leopold's in the Congo. Before the year was out they became partners.

Back in Washington after the Exposition was closed, the Walshes built a home of staggering proportions, considering that the family still numbered but four people. The house had sixty rooms including a theater, a ballroom, a roof garden, and hydraulic elevators. Once ensconced, the Walshes settled down to some real spending.

Evalyn, now seventeen, still hankered after Paris and Tom obligingly sent her back with a $10,000 letter of credit when she beguiled him with a soul-felt desire to study voice. But once in the City of Lights, it was too soon excruciatingly apparent that Evalyn's singing capabilities were closer to those of the bray of the Rocky Mountain canary than to the golden tones of a Mary Garden. The daughter of Tom Walsh was delighted at this verdict. Now she could devote her time uninterruptedly to spending money on her dearest desires: jewels, furs, and handsome, struggling young artists. When a friend tipped off the family that Evalyn was leading a life a touch too dissolute, Tom called her home. After this episode, concerned not so much about their daughter's spendthrift ways but more about her reputation, the Walshes considered moving the family seat from Washington back to Colorado where the pace if not more conventional was at least not so noticeable. But before they left the East, they spent one more summer in Newport where Evalyn and her brother Vinson were in an automobile accident in which Vinson was killed and Evalyn was badly injured. That event clinched Tom Walsh's resolve. Life on the Atlantic seaboard, both in America and Europe, was too fast, and as soon as Evalyn recovered, the family moved to Wolhurst, a modest five-hundred-acre estate on the outskirts of Denver, formerly owned by the flamboyant Republican senator Ed Wolcott. Here, Carrie Bell continued to receive distinguished guests at her tea table, including President Taft, and Tom continued to pick up mining properties that added neatly ledgered sums to the millions he got for the sale of the Camp Bird to English interests. Nor did Evalyn languish in the high dry plains of Colorado.

Despite the fact that Colorado was not exactly a mecca on the itinerary of the nation's social set, the glitter of Walsh gold and Evalyn's jewels drew a steady and wide assortment of beaux, both barefoot and well healed, to Wolhurst. Tom Walsh carefully screened them all. And when Evalyn and her childhood chum of Washington Days, Ned McLean, decided to elope, with the help of Nathaniel Hill's son, Crawford, who arranged for the parson and the place, Tom Walsh looked delightedly the other way. Ned's parents were *the* John R. McLeans, a three-generation storehouse of wealth gleaned from newspaper publishing, notably the Washington *Post* and the Cincinnati *Inquirer*. For their honeymoon on the Continent, both families tendered the newlyweds matching letters of credit totaling $200,000, with the assurance that if they needed more they had only to cable home. In two months of hard and diligent spending Evalyn and Ned managed to dispose of every cent of their play money, plus a handsome bonus contributed by the bride's father. The tangible evidence of their entry into the market place was two Mercedes saloon cars, two sets of Arab burnous, one chinchilla wrap, and the 94.8-karat pear-cut diamond known among collectors as the Star of the East.

A few years later Evalyn and Ned journeyed to Paris again where a dismal morning-after was suddenly brightened by the arrival of Monsieur Pierre Cartier bearing in his hand a blue-white jewel of such singular cast that Evalyn recognized it instantly. She had seen it once before suspended from the lissome throat of a sensuous beauty in the harem of Sultan Abdul-Hamid when she and Ned visited Turkey on their honeymoon, and the memory of it was indelibly fixed in her mind. After a decent interval of courtly backing and filling that accompanies a transaction involving precious gems, Evalyn came away the owner of the Hope Diamond, complete with the long history of malevolence attending its many possessors.

The daughter of Tom Walsh had breathed enough of the free atmosphere of the frontier to lose any Irish inheritance of superstition and cared not a whit about a jinx. She might well have cared, however, for her life, too, was touched by tragedy that many would say was the curse of the Hope Diamond. Charming, debonaire, dissolute Ned

McLean died of acute alcoholism in his mid-fifties. Their daughter committed suicide shortly after marrying Senator Robert Reynolds of North Carolina, a man thirty-six years her senior. And their son, Vinson, like his namesake uncle was killed in an automobile accident. Tom Walsh died in 1910 of lung cancer, a malady that afflicted scores of hard-rock miners who day after day for years breathed the fine rock dust of the mines. Evalyn herself died unobtrusively of pneumonia in 1947, and her cherished jewel was sold to the famed New York buyer and seller of gems, Harry Winston, for $1,500,000. When he could find no one who was willing to put down the required cash or put up with the stone's bad reputation, Winston presented the Hope Diamond to the Smithsonian Institution where it is on display for the vicarious enjoyment of those whose tastes run to rare jewels but who have not had the luck of the Irish to stumble onto a bonanza like the Camp Bird to pay for them.

Until a few years ago, the fabulous Camp Bird mine was still in production, not to the degree that it was in the halcyon days at the turn of the century but still disgorging thousands of dollars of valuable ore annually and providing a memorable training ground for graduates of mining schools.

✧◈✧

Thomas Walsh was not the only man to strike it rich in the San Juans. He merely struck it the richest. Hundreds of others rose from indigency to affluence on the strength of good eyesight and gut-feeling geology. From Silverton and Ouray, dogged prospectors, undaunted by the formidable terrain or the likelihood of calamitous storms even in the middle of summer, climbed up and over the mountain passes to the west to follow their hunches that the veins of ore that outcropped along the east side of the Uncompahgre Mountains also ran through to the west side. Their deduction was correct. On the west side in 1875, in an alpine bowl named Marshall Basin, John Fallon in quick succession laid claim to three parcels of land that became the famed Sheridan, Mendota, and Union mines. Not far away, W. L. Cornett set stakes around a promising vein he called the Liberty Bell. A mile to the southeast, as the crow flies, two other lucky miners came onto the Tomboy lode set in a glacial cirque. J. B. Ingram, who came along after John Fallon, decided that the stakes for the Sheridan and Union claims encompassed more land than was legal. When his measurements proved him right by a comfortable five hundred-foot margin, he cannily stepped off his own claim between the two, calling it the Smuggler.

Soon the slopes and basins of the Uncompahgres were crisscrossed with trails. The braying of burros, laden with ore on the down trip and supplies on the up trip, combined with the wind swirling over timberless reaches to startle small animals in their lairs, and the booming voices of teamsters bounced off rock walls in a day-long cantata of transport. Everything from tins of snoose to gigantic iron shaft wheels was delivered by mules or burros. Heavy machinery was loaded onto "go-devils," the great stone sleds with iron rollers fitted to their undersides. Teams of burros or mules in tandem were then harnessed to the sled to drag it up the tortuous trails to the mines. Cabling for hoists, some of which reached 2,000-and 3,000-feet-in-length, destined to reach down into the heart of the mountain mines, was coiled onto pack saddle horns and carried up the mountains in one continuous piece. The record was 4,000 feet of unbroken cable hung on the saddles of fifty-two burros.

Down at the foot of the mountains in a narrow reach of the San Miguel River Valley grew Columbia, a camp to supply the sky-high workings of the mines above it. The main street, fronted by rude clapboard buildings, was crowded with wagon trains transshipping ore from the backs of burros to the railhead thirty-seven miles away, around Dallas divide to a terminal a short distance above Ouray. As the camp's population swelled and it took on the rudimentary features of a town, someone suggested its name be changed to identify it more closely with the reason for its being. Consequently, Columbia became Telluride, in recognition of the fact that the streaks of silver laced through the mountains were cached in tellurium ore. Despite the high cost of transport— $60 a ton to the railhead, $35 a ton for smelting—the ores of Telluride brought enough profit to keep its miners employed, its mine-owners in the black, and investors interested. By 1882, the town could claim ninety business houses. The busiest, at least on a Saturday night, were the Pick and Gad, the Senate, and the Silver Bell, three of the twenty-six parlor houses and lace-curtained cribs called home by no less than 175 ready and willing girls. Telluride's madams were proud to say that no man had to wait in line.

About 1890, Max Hippler and Gus Brickson, whose pockets jingled with coins from some lucky grubstakes, decided that Telluride lacked tone. To offset the notorious establishments run by Jew Fanny and Diamond Tooth Leona, they put up the Sheridan Hotel, a three-story brick manse of high ceilings, broad vertical staircases, and canyon-like hallways. Hippler and Brickson spared nothing in the construction of the hotel. They even sent to London for the cherrywood bar. But at the same time they kept a weather eye out to keep the cost of operation low. When it came to providing for public rooms for the entertainment of the guests, they built a musician's loft that opened onto the bar, the dining room, and the ballroom, individually or all at once by a slick arrangement of sliding panels, thus making it possible to entertain a maximum of people at a minimum of cost. In the front of the hotel were three handsomely decorated suites, each with a private bath, a bedroom, and a sitting room, in which such distinguished guests as William Jennings Bryan, Lillian Gish, and Sarah Bernhardt relaxed before they met their public. The ladies of the theater came to play in Telluride's new opera house, which was conveniently connected to the Sheridan by a passageway on the hotel's second floor. The gentleman of the Congress came to exhort the silver-miners of Telluride to fight for the maintenance of the silver subsidy.

Bryan and his mining friends, however, did not prevail, and in 1893, along with the rest of the silver towns in Colorado, Telluride hit the skids. But not for long. Under a new wave of prospecting, sparked by Tom Walsh's find over in Ouray, the camp came alive again as good deposits of gold tellurium took the place of the now worthless silver ore. The gold boom lasted until 1901 when the miners, under the impetus of a growing labor movement in the West, struck to abolish the fathom system and in the process set the camp on the road to a permanent decline.

When the crowds of Cornish miners came to Colorado in the late 1860s and early 1870s, many of them leased mines and paid the men they hired according to the ancient Welsh custom by which a miner received a fixed price for mining one "fathom" of earth. One fathom was arbitrarily six feet long and six feet high and as wide as the vein being worked. The custom soon spread throughout Colorado when it became clear that by this system a mine-owner would suffer no penalty in the cost of mining if the vein varied sharply in width since a miner's wages stayed the same whether the vein was wide or narrow, despite the fact that it took a man more hours to mine a fathom if the vein were wide than if it were narrow.

The miners of Telluride walked out on May 2, 1901, calling for the abolition of the fathom system and a return to a daily wage. The newly organized Western Federation of Miners sought out Arthur Collins, manager of the combined Smuggler-Union mines, and demanded that he submit the matter to arbitration. Collins refused. Instead, he promptly hired nonunion men from out of town and reopened his beleaguered mines. Enraged, some 250 strikers strapped on pistols and clambered up the side of the mountain to lay siege to the Smuggler-Union. When the day shift ended, the scabs were met at the portal by a deputation of strikers demanding that they quit. The imported miners, all of whom were also armed, shoved the strikers aside. In an instant there was the stutter of gunfire as men leaped for cover among the crags of rock. When the shooting stopped, three scabs lay dead and six were badly wounded. The striking miners then triumphantly rounded up the rest of the out-of-town men and conducted them to Telluride's town limits urging them to get out and stay out or run the risk of being shot to death. The scabs left. With the strikers in the driver's seat, the dispute was settled with an agreement that allowed the use of the fathom system as long as the miner's daily take-home pay did not drop below $3 for an eight-hour shift, the going pay rate among mines in other districts that paid a daily wage.

It was a fair settlement, but it did not bring peace to the mines. Throughout Colorado, as in the nation, the growing force of unions pitted against unyielding management brought violence. Arthur Collins was assassinated the following year by a sniper's bullet as he sat at his desk in his office. And again, in 1903, along with a general strike of mill and smelter workers, Telluride's mines were struck. This time management won after the state militia instituted martial law and deported the strikers at the point of bared bayonets. The breach between the miner and mine-owner grew ever wider. In the first fourteen years of the twentieth century, Colorado would see labor wars in which white man fought against white man in a conflict as bitter and bloody as were the clashes between the red and white men of the previous century.

❧❧❧

South of Telluride, over the mountains to Howard's Fork on the San Miguel River, a passel of prospectors turned up the Silver Bell, the Gold King, the Alta, and the Butterfly-Terrible lodes, all names to conjure with in the seemingly unending production of silver. The camp that came to life in this circle of mines proudly gave itself the biblical name of Ophir. Typical of the treasure of Ophir's mines were the ten sacks of surface ore from the Gold King that netted Jack Nunn a round $5,000.

Nunn was a Telluride lawyer who, with his engineer brother R. P. Nunn, did a little fortune hunting on the side. After finding the Gold King, the brothers spent all of their free time working in the mine alongside their hired miners and they soon discovered what a costly and backbreaking job it was to operate a mine on manpower alone. Setting about to figure out a better way, the Nunns, after a few fits and starts, created the first long-distance alternating current system in the world to bring electric power to their mine. When they showed George Westinghouse their design for a generator to do the job, he laughed at them. But under their entreaties and assurances he at last consented to build the contraption if the brothers would first plunk down a $50,000 deposit of Gold King cash. They did, and Westinghouse had to eat crow. From Trout Lake, which lies in a basin formed by three mountains, the Nunns built long flumes down to the power plant at Ophir that provided the necessary high pressure water to operate their generator. By 1891, their high voltage transmission lines sent alternating current to all of the major mines of Ophir and Telluride. Out of their efforts came the Telluride Institute, a branch school used for years by Cornell and other universities to give their embryo electrical engineers some first-hand experience in the practicalities and ingenuities involved in the transmission of light and power.

Over a ten-thousand-foot pass to the south of Ophir, marked by a formation of peaks early prospectors called Lizard Head, came the silver-hunters. Led by ex-Army Colonel Hagerty, one party turned up lead carbonate shot through with high-grade silver. Within one month after Hagerty's group hit the rocky, mountain-lion-infested gulches, the camp of Rico flowered into a bouquet of six-hundred residents, twenty-nine buildings, seven of them saloons, and four of them bearing signs reading "ASSAY OFFICE." Trumpeted the Ouray *Times* of November 29, 1879, regarding its new neighbor over the mountains:

> ...everything is there, wine and women, cards and caterers, houses and horses, men and burros, moneys and mines, busy boys and blowing bummers, working men and working women, pack trains and bull trains, carpenters and sign painters, assay offices and bunco steerers, Sunday schools and keno chambers.

As shafts went deeper, vein after vein was exposed and profits steadily rose. Wagon trains heaped with pay dirt hit the trail south to Durango every hour of the day. Profits had to be high with eggs selling for $3 a dozen and flour at $50 for a hundred pounds. But Rico was an easy come, easy go camp. Even the local parson indulged himself in the pleasures of the layman. Dice and card games came to a halt during the hour he preached on Sunday mornings in his favorite saloon. But after his friend the faro dealer passed the hat for the collection, the minister was not loath to dip his hand into the offertory to buy a stack of chips for a round at the tables. Along with gambling, Rico residents favored musical entertainment, and when one miner spent some of his silver horde on a piano the whole town turned out to help him assemble it and then respectfully listened to matronly Helen Wixson "paw hell out of the ivories at the same time she sang like an angel."

With the birth of every new silver camp to the west of Silverton, Otto Mears's itch to link them together with his narrow-gauge rails grew worse. Once in Ouray, Mears knew he could build north and west over the Dallas divide to the valley of the San Miguel to tap the rich commerce waiting at Telluride, Ophir, and Rico. But there his Silverton Railroad sat, dead-ended at Ironton, six impassable miles short of Ouray. No matter how he surveyed the route ahead, it always revealed insurmountable grades of 7 percent and for once Mears was stymied. But never did his mind accept as permanent the barrier of the impossible. After a few weeks of hard cogitating, Mears came upon the solution. He would simply go around the other way. Incorporating a new road to be called the Rio Grande Southern, Mears started at Durango and thrust westward to Dolores, a farming community on the west side of the La Plata Mountains. From there the Rio Grande Southern wound along the Dolores River, its "thirty-pound rails laid without ballast on elemental earth," over rock ledges projecting over sheer-sided canyons hundreds of feet deep, through countless snowsheds and around sweeping loops as thrilling as Georgetown's, and over Lizard Head Pass to reach Telluride on a snowy day in December, 1891. Tracks from Ouray to Telluride via the Dallas divide completed the link. The incomparable pathfinder had won over uncompromising nature, spending a mere $9 million on 162 miles of railroad to get around six.

With the coming of the railroad, ore transport costs dropped abruptly in the familiar pattern and the large producers kept the little narrow gauge uniformly busy. The Liberty Bell alone regularly shipped 150 cars of concentrates a month. Nor did the line want for passengers. Mining magnates and presidents of rival railroads, sporting passes of handsomely engraved local silver, rode in the plush splendor of diminutive palace cars while lesser, paying customers crowded aboard unadorned wooden-seated coaches, taking "front seats on the way up the mountain and rear space on the way down to avoid the pools of tobacco juice which flooded the floor at whichever end of the conveyance was inclined downward at the moment."

The Rio Grande Southern came on hard times along with the rest of Colorado in 1893 and ultimately became the property of the Denver and Rio Grande. It was eventually abandoned in 1957 after sixty-five years of loyal and precarious service to a rabidly devoted public.

For all of its rash of boom towns and rich strikes, the silvery San Juans were but an *entr'acte* to the real silver bonanza. In the center of the state on the headwaters of the Arkansas, in California Gulch already panned clean of gold, prospectors would find the richest silver lode in all the Rockies.

10

Leadville—
The Biggest Bonanza of Them All

CALIFORNIA GULCH, ON THE WESTERN SLOPE OF THE MOSQUITO RANGE, HAD ALREADY been the scene of one boom and bust. Abe Lee had seen to that back in 1860 when his pan turned up mint-rich gold on one of the streams coursing down the mountainside. But it was a short-lived rush, and within two years it was all over. All over for gold, that is. Unknown to Lee and his compatriots, another money metal lay for the taking just beneath the surface of the gulches draining westward into the Arkansas River as it cut its way southward through the pine-studded flat lying between the Sawatch and Mosquito ranges.

Will Stevens had been knocking around Colorado since 1864, first trying his luck without much success in Gilpin County, then meandering over the mountains to California Gulch. He arrived in the gulch on the ebb of the rush started by Abe Lee. Some of the people he talked with had made their fortunes and were moving on. A few remained to eke out a living providing services to those who still worked some moderately paying placers. Others, whose claims all failed to materialize ore in paying quantities, were leaving to follow new leads. Stevens, after a few desultory months of panning, was one of the latter. However, instead of chasing after gold again he left to take a job surveying in Utah. But as the years passed he could not shake the hunch that there were still buckets of hidden riches in California Gulch. Finally, in early 1874, he packed up his transit and his plumb bob and headed back to the headwaters of the Arkansas River. When he got there, he dropped into H. A. W. Tabor's little general store, the only one left in what was once booming Oro City, and caught up on what, if anything, was going on in camp. Tabor told Stevens that there was some fitful gold mining still being done and that about twenty men were in the process of digging a ditch to bring more water into the gulch to improve their sluicing operations. Stevens

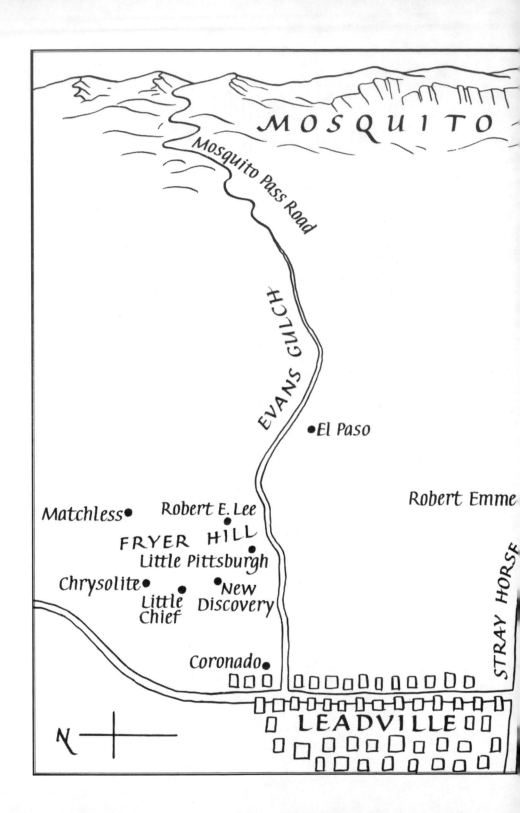

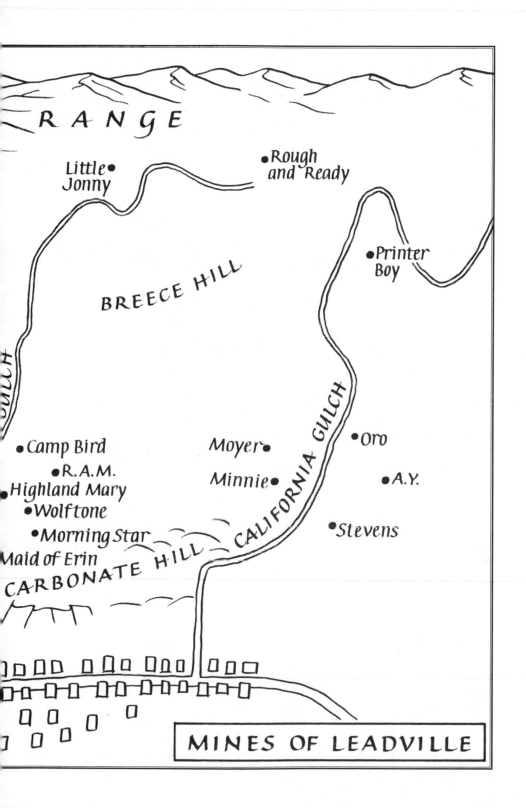

RANGE

Little Jonny

Rough and Ready

Printer Boy

BREECE HILL

GULCH

Camp Bird

R.A.M.

Highland Mary

Wolftone

Morning Star

Maid of Erin

Moyer

Minnie

Oro

A.Y.

Stevens

CALIFORNIA GULCH

CARBONATE HILL

MINES OF LEADVILLE

then roamed the area looking over the claims owned by the men digging the ditch. What he saw prompted him to get in touch with his friend, geologist Alvinus B. Wood in Ann Arbor, Michigan, and persuade him to come to California Gulch. When Wood arrived, the two of them rounded up $50,000, formed the Oro Ditch and Fluming Company, and bought out the twenty miners who dug the ditch.

For the first year of operation, despite the heavy black sands that clogged their sluices, their placering operations paid well, averaging a 30 percent return on their $50,000 investment. The troublesome sands were the same as those that Augusta Tabor had complained made working her husband's claim so difficult. But to Alvinus Wood, they were more than troublesome. They piqued his curiosity. One day, he gathered up a packet of samples and performed an assay on them. When he finished, he rushed down the hill to Stevens who was working on the sluices and took him aside to tell him the results. The black sand turned out to be carbonate of lead carrying forty ounces of silver to the ton. Quickly and quietly, the two men traced the source of the sand up the west side of California Gulch to its source, a vein ten-feet thick. Here they staked out claims and set a batch of miners to work. After a few weeks of mining the ore, the crew foreman, an Irishman by the name of Walls, came to Stevens.

"Beg pardon, sir," Walls began, "but me and the boys, we was wondering what it is you be digging for. There's not a smidgen of color in all them black dirty rocks we've taked outta this pit."

"So there isn't, Mr. Walls," answered Stevens. "There is not an ounce of gold to be had in this claim. But these black rocks are carbonate of lead, and they are full of silver."

"I see," said Irishman Walls. Two days later he quit the Oro Ditch and Fluming Company to stake out his own claims nearby. He also told his son-in-law of what he had heard, who told his father, who told his wife, who told her neighbor, who told her husband, who wrote his brother, who gave it to the newspapers, and the rush was on.

It was Pikes Peak and Clear Creek all over again. They came by the hundreds. They came by wagon, on foot, sometimes dragging a handcart, sometimes leading a droop-eared mule, swarming into California Gulch and spilling over into two more inviting gulches to the north, Stray Horse and Evans. By the summer of 1877 the hillsides on each of these three gulches were crawling with silver-seekers. Rich men, poor men, famous and infamous, side by side, they wielded their picks and shovels in a frantic contest to see who could make the biggest strike.

One of the first to cleave open a rich body of carbonate was George Fryer, working the hill on the north side of Evans Gulch. His New Discovery mine atop what was promptly named Fryer Hill brought a wave of prospectors to comb the adjacent property. John Campion, a Canadian who answered the call of the silver rush, grabbed up a shiny new pick and followed the crowd into the scrub-pine hills. Everyone but John, however, seemed to know just where to dig. He was at a loss; the terrain looked all the same to him. Spying a man intently gathering rock samples on a hillock overlooking Stray Horse Gulch, Campion rushed up to him to ask where he should begin. Annoyed at being interrupted, the prospector looked around hurriedly, picked out a lone pine as far away from where he was working as he could see and pointed to it. "Over there, under that tree," he said gruffly and went back to his rock-gathering. Delightedly, the Canadian dashed over to the tree and swung his pick into the yielding ground. In a matter of hours, John Campion was a wealthy man. His Little Jonny held one of the richest lodes in the gulch.

So contagious was the new frenzy that even Governor John L. Routt left his aides to mind the state while he scuttled over to Stray Horse Gulch to roll up his executive sleeves, don miners' boots, and personally muck out the silver-rich rock of his Morning Star mine, returning to Denver a millionaire.

Over in California Gulch, Alvinus Wood sold his interest for $40,000 to Levi Leiter of the Marshall Field Company in Chicago. By the spring of 1879, the new partnership of Stevens and Leiter capitalized their holdings at $2 million and had a dozen claims in full production under the name of the Iron Silver Mining Company. One of their mines, its silver ore overlaid by a blanket of rust-colored, iron-containing rock, Leiter named the Iron Hat from an old German proverb: "A lode will never cut rich and fat/ Unless it wears an iron hat." In their day, the Iron-Silver mines produced more than $15 million.

Over the hill from Stevens' and Leiter's property, on Carbonate Hill, Charles, Patrick, and John Gallagher opened the Camp Bird. After taking out a comfortable fortune, the brothers sold it to the St. Louis Smelting and Refining Company for a tidy $250,000. Across the gulch from the Iron Hat, A. Y. Corman found the A.Y. claim that assayed high in silver. Corman, who liked prospecting better than profit taking, looked around for buyers. His search led to Philadelphia and Charles Graham, a moderately well-to-do grocer. Graham bought Corman's A.Y. and another promising claim of Corman's called the Minnie on speculation. After a few months Graham was ready to unload his silver stock and he chose the likeliest candidate from among his business friends on which to do it. Meyer Guggenheim was a Swiss emigrant who with his seven sons had parlayed a peddler's route into a fortune in lace and embroidery imports. But wealthy as he was, Guggenheim still had not reached the goal he set for himself—to leave a million dollars to each of his sons. So when Charles Graham came to him with some Colorado mining shares to sell, Guggenheim, never a man to hold back when faced with an opportunity to turn a dollar, peeled off a few thousand-dollar notes and found himself the owner of the A.Y. and the Minnie.

Shortly after his purchase, Guggenheim went west to inspect his new property. By the time he got there the shafts on both mines had been sunk to a depth of about two hundred feet, but there seemed to be little production. Guessing the reason, Guggenheim proceeded to drop a stone down each shaft, and to his dismay he found what he had feared—they were half full of water from seepage of underground springs. Nevertheless, the persevering ex-peddler bought pumps and set a crew to unwatering the A.Y. and the Minnie. And like everything else he touched, his mines returned him vast quantities of money once they were dried out and in production—$20 a ton was the average, with one stope alone giving up $180,000. But along with silver, there were hundreds of tons of copper in the A.Y. and the Minnie and it gnawed at Guggenheim that the smelting process then in use for copper was too costly to make his ore pay. Calling in his troop of sons he instructed them to look for a cheaper way to smelt copper. As luck would have it, a friend of the family, Bernard Baruch, had heard of a newly developed process that he believed would answer the Guggenheim's needs. It did, and the cost of copper smelting dropped to a highly profitable $8 a ton. With their new reservoir of riches turning out millions, M. Guggenheim & Sons phased out their fastidious lace and embroidery business to concentrate on the grimy commerce of smelting. The purchase of the Guggenheims' first smelter in Pueblo in 1888 began the mining and metal empire that gave Meyer Guggenheim many times more than the $7 million he hoped to leave to his sons.

As more and more strikes were made in Stray Horse and Evans gulches, the population center shifted from the old site of Oro City in California Gulch a few miles north to the flat between the Arkansas River and the two other gulches. One of the first people to follow the move north was H. A. W. Tabor. In mid-1877, he rented a cabin along the roughly parallel rows of tents and cabins running east and west that sprang up to house the five hundred silver-struck residents. One of the early comers euphemistically named the row of dwellings and stores Chestnut Street after Philadelphia's famous artery. In time the name faded to an anonymous "Second Street" as the business district pivoted to Harrison Avenue, the broad thoroughfare cutting north and south along the foot of the gulches pocked with a growing number of what appeared to be gargantuan gopher hills as mine tailings rose in mounds all over the hillsides.

But in July, 1877, when Billy Nye opened his saloon and lodging house that called itself the City Hotel and took in paying guests, Chestnut Street was still the center of commerce. All through the summer, the camp churned in the excitement of the new rush. When winter came, however, and with it the prospect of the three-month isolation and inactivity that the heavy snows would inevitably bring, the spirits of even the most enthusiastic of the men were dampened.

It was left to the women in camp to sustain the population through the winter doldrums. In this, Augusta Tabor led the way. When she heard that Susan B. Anthony was lecturing in Colorado, she got Billy Nye to turn over his saloon for one evening to the suffragette. Augusta appropriated fifty yards of bright calico from Tabor's stock of yard goods to hide the bar, and Billy agreed to keep the drunks out during the lecture. On the appointed evening, mackinawed miners and their womenfolk crowded into the saloon, the men attending more in the interest of having something to do than in the interest of the speaker's views. It was, however, a respectful audience, and Miss Anthony's ringing phrases fell on attentive ears. Afterward when the lady lecturer allowed as how her expenses in making the trip up to the silver camp had been simply "magnanimous," the miners were impressed with her pluck if not her malapropism and they forthwith showered her with hard cash.

The generous miners willingly came forward with cash again at Christmas when Augusta made the rounds for funds for a community holiday party. The festivities took place on Christmas Eve around one of Billy Nye's beer barrels filled to the brim with freshly popped corn. Every man, woman, and child got a gift—perhaps a plug of tobacco, a frying pan, or a pair of mittens, and the bar was laden with dried apple pies and sugar cookies to be washed down with ginger beer and coffee.

With Christmas and New Year's over, H. A. W. Tabor, merchant Charlie Mater, ore-buyer August Meyer, and several other men met on January 14, 1878, in Gilbert's wagon shop to give the new camp a name and a municipal government. At the meeting a half dozen names were suggested, among them Carbonateville and Agassiz, (for Louis Agassiz, the renowned geologist and naturalist). But none of these struck the fancy of the town planners so much as J. C. Cramer's homely but accurate offering of Leadville. By the end of the meeting the new camp that owed its origin to carbonate of lead had a name, a mayor, a town treasurer, and a postmaster, all of which fell to H. A. W. Tabor, not because of his capability so much as his availability. The Tabor store was the clearinghouse of the camp, and it seemed reasonable to let it become, for the time being, the town hall and the post office.

Later, in April, Governor Routt left off silver-mining long enough to grant Leadvle's petition for a special election. Again, Tabor won the job of mayor.

In the meantime, the camp swelled with humanity. They came with the spring as soon as the Mosquito, Weston, and Trout Creek passes were free enough of snow to allow foot and wagon travel. Pouring onto the wild pine flat they quickly felled trees and tore down wagons to build makeshift tents and cabins and then they headed for the gulches to find their fortunes, learning to step nimbly aside as ore wagons from mines already in production clattered down the narrow roads en route to the railhead at Salida. Among the timber cutters, prospectors darted like ants, hacking away at a rock here, a rock there, storing away deep in the pockets of their canvas trousers any likely looking specimens. To the fanatical silver-seekers, almost every stone was likely looking, and assayers were besieged with countless samples to analyze. So rich were most of the specimens that in their own excitement assayers fell prey to jokers. Presented with a chunk of broken grindstone, one assayer confidentially reported that it ran nearly a hundred ounces of silver to the ton.

The frenzy of activity in the gulches was mirrored ten times over in town. A dozen new streets were laid out. Just north of Chestnut came State Street where saloons, gambling halls, and parlor houses opened so quickly that patrons had to step over carpenters and tin-smiths still adding finishing touches to the buildings. On Harrison Avenue, lots sold for the incredible price of $250 a front foot. Men waited in line at the hastily built saw mill to grab up boards as they came off the planer. So scarce was wood and so prevalent was hijacking that no man with a wagonload of lumber traveled far without a loaded shotgun.

Next to hijacking, lot-jumping was a favorite pastime of some of Leadville's newcomers. Parson Tom Uzzell staked out a lot and preached every night in a different saloon to raise money to build his Methodist church. He got the money within a week, but in the meantime a trio of hoodlums jumped his lot. The parson unceremoniously whipped out his pistol and vowed to send every one of the lot-jumpers to join their Maker if they did not get off his property. A well-aimed bullet that burrowed into the sod at the foot of one of the defiant culprits convinced the threesome to withdraw. The little church went up in record time, and when the bell for the steeple was in place, "Fighting Tom" Uzzell, as he was now known, grabbed the rope and sent the bell clanging over the gulches and flats in a paean of triumph. On Fryer Hill, a startled miner stopped his pick in mid-swing at the sound. "Well, I'll be damned!" he exclaimed, "if Jesus Christ hasn't come to Leadville, too!"

Land speculators bought up lots and threw together false-fronted structures that rented for $400 a month. Throughout the summer of 1878, the price of staples soared under the ever-increasing demand until flour and bacon were ten to twelve times higher than in Denver. Nobody seemed to mind, and there was enough money around so that a barrel of whiskey, for example, properly watered, could net Billy Nye a clear profit of $1,500. Beds in town were at a premium. An enormous shed, called the Mammoth Palace, fitted with five hundred bunks in two tiers sold space around the clock at 50 cents for eight hours. Those who could not find rooms fought for a corner on a saloon floor, and some rolled up in their saddle blankets outside in the chill night air. Pneumonia and then smallpox swept the camp to put a temporary stop to the growing population, but by the end of 1878, Leadville harbored five thousand frenzied people.

Night and day, heavy-booted miners, mainly Irish, clomped the crowded boardwalks on the way to the assay office jostling derbied dandies who idled their way past the decorated windows of David May's fancy new clothing store, swinging silver- and pearl-handled walking sticks. Swarms of small boys trailed behind flush and free-handed miners hoping for a handout. Gaunt-faced women in modest grays and blacks, with strings of children in tow, hurried up State Street to Harrison's markets, their eyes cast down to avoid the curl-ing-lipped smiles of the camp followers leaning provocatively in the doorways and windows of the rows of cribs strung along the street. Iron-clad wheels of carriages and freight wagons churned the streets to mud, the clop and squish of hooves mingling with the braying of mules, and the strident voices of hawkers for the camp's rival saloons and variety houses added to the continuous din of steam whistles and squeaking windlasses. Word of every new strike rolled over town like a tumbleweed in the wind.

To report the fast-moving events of the budding town came John Arkins, who left his job as foreman of the composing room of the Denver *Tribune*, invested all of his savings in a press and a 20-by-30-foot clapboard shanty, enlisted Carlyle Channing Davis to help him, and inaugurated Leadville's first newspaper, the *Chronicle*. His first feature story was very nearly his last. Spying a crowd of people hovering around a cabin on the edge of town, Arkins, with the nose-for-news technique of a cub reporter, pushed his way through the crowd and to the door. The odor of chloroform filled his nostrils as he peered into the dimly lighted room to see a woman lying on a bed, and a doctor working over her with a stomach pump. Assuming that the woman was one of the town prostitutes who had attempted suicide, a fairly common news item in the silver camp, Arkins printed the story that very afternoon. The next morning a brutish man, bewhiskered and reeking of stale tobacco, stormed into the *Chronicle* office pull-ing the unfortunate woman along behind him. Confronting Arkins, he vowed to kill the man who suggested his wife was a woman of low repute. With a swift and facile rejoinder Arkins explained that the originator of the story was out but that obviously there had been a terrible error. Bowing ever so slightly toward the woman, Arkins gravely declared, "Why, it is plain, madam, that you are the flour of sulphur and the cream of tartar!" The pair looked hard at Arkins' solemn face for a moment, and then the woman blushed, preened the folds of her tattered skirt, and let a bashful smile crease her worn face. Her husband digested the editor's words by working his tongue over the cud of tobacco in his mouth. Then he hit the cuspidor with a ring-ing bullseye and grabbed Arkins by the hand. "By Heaven, sir, you're an honest man!" he exclaimed as he dragged Arkins along by one powerful arm and his wife by the other to the nearest saloon where he solemnized his new regard for newsmen by drinking the captive Arkins under the table.

Arkins' reporting improved overnight, and within a year the *Chronicle* was a three-edition, nine-column afternoon paper, with a battery of newsboys turned out in spiffy blue uniforms with shiny brass buttons. Even its second-running rivals, the *Herald* and W. A. H. Loveland's *Democrat,* volunteered that the *Chronicle* was a competent and stylish sheet whose campaigns for town improvements, fairness in politics, and a reduction in materialistic considerations served as constructive rallying points for the civic endeavors of the leaders of Leadville's self-conscious new society.

Not only newspapers spread the story of Leadville's fantastic orgy of carbonate discoveries. Western Union's newly installed line strung from Alma in South Park carried hour to hour quotations as silver shares multiplied by the thousands. Leadville's mineral production in 1878, one year after Stevens and Wood made their find, reached a staggering $2,490,000. And it was only the beginning.

Just above George Fryer's New Discovery was the Robert E. Lee, a claim Jim Dexter picked up in early '79 for $16,000. He sunk a shaft to a depth of one hundred feet without turning up so much as a trace of silver. Disgusted, he called off his crew just as they were preparing to blast out another section of the shaft, found a buyer for his property, and unloaded it for $30,000, thinking himself lucky to have made a few dollars on his original investment. The day after the sale was final, the new owners put the crew back to work on their blasting. With one charge they blew open a vein of nearly pure silver that in one twenty-four-hour period produced ninety-five tons of ore worth $118,500.

Richard Dillon and his partners gophering among the outcrops on Carbonate Hill uncovered a mine they named the Robert Emmett, for the Irish patriot. Scampering over to Fryer Hill, they spotted a section of hillside that showed a "contact," a carbonate vein bedded between limestone and porphryry. Dillon staked out claims covering eight acres, a small parcel for that area, and hired a crew to start digging. The miners found the vein to be eighty feet thick in some places and Dillon's Little Chief got the reputation of being the smallest and richest mine in Leadville. Meanwhile, Ezra Dickeman's El Paso over in Stray Horse Gulch was well on its way to giving up ore that would reduce to $3 million in silver. On the west side of Carbonate Hill the peripatetic John McCombe opened in quick succession the high-grade ore bodies on his Crescent, Big Chief, Castle View, and Evening Star claims. The twenty-six-year-old McCombe, his pockets bulging with silver dollars, then returned to Ireland to marry his childhood sweetheart. He brought his bride to Leadville where the compulsion to prospect seized him once more and he proceeded to uncover two more bountiful lodes, the Parnell and the McCombe.

Another Irish immigrant, Black Jack Morrissey, did as well. His Ready Cash mine grossed $480,000, and he sold his equally profitable Highland Mary to the Guggenheims for a reported $3 million. Like Pat Casey in Central City, Black Jack Morrissey was illiterate, but he found it no handicap. His enormous, bulky figure was a familiar sight as he made his daily way from saloon to saloon on State Street, dispensing coins to bootblacks and beggars in moments of spontaneous expansiveness. Across his ample front hung a glittering gold watch chain. Town wags took perverse pleasure in trying to embarrass Morrissey by stopping him and asking the time of day, knowing full well the prospector could not even read numbers. But the canny Morrissey foiled his foes. Pulling out the enormous golden disk that was his watch, Black Jack flashed it in front of his grinning tormentors. "Here," he boomed, "look for yer self and then there'll be no lies told."

It took no time at all before Morrissey's reputation as a soft touch got around Leadville. And when a town committee asked him to donate money toward the purchase of half a dozen gondolas to decorate nearby Turquoise Lake, Morrissey was quick to say yes. But as he shelled out the gold dollars, a puzzled expression crossed his wide, friendly face. "I'm glad to do it," said he, "but why get six? Get two and let 'em breed, and by summer there'll be flock of 'em." Like many another rags to riches millionaire, Black Jack's fortune slipped through his fingers in good works and bad investments and he ended his days in the Denver poor-house, steeped in a bewilderment of how it had all happened.

Relief stagecoach driver Daniel Ellis, better known as "Broken Nose Scotty" after he drove his coach off the road coming over Weston Pass, was a rowdy who spent his time stagecoaching, prospecting, and sleeping off drunk and disorderly charges in the camp jail. During one of Scotty's sojourns in the pokey, a local agent for a syndicated mining company came to him and offered him $30,000 for one of his claims on Breece Hill. To Scotty that $30,000 was more money than he had ever hoped to see. He quickly petitioned the sheriff for an hour of freedom to conclude the deal. The sheriff granted his request and off went the stagedriver in the custody of the claim agent. Within half an hour Scotty was back with $30,000 in a buckskin bag and a broad smile on his face. Before an incredulous judge and sheriff, Scotty laid down enough money to cover his fine and that of all his fellow prisoners. He then marched his cellmates to a clothiers where he bought each man a new suit, and after that a full meal at the Saddle Rock Cafe, one of Leadville's fanciest restaurants. When his pals were full of food and champagne he gave each one a ten-dollar gold piece and sent him on his way. Before midnight every man jack of them was back in jail for disturbing the peace, including Scotty. Scotty bailed himself and his friends out again—and again—and again. Finally, he left Leadville, making the long land and sea journey to Scotland to see his aging mother on whom he settled $10,000. With the remainder of his windfall, Scotty came back to Leadville to smoke large cigars, play the tables, drop dollars into the open hands of his fairweather friends, and take up his old habit of alternately prospecting the gulches and getting drunk as a lord. But never again did Scotty make a strike, and in a few years, the $30,000 dribbled away so that at his death there was no money with which to bury the colorful spendthrift. But the county came to the rescue, dipping into the coffers Scotty had so amply lined in his misdemeanoring days, to buy a plain pine coffin and lay him to rest in potter's field.

Scotty at least had the good manners to keep his shade from inhabiting his mine after his death. Others, according to the superstitious Irish miners, were not so well mannered. Persistent stories of ghosts roaming the sepulchral underground workings kept many a youngster wide-eyed and rigid far into the long Leadville nights. In Fryer's New Discovery, miners swore they were often startled by a shadowy figure of a woman gliding through the drifts as if looking for someone. Over on Iron Hill, not far from the first carbonate discovery by Stevens and Wood, stood the Moyer, the most haunted of all the silver camp's mines. According to the tales of the men who worked in its bowels, the Moyer housed the restless spirits of twelve men killed in a cave-in and one particularly upsetting lone spirit who would suddenly appear near a tram of ore, its face "intense with woe and agony" and its fingers clutching air as if to find a handhold to keep from plummeting down a shaft. But the most celebrated of the Moyer's ghosts was that of State Senator Gallagher. Gallagher fell to his death down the mine's shaft when the bucket in which he was riding broke loose. His spirit, the face clearly bearing the senator's image, was often glimpsed irresolutely wandering through first one drift and then another, with a pick in hand, as if searching for a way to the surface.

꧁ꕥ꧂

Of all those who struck a bonanza in the black sands of Leadville no one struck it so easily or exhibited it so flamboyantly as Horace Austen Warner Tabor. For eighteen years he had dragged the uncomplaining Augusta from one Colorado boom camp to another—from Idaho Springs on Clear Creek to Buckskin Joe in South Park, to Oro City in California Gulch, and finally to Leadville. Between his storekeeping and her pie baking they had eked out enough to live on with a little bonus now and again from flash-in-the-pan strikes. And when Tabor moved to Chestnut Street in the early months of Leadville's beginning, he appeared to achieve what would be the apex of his career. With Augusta to help him, he built a thriving trade, and to go with his business success there was the prestige of being Mayor Tabor. He had altogether given up prospecting, convinced that he would never strike it rich. Besides, prospecting was arduous work and full of frustrations, and at forty-five-years-old, Tabor was ready to take it easy. The comfortable living his store brought left plenty of time for his two favorite pastimes, all-night poker games and hours of chewing the fat with his cronies. And there was enough stock on the shelves so that he could at least vicariously enjoy the excitement of the chase after sudden riches by now and then grubstaking his friends.

As the sun went down on Saturday evening, April 20, 1878, Tabor and a half dozen others sat on wooden crates around the iron stove in his store puffing on their pipes and cigars and discussing their favorite topic: the future of Leadville. Only that morning, word of George Fryer's strike had made its way down to the camp and sent new shivers of craving through the populace. Among the tousle-headed and flannel-shirted men clustered in Tabor's store were two prospectors whom Tabor had known from South Park days, George Hook and August Rische. Hook had left off steelmaking in Pittsburgh at the first news of Pikes Peak gold to join the rush west. On the way he met Rische, who had thrown aside his cobbler's tools to do the same thing. They formed a partnership and headed over the plains as fast as their feet would take them. But when they reached the new Eldorado their luck turned out to be as desultory as Tabor's and all they had to show for years of hard work and vain hope were calloused hands and the clothes on their backs. Even the tobacco they smoked they had cadged from the generous Tabor. As Rische and Hook listened to the talk around them on that Saturday evening, the old excitement crept into their consciousness, and on Sunday morning, the two grizzled prospectors, stiff from a cold night's fitful sleep in their drafty tent-cabin, trudged up Fryer Hill to size up the site of the New Discovery. After staring at the contact that held the rich ore and poking around the surrounding terrain for a while, both men concluded "the stuff had come off the hill" above Fryer's claim. To guess where the continuation of the vein lay was one thing; financing an exploratory venture was quite another, especially since they were stone broke and bereft of assets. After a few minutes of thought they turned to the one man whose generosity they already knew well.

"Tabor," said Hook, "we think we know where Fryer's vein lies up the hill. If you stake us we think we can find it."

"And if we do, we'll go thirds with you," offered Rische.

The genial storekeeper grinned broadly. "Sure, take what you want," he said, waving his hand at the neatly stacked shelves of goods.

"Even a jug of whiskey?" asked Hook, winking at Rische. "Why not," said Tabor as he sorted the mail for the morning stage to Alma. "You can have the whole store if you bring in a bonanza."

Tabor's new partners gathered up $17 worth of tools, blasting powder, fuses, beans, bacon, flour, and a stone jug of rotgut, and set off for Fryer Hill. They climbed the bracken-covered hillside until they reached a rise well above the New Discovery and somewhat to the south of it. Here in the shade of a scrub pine, they sat down their packs, wiped the sweat from their weathered faces, took a slug of whiskey from their jug, and discussed where they should begin to dig.

"What's the matter with right here?" asked Hook.

Rische looked around for a minute, and then shrugged, "As good a place as any, I guess."

Thirty days and thirty feet of shaft later, August Rische and George Hook struck the vein of high-grade carbonate ore that started their partner H. A. W. Tabor on the road to becoming the richest and most notorious man in Colorado. If the two German prospectors had dug in their picks ten feet farther to the south, they would have missed the vein altogether.

Hook named their claim the Little Pittsburgh, and within a month the mine was producing $8,000 a week. But the sudden input of wealth unnerved the ex-steelmaker, and he sold out to Rische and Tabor five months after the discovery of the bonanza. Not long after Hook liquidated his share, Rische sold his interest to David Moffat, whose enterprises had widened from bookselling to banking, railroading, and mining. For his share Rische received $262,000. With it, he took the stage to Denver, opened a saloon, went bankrupt, and eventually disappeared.

Not so Tabor. Once he had the opportunity, Tabor bulldozed his way forward, turning his sudden wealth into a string of monetary successes, some of them incredible by any standards. The first step taken by Tabor and his new partner Moffat was the capitalization of the Little Pittsburgh at $20 million, and for ten successive months the new company returned dividends to its shareholders of $100,000 a month. With this potboiler, Tabor came from behind the counter of his store and bought up very nearly every silver claim in Leadville that he could get his hands on. Two of his more spectacular buys were the Chrysolite and the Matchless.

A teamster with the unlovely name of "Chicken Bill" Lovell, hung on him by fellow freighters after Lovell survived a blizzard on Mosquito Pass by eating raw a load of chickens he was transporting to Leadville, poked around Fryer Hill and laid claim to a parcel of hillside to the north and below the Little Pittsburgh that he called the Chrysolite. In the few feet of shaft that he dug Chicken Bill found no ore, but he was determined to get paid for his work. Well aware of H. A. W. Tabor's reputation for gullibility, Lovell stole some high-grade ore from the Little Pittsburgh and carefully salted his own claim with it. Then he invited Tabor over to inspect the Chrysolite, telling him it was for sale. The credulous storekeeper eyed what looked like ore as rich as that in his Little Pittsburgh and bought Lovell's claim on the spot. In no time, Chicken Bill was cackling all over Leadville about how Mayor Tabor had fallen for one of the oldest tricks in the history of mining. But the snickers, guffaws, and ribbing did not fluster the easy-going butt of the joke. Tabor merely hired some miners to deepen the shaft of the Chrysolite and with what soon came to be known as "Tabor Luck" the miners turned up a vein of carbonate within eight feet of where they began digging that netted their boss a neat $300,000.

When it got around town that Tabor gave Peter Hughes and six other grubbing prospectors $117,000 for their Matchless, a worthless hole on Fryer Hill named for a popular chewing tobacco, and shelled out another $30,000 to settle pending boundary disputes over the mine, Leadville was convinced there was no bigger fool than Horace Austin Warner Tabor. But the hooting and finger pointing were again quickly cut short when the Matchless began giving up ore that was nearly pure silver, containing ten thousand ounces to the ton, and returning to its owner another $100,000 a month in take-home pay. Shortly afterward Tabor sold his interest in the Little Pittsburgh to Dave Moffat and Jerome Chaffee for a round million dollars, and, fifteen minutes later embarking on a spree of investing that would last fifteen years, the ex-stonecutter bought 80 shares of stock in the First National Bank of Denver.

From little more than enough for subsistence Tabor's income soared to superabundance. He was ecstatic. Immediately, he saw himself as the benefactor of Leadville, the town that made him rich. He plunged thousands of dollars into the building of the first brick building in town. When that project was well underway, he organized a company of volunteer firemen, fitted them out in scarlet uniforms with the name of Tabor emblazoned on their chests in gleaming silver, ordered from San Francisco the most expensive fire hose cart he could find, and built a fancy firehouse to keep it in. And then he put up funds for a public hall, the Wigwam, to seat five thousand people, and financed a horse-drawn street railway for his town. Determined to have Leadville in the forefront of modern cities, Tabor organized a telephone company with 150 subscribers and long-distance lines to Denver. When the hookup to Denver was completed Tabor bellowed loudly into the mouthpiece to an executive of the First National Bank. Then, stated the *Chronicle*, someone in Denver sang "'In the Sweet Bye and Bye' which could be distinctly heard."

Distinctly seen and heard all over Colorado was the stonecutter-turned-Croesus. With Dave Moffat's private railway car, enamel-bodied stagecoaches, and spirited teams at his beck and call, Tabor flitted from one boom town to the next, expanding his silver empire into the remote San Juans and to the other side of the Sawatch range, to Aspen, where he gathered up promising claims with the nonchalance of a beachcomber gathering shells at the shore.

Between epic episodes of wheeling and dealing in mining, real estate, insurance, and in a portfolio full of other enterprises, Tabor moved Augusta and young Nathaniel Maxcy into a plain, clean-lined, red clapboard house just off Harrison Avenue. Here, for the first time in their married life, Tabor's wife had a room whose sole purpose was to be cooked in; for the first time she had a parlor and bedrooms that were units unto themselves with walls and doors to make them so. In the beginning she was elated, but as Tabor's wealth grew and with it his unbridled spending, Augusta became resistive. She refused to attend the lavish champagne suppers her husband threw at the Saddle Rock and the Tontine restaurants for eastern investors, Denver businessmen, and for politicians, from the White House down to the county level. She was tight-lipped and livid when Tabor arrived home in the early hours of the morning after spending all night in the upstairs parlor of the Texas Club over poker hands whose values ran into thousands of dollars. When she cautioned against extravagant expenditures, Tabor sent to Denver for a sapphire pendant and a pair of crystal and diamond earrings to assuage her, and showed her the bill of sale stamped "paid" to prove to her that their affluence was real and that she had only to accept the fact and enjoy it with him. But the hard-stemmed frugality of Augusta's New England heritage could not be bent. She wallowed into an unremitting and hopeless fear of impending disaster that turned her from a placid helpmate into a harpy. Tabor was exasperated and perplexed, and before long, with his ever-cheerful gregarious spirit propelling him, he gravitated to others to share with him his stupendous, preposterous, and entrancing good fortune. With the same indiscriminate enthusiasm with which he picked up mining claims, Tabor gathered around him troupes of women—hurdy-gurdy girls, variety hall performers, and harlots, treating them to champagne and fine clothes, in exchange for which they had to endure his mawkish displays of affection. One of his favorite girls, according to her own account, was Willie Deville, a bride of the multitude, whom Tabor spirited away, without much difficulty, from a Chicago house of ill-fame to travel with him and give him the comfort he required in the face of the distressing problem of how to spend his money. Augusta bore the snide hints and open publicity about her husband's transgressions with some grace. But there is no record of her trying to temper his new exuberance with the warm persuasion of understanding. Uptight and blind with fear and shame, she withdrew from the sight of their friends.

As H. A. W. Tabor's horizons expanded, so did Leadville's. Three stage lines, Barlow and Sanderson out of Canon City, McClelland and Spotsworth from Denver, and Wall and Witter from South Park, in fierce competition for passengers, brought twelve fully loaded coaches a day clattering into town with people jammed inside and hanging on the boot. By 1879, two thousand men were engaged in freighting and on one day in May alone, 146 freight wagons rumbled into the dusty corrals at the end of Chestnut Street loaded with typical stores of a booming mining camp—bacon and sealskin coats, flour and dia-monds, champagne and sticks of dynamite. Shops of dressmakers, milliners, and photographers opened their doors for business as fast as buildings to house them could be completed. Over from Central City came Bill Bush to build a three-story hotel on Harrison Avenue, which he called the Clarendon, a name reminiscent of English nobil-ity. And for its patrons, Bush's hotel drew from the ranks of those who fitted the American concept of nobility. On opening day most of its eighty sleeping rooms were reserved by senators, congressmen, Wall Street financiers, and Colorado mining moguls, including Dave Moffat and ex-Governor McCook. To sample the fare of Monsieur L. Lapierce, the Clarendon's chef, lately of Delmonico's in New York, the carbonate kings invited such notables as General Sherman, the Duke of Cumberland, Jay Gould, Commodore Vanderbilt, and of all incon-gruities, esthete Oscar Wilde.

On a lecture tour of America, Wilde was booked into Denver and Leadville by his world-renowned manager D'Oyly Carte. In Leadville, Wilde, whom the press pictured wearing his hair long and flowing over his shoulders, his legs encased in silken knee breeches, and carrying in his hand a single lily, was greeted with some deri-sion by the roughnecks of the silver camp. But the undercurrent of lewd remarks changed to handshakes and walloping slaps on his velveteen-jacketed back when he was invited by Tabor to view the underground workings of the Matchless and succeeded before a host of hard-rock, hard-liquor miners in putting away twelve straight and successive shots of venomous whiskey without so much as a hiccup or a stagger afterward.

The poet spoke from the stage of the three-story Tabor Opera House, another of the silver king's munificences to his town. It was located across the alley from the Clarendon and was connected to it by a covered passageway that allowed hotel guests and theater patrons to avoid the ankle deep mud of the alley and the slicing winter blizzards in passing to and from the theater. Its interior was the last word in theater de-cor, with frescoed ceilings, gilded moldings, and a view of the Royal Gorge hand painted on the drop curtain. For the opening, Tabor persuaded Jack Langrishe to return to Colorado from Chicago to begin a thirteen-week engagement of *The Serious Family* and a farce called *Who's Who*.

When Oscar Wilde was scheduled to appear on the stage of the Tabor Opera House he decided that since Leadville was populated by "miners, men working in metals" he would lecture them on the "Ethics of Art." As he wrote in his memoirs, "I read them passages from the autobiography of Benvenuto Cellini and they seemed much delighted. I was reproved by my hearers for not having brought him with me. I explained that he had been dead for some little time which elicited the enquiry, 'Who shot him?'"

True or not, Wilde's story was a fair comment on law and order in Leadville. As the silver camp's prosperity heightened, so did its crime rate. The Leadville *Chronicle* was proud to boast on February 25, a month after its first edition appeared, that "There is no city in the country of equal population to her own that can today show a lower criminal record than Leadville." So rapid was the acceleration of crime, however, that by April 24 of the same year, the editors were indignantly demanding, "Is there no law in Leadville?"

On the heels of the prospectors, there poured into town the usual regiment of gamblers, prostitutes, con men, and footpads. Flush with a week's wages and thirsty, many an unsuspecting miner betook himself to the nearest saloon on State Street where he found a girl of the line ready to pour him a drink. What he didn't know was that the bottle held a mixture of whiskey and snuff, guaranteed to cause racking nausea. Later, out back in Tiger Alley, the hapless miner, sodden and sick, was relieved of his pocketful of coin by his benefactress and left to sober up in the bleak cold of a Leadville night. Brawls and shootouts over the slightest issue were as common as the rats in the hay of the livery stable. Among the miners, who were mainly Irish and Cornish with a scattering of other nationalities represented, hereditary emnities burst into violence under the free and continuous flow of distilled spirits.

Red Dan Sullivan and his gang dressed to the nines and swaggered downtown after church on St. Patrick's Day, their derbies tipped over their foreheads, their thumbs hooked into vest pockets. When they got to their favorite intersection, the one with a saloon on each corner, Red Dan paused, while he decided which establishment to grace with his presence on this festive day. As he pondered the question, there lurched up to him Swede Berg, emboldened by a morning's head start at the tap and leeringly ready for a fight.

Berg glared at Red Dan. "St. Patrick was a son of a bitch," said he slowly and clearly.

Red Dan contemplated the weaving, heavy-lidded figure before him. "Berg, you can't blaspheme my patron saint," said the Irishman calmly. "I'll ask you to take back your words."

Berg took a deep breath and said again, "St. Patrick was a son of a bitch."

In a flash Red Dan's Sunday suit coat was off and the two men, roaring in fury, tore into each other with fists and feet and teeth. When it was over, and it was over soon, Berg lay gasping on the rutted street, his diaphragm ruptured. "Broke his wind," one of Red Dan's cronies announced with satisfaction. Berg recovered sufficiently to leave Leadville under his own power to seek a lower altitude where his permanently impaired breathing apparatus would be less taxed.

Often what started as a diversion led to death. On a nippy October day a gang of miners got together for an afternoon of boozing in a saloon up in Stray Horse Gulch. After a few belts one brawny mucker challenged another to a wrestling match. The miners formed a weaving ring of humanity around the pair and they went to it with snarls, bellows, and gouges. Who won was not clear-cut, and the barroom filled with the sound of agitated voices and scuffling as the match grew from a two-man encounter into a free-for-all. Soon knives flashed and gunsmoke clouded the room. After the smoke cleared, Michael Higgins, who had taken no part in the fracas, lay dead, struck by a wild shot.

Sneak thieves took everything they could lay their hands on from oysters to shrouds off the dead. Holdups and murders in broad daylight became an everyday occurrence. Si Minick, with a bellyful of 40-rod joy juice, blasted the head off a stranger and robbed his victim to buy one of the new sealskin coats he had seen displayed in Daniels and Fisher's store window. Roving gangs threw up barricades on the main roads into town and collected a toll at the point of a Colt revolver. The contagion of violence caught up even the most respectable Jim Bush, brother of the Clarendon's Bill Bush, discovered that Mortimer Arbuckle had jumped a lot the Bush brothers had sold to another man. Jim Bush peremptorily sought out Arbuckle and shot and killed the unarmed trespasser and was indicted for murder. So rife were murder and robbery that the *Chronicle* in a macabre vein ran a daily column called "Breakfast Bullets," tabulating the night's crimes. On one particularly busy morning-after, Davis and Arkins headlined a deadly catalog: "Murderous attack upon Kokomo frieghter," "Assault and robbery on Harrison Avenue," "A tenderfoot carroted on Capitol Hill," "Daring robbery of a man at the Comique," "Arrest of a notorious confidence man."

The climate was right for a vigilance committee, and it was not long in coming. The first act of the self-appointed custodians of Leadville's morality was to break into jail on November 20, 1879, and, holding Sheriff Watson at bay, drag out gang leader Frodsham and footpad Stewart and lynch them. The committee hung a note on the bodies, saying: "Notice to all lot thieves, bunko steerers, footpads, thieves, and chronic bondsmen for same...We mean business and let this be your last warning..."

Anti-vigilante townsmen sent lightly veiled threats of dire retribution to those they suspected of being among the lynch gang. Men of all stations strapped pistols on their hips before setting foot on Leadville streets, day or night. Miners going to and from work walked in groups of eight or more. And the camp slid into the uneasiness marked by armed guards posted at the main buildings—the post office, the telegraph office, and the jail. Traffic to Leadville came to a standstill.

After a hiatus of several days, Tabor and some of the other carbonate nabobs called in notorious gunfighter Mart Duggan, pinned a marshal's badge on his chest, and watched with grim satisfaction as he used his inimitably strong arm and a covey of henchmen to bring some semblance of order to the crime-ridden silver camp. Occasionally, Duggan's tactics got him the wrong man—once he mistakenly laid hands on Tabor himself and threatened to throw him into the pokey. And occasionally the mayhem he was hired to combat took a seriocomic turn—such as on the Saturday night when Johnny Boadman and his dance-hall girl were whooping it up in a State Street saloon. Between dances, John Appleby sauntered up to speak to Johnny's girl. At this point Boadman's normally sluggish senses were sufficiently primed with liquid fuel to ignite in a flash of fury. Grabbing the startled Appleby by the shoulders, Boadman proceeded to bite off the end of his antagonist's nose.

At the peak of Leadville's productivity, nearly a thousand roustabouts were kept busy around the clock hauling the ever-increasing tons of ore to the smelters ringed around the town. Like seemingly all else in Leadville's rise, it was pure luck that the ores of its silver mines lent themselves to comparatively easy and straightforward smelting in a blast furnace. The Leadville story was a far different one from the trial and error, boom and bust of an earlier day in Colorado when the lack of scientific knowledge thwarted miners' attempts to reduce the refractory gold ores of Gregory Gulch until Nathaniel Hill's research opened the door to success. Leadville's rapid rise to prosperity and sustained success was in no small part due to the arrival in camp, almost coincidentally with the first big silver discoveries, of two Freiburg-trained metallurgists, August R. Meyer and James B. Grant.

Meyer, an ore buyer who operated a sampling works in Alma, was hired by Edwin Harrison, president of the St. Louis Smelting and Refining Company, to make the taxing journey over thirteen-thousand-foot Mosquito Pass and report on the truth of the Stevens and Wood lead carbonate bonanza. Meyer got hold of some ox teams in New Mexico and dutifully sent a load of carbonate to Colorado Springs for shipment by rail to St. Louis. Harrison was unimpressed by the results of the first tests on the ore—the amount of silver in the samples did not pay even for the cost of freighting the ore. But Meyer was not discouraged. In the meantime, Fryer's New Discovery and the Morning Star and several other higher-grade lodes were uncovered, and the second batch of ore Meyer sent to St. Louis was sufficiently rich to prompt Harrison to put up the money for a smelter in Leadville. Meyer named it the Harrison Reduction Works, and in the spring of 1878 he blew in his first furnace. By fall another was in operation.

In October, just after Meyer blew in his second furnace, thirty-year-old James Grant arrived in camp armed with a first-class technical education and a $300,000 grubstake from a doting namesake uncle. Grant, who had studied the mining and ore-refining centers in Australia and New Zealand after taking post-graduate work at Freiburg, saw that Meyer's smelter would soon be swamped by the burgeoning production of the Leadville mines. So he used his uncle's money to build his own smelter, an eight-furnace giant, that within a few months after its opening was smelting the entire production of Tabor's Little Pittsburgh and a number of other steady producers. Within another year, fourteen smelters belched layers of sulphurous smoke over Leadville twenty-four hours a day.

Both Meyer and Grant took fortunes out of their smelters. With his, Meyer built a handsome three-story frame house on the Hill at the north end of Harrison Avenue. He furnished it in black walnut, diamond-dust mirrors, and Tiffany glass, and brought the bride he married in 1878 in the home of H. A. W. Tabor to live in it. After that the lives of the Meyers were notably uneventful. Eventually, with Leadville's decline Meyer shut down his smelter and moved his family to Kansas City where he spent a happy score of years before his death dealing out in philanthropies some of his Leadville wealth.

James Grant became a pillar of Leadville society and, at the same time, by virtue of his fair dealing and unassuming manner, he found many a friend among the rough-and-tumble miners. His plebeian ways paid off—in 1882, Grant became Colorado's first Democratic governor. In that same year Grant's smelter burned and he moved his operations to Denver where he erected a new and larger smelter. Later, in the days when gigantic trusts became the game of industrialists, Grant's company was one of twenty-three that were tied together to become the American Smelting and Refining Company. When the Guggenheims were asked to join the trust they demurred, at the same time promising to act "in harmony." They did the opposite, and in 1901 at the end of a Wall Street gavotte involving market glutting, the forcing of below-cost sales on the trust, and stock acquisitions, the Guggenheims came off the dance floor in control of the American Smelting and Refining Company. Once in the driver's seat of the juggernaut, the Guggenheims streamlined their new operations by shutting down a number of (comparatively) small smelters in the combine, including Grant's in Colorado, leaving to their own Pueblo plant a market in ore smelting free of any local competitors.

But long before that happened, while Meyer and Grant were still riding the crest of Leadville's ascendency, it became clear to them and the rest of the camp's businessmen that Leadville needed rail service as soon as possible. There were simply not enough local smelters to take care of all the ore coming out of Leadville's mines. Besides that, the maws of the blast furnaces required a source of fuel larger and more efficient than the charcoal made from pine trees logged in the gulches and hillsides around town. And despite their number, there were not enough freight wagons to haul the ore and ingots to distant smelters and refineries and keep the camp supplied with necessities to sell to clamoring newcomers and old-timers. Nor did the recognition of the need for a railroad lie exclusively with Leadvilleites.

Down in Denver, ex-Governor John Evans, an avowed railroad bug from the moment of his arrival in Colorado in 1862, tapped his pencil on his desk in anxious anticipation when the first assays of Leadville's high-grade ores were published. His Denver and South Park Railroad, chartered in early 1877, was already approaching Kenosha Pass on the northeastern brink of South Park. The planned route to the new silver town was across South Park over Trout Creek Pass to the Arkansas and up the river to Leadville. Evans' little narrow gauge road leaped forward as business boomed—people and products, all destined for Leadville rode the South Park to each new westward station and were relayed the rest of the way by stagecoach and freight wagon. But the real prize was the mountains of ore and ingots to be brought out of Tabor's town for smelting and refining. And the reason for Evans' impatience lay in the acquisitive maneuverings going on between his two rivals for the same prize, the Denver and Rio Grande and the Santa Fe, one or both of which could snatch the trophy from in front of the cowcatcher on Evans' straining locomotives.

The Denver and Rio Grande was the inspiration of William J. Palmer, a trim, small-statured, tightly wound dynamo of unassailable Philadelphia heritage whose teen years were spent studying railroading in England and in his home state. When the war between the states broke out, Palmer, despite his Quaker-fostered pacifism, set himself to raising and leading into the fray the Fifteenth Pennsylvania Volunteer Cavalry. Except for one blot on their scutcheon in the form of a spontaneous mutiny, quickly put down, Palmer's troops acquitted themselves with medal-catching ability—even to harrying Jefferson Davis through the Appalachian Mountains into the hands of federal forces in Georgia. The result was that Will Palmer returned from the service a brevet brigadier general at the buoyant age of thirty-two.

Once shed of his Union blues, the youthful hawk turned again to railroading, becoming the pathfinder for the westward-building Kansas Pacific which after fits and starts he brought into Denver in August of 1870.

When that job was finished Palmer decided to build his own railroad, a narrow-gauged one running north and south. Legend has it that his predilection for a narrow gauge was based on his profound conviction that the common practice of selling space in a berth for two occupants led to an immorality that could be avoided by using narrower cars in which only one person could fit into a berth. Having decided the size and the axis of his road, Palmer then turned to the last decision, one that was the easiest to make—where to locate his railroad. On his surveys for the KP and his own road, Palmer, like William Gilpin and many another, succumbed to the spell of the mountain West, recording his reaction to the stunning eminence of the Rockies and the clarity of the air in letters to Queen Mellon, his cultivated and fashionable fiancée on Long Island.

For 120 miles along the Arkansas river we drove along...generally within full view of the Rocky Mountains. The long ride was by moonlight. I spread out my blankets on the top of the coach back of the sociable driver and slept soundly in the fresh, keen air until awakened, perhaps by the round moon looking steadily in my face, when I found the magnificent Pike's Peak towering immediately above at an elevation of 14,000 feet and topped with a little snow. I could not sleep any more with all the panorama of the mountains gradually unrolling as the moon faded and the sun began to rise; but sleepy as I was, sat up and drank in, along with the drafts of pure air, the full exhilaration of that early morning ride.

At Colorado City...we stopped for breakfast. Near here are the finest soda springs and the most enticing scenery. I am sure there will be a famous resort here soon...

The reason for Palmer's certainty about a resort at the foot of Pikes Peak was because he would build it. As soon as he financed his railroad, most of it with English pounds, Palmer selected a site near the soda springs and the evocative spires of red sandstone known to travelers since the first days of the gold rush as the Garden of Gods, and formed the Colorado Springs Company. The express purpose of the company was to create for the passengers on the D&RG and, more importantly, for Palmer's bride-to-be, a Newport in the West, an oasis of civilized amenities in the midst of the savage reality of the raw frontier. The red-headed and determined little developer had far and away more success carrying out the original plan of his town than he did of his railroad, although both ventures turned out to be abundantly successful.

Palmer's narrow gauge had as its projected route a continuous rail link between Denver, El Paso, and Mexico City, and even, some said, there was floating around in the back of Palmer's agile mind the dream of an unbroken line from Denver to Mexico City, and through Central and South America to Buenos Aires. The first leg of this elaborate notion, from Denver to Colorado Springs, a distance of seventy miles, was completed in 1871. The next leg south reached Pueblo in 1872 and from then on the D&RG was, so to speak, sidetracked. Its ultimate destiny, ironically, was prescribed not by its board of directors, or the federal government, but by another railroad.

Palmer's north-south narrow gauge was not a real competitor to the east-west inclined broad-gauged Atchison, Topeka, and Sante Fe, but the directors of the latter had recurring nightmares over one large "What if." What if Palmer's track layers, once in New Mexico, suddenly veered westward and copped the Santa Fe's cherished route to the Pacific?

It was clear that the only way to preclude this eventuality was to slam shut in Palmer's face the only door through which his road could

proceed southward. That door was Raton Pass down on the Purga-
toire River on the border between Colorado and New Mexico. And
who held the key to the door? None other than that wily old trader,
Richens Lacey Wootton.

Uncle Dick Wootton had scuttled out of Denver in 1869 to spend
his retirement as the sole beneficiary of a toll road he built over the
pass and an adobe tavern he put up beside it. In 1878 as the two
railroad lines converged on Raton Pass, the D&RG from Denver and
the Santa Fe slanting westward through southeastern Colorado from
Topeka, Kansas, Uncle Dick found himself clinking glasses and crying *Salud!* more and more often with a slouch-hatted, black-draped
figure of a man who claimed to be a Basque sheep-herder. The two
got to be great pals as Uncle Dick spent countless hours exuberantly
banging his moccasined feet on the floor to the catchy fandango
rhythms of the herder's violin. And then there came a night when the
would-be herder threw off his costume and introduced himself as Ray
Morley, advance agent for the Santa Fe. Before Wootten finished emit-
ting bilingual oaths of astonishment, Morley asked him for the right-
of-way over his toll road in exchange for a hefty remuneration. The
agent explained that his abrupt unveiling and request was prompted
by word intercepted from coded telegrams that Palmer in the dead
of the moonless night had secretly dispatched a crew from Pueblo to
race down to Raton and take possession of the vital pass.

Uncle Dick was quick to make the choice between being bought
out and being run out. Without hesitation, he saw it Santa Fe's way.

The time was now four in the morning and in the cast stillness
that pervades the high desert just before the dawn, Morley, with a half
dozen of Wootton's boozy compadres, clambered up the caliche toll
road and in the light of a lantern drove the stakes that gave Santa Fe
control of the pass. And none too soon. Within a half hour Palmer's
grading crew arrived on the scene to find their southern exit closed
and locked and the key irretrievably secure in Santa Fe's pocket.

But Palmer's troubles with the cocky Santa Fe were not yet over.
With the way south apparently blocked, it looked for a time as if
the D&RG would live out its life as a short line between Denver and
Pueblo. And then the Leadville boom hit the headlines. Palmer, sitting
on the silver camp's doorstep with his three-foot track ideally suited
to mountain railroading, immediately set his gandy dancers heading
northwestward toward the headwaters of the Arkansas. But the Santa
Fe, drunk with its Raton Pass victory, had the gall to announce that
it, too, would build to Leadville posthaste.

Both lines logically chose to follow the cleft of the Arkansas River
to their destination. It was a feasible path with only one obstacle
lying in the way of parallel tracks—the canyon of the Royal Gorge,
so declivitous and narrow that there was room, and only barely, for
but one set of rails. Whose would it be? All Colorado, but especially

Leadville, watched to see which of the roads would win this battle. Particularly pleased at the turn of events was the South Park's John Evans. While his rivals fought it out among the boulders of the Royal Gorge and in the courts, his line pushed steadily closer to its goal.

While attorneys for both the Santa Fe and the D&RG marshaled their writs, subpoenas, and briefs to lay siege to the courts, down on the Arkansas River in the midst of the breathtaking scenery of the Royal Gorge the battle was joined in strict accordance with the precepts of warmaking. Chief Engineer De Remer of the D&RG led his armed track layers into the breach held by Ray Morley, again acting as Santa Fe's general, and his determined section crews. Behind hastily thrown up breastworks of river rock and mud, snarling groups of partisans kept oncoming track layers at bay with Colt repeating revolvers, while D&RG demolition teams made their way behind the lines to blow up track already in place. Snipers' bullets sent advance surveying parties diving for cover, and the nights were alive with hijackers who made off with blasting powder, tools, and grading equipment; arsonists who set storehouses and labor camps ablaze; and special forces men who deftly relocated survey stakes to confuse the enemy. Telegraph wires sang with fraudulent orders and counterorders as each side succeeded in breaking the code of the other.

To repel the D&RG encroachment, Santa Fe hired Bat Masterson and a motley gang of roughs recruited in the border towns of Kansas and Texas. Masterson's mercenaries, borne on an armored train, arrived in Pueblo bristling with arms and braggadocio and, to the Santa Fe's exasperation, promptly fell victim to the pleasures and pains of the myriad gambling and watering houses in town and were rendered somewhat less than useless in record time with the help of dealers, shills, bartenders, and soiled doves, all in the pay of the D&RG.

Topping off Santa Fe's keen disappointment at the demoralization of its troops, Masterson sold out to the D&RG in a clandestine exchange of money for loyalty, and the Battle of the Royal Gorge was over. It was not a military victory, but a negotiated peace. The courts, dealing in decisions and reversals with some resemblance to a switch engine shuttling cars in a railyard, finally declared the D&RG the winner of the right-of-way. Santa Fe might have accepted the decision with less grace had it not been that during the fracas, that master railroader of the period, Jay Gould, had quietly bought controlling interest in the D&RG. As it was, rather than risk the kind of retribution Gould could levy against their strongboxes the Santa Fe cheerfully accepted from the D&RG $1,400,000 for the track it had already laid in the Royal Gorge, and in return promised to forever forego Leadville. The D&RG did not get out scot-free, however. In addition to the cash, its penalty was to put aside Palmer's dream of a southern line to Mexico and agree not to build south of Colorado.

With the way cleared to Leadville, the D&RG wasted no time getting there, putting the first train, filled with dignitaries, including former president Grant, into the wildly expectant silver camp on July 23, 1880.

Throughout the battle taking place on the Arkansas, the South Park inched ever closer to its last barrier, Trout Creek Pass, before descending to the Arkansas and wheeling north to Leadville. But despite its rapid progress, the D&RG scotched the South Park's plans by attaining the prior right-of-way on the Arkansas north of Trout Creek Pass, and when the South Park got there in the fall of 1880, Evans and company had to eat crow and negotiate permission to run its cars into Leadville over the D&RG track.

Four years later the South Park, now also under the controlling thumb of Jay Gould, built into Leadville from the north, after Gould had a falling-out with Palmer and his D&RG friends. The Highline, as it was called, branched at Como to pass through Breckenridge, Kokomo, Robinson, and into the silver town. Although shorter and faster than the southern route it was still an all-day trip from Denver to Leadville. In the annals of Cloud City, the Highline is remembered for the sleety winter night when the little narrow gauge engine pulling a Barnum and Bailey Circus train suffered a steam-line break on a grade ten miles north of town. While the muffllered trainmen scratched their heads under their ear-flapped caps deciding what to do, the traveling manager of the Greatest Show on Earth rousted groggy mahouts out of their berths to unload the elephants and push the train over the rise so it could coast into Leadville.

In the spring of 1880, one of the South Park's more attractive passengers entrained for Leadville with high hopes and few regrets. Elizabeth McCourt Doe was free, smolderingly beautiful, and bursting with the vigor of her twenty-four years. The trip was maddeningly slow and there were long stops for water and fuel at Como and Fairplay, the once booming gold camps now suspended in mesmeric indolence. Not even the gorgeous mountain scenery and the wildflowers carpeting South Park could hold Baby Doe's attention for long. She was fussily impatient to get to the silver town and begin her new life, whatever form it would take.

After a day and a half, the train reached Buena Vista on the Arkansas River thirty-two miles south of Leadville, the end of the South Park's tracks. Here, Baby Doe's trunk and leather hat boxes were piled atop a stage, and along with the other road-weary, soot-covered train travelers she climbed into the coach for the rattling, jolting last miles to her destination.

Waiting for her at the stage depot was Jake Sandelowsky. As she stepped down from the stagecoach Jake's hand reached for her arm and they walked down the wooden sidewalk to his waiting buggy. Jake drove Baby Doe up and down the busy streets to Harrison

Avenue where he proudly showed her his new store, occupying the left ground-floor space of the grandiose and new Tabor Opera House. Baby Doe stared at the colorful awning.

"But, Jake," she said, "it says 'Sands, Pelton and Company'! What happened to my old friend Sandelowsky?"

Jake laughed as he expertly steered the buggy through the street crowded with delivery carts, ore wagons, and men on horseback. "Listen, Baby," he said, "everything happens so fast here no one would wait for me to spell out Sandelowsky, so I shortened it to Sands."

It was Baby Doe's turn to laugh and laugh she did, for the first time in many months, caught up by the lively atmosphere that greeted her on all sides of the boomtown. Here she saw what she had expected to find in Central City, but she and Harvey had arrived there too late to experience the scintillating life of a mine town in its heyday. Baby made up her mind not to miss her chance in Leadville.

Jake established her in a cheery room in a boarding house not far from the center of town. From the hall window she could see the snow-covered peaks of thirteen-thousand-foot Mt. Massive and Mt. Elbert standing high above the already towering Sawatch Range to the west. Central City's eight-thousand-foot altitude had been enough to make her giddy but Leadville's ten thousand feet combined with the rapid change in her fortunes made Baby Doe's first few weeks in town heady ones.

One of her first evenings with Jake was spent in a glorious whirl of color and gaiety at a ball given by the Tabor Light Cavalry. Nearly every mining magnate and syndicate in Leadville had, as a part of their operations, a private army. There were the Wolfe Tone Guards, the Carbonate Rifles, the Leadville Guards, and the Pitkin Cavalry, among others. How they came to be formed and why was manifold. Some said they were necessary to protect the populace and the mineral properties from the marauding Ute. But the likelihood of the Ute attacking a settlement as large as Leadville was extremely remote when the first of the mining militias was formed in July, 1879. There was some talk that they were insurance against lawlessness since Leadville's small custodial police force was woefully inadequate should a major civil uprising occur.

However the real and unspoken rationale for their existence was the constant threat of ownership disputes that could quickly fulminate into physical attack and the growing power and militancy of the miners' unions. But no matter what their purpose, it was a mark of achievement in an up and coming young man in Leadville to be invited to join one of these spirited companies.

Jake, perhaps through the auspices of his old friend from Central days, Bill Bush, was tapped for membership to Tabor's Light Cavalry soon after his arrival in Leadville. The Light Cavalry was one of two such armies Tabor maintained. His Highland Guards were the talk of the town with their frequent parades, sometimes with their leader at their head, resplendent in black doublets, Royal Stewart kilts, complete with Bonnie Prince Charlie bonnets on their heads and flopping goat hair sporrans hung from their waists. Somewhat more sedate but no less colorful were the uniforms of the Tabor Light Cavalry—scarlet trousers seamed with a gold stripe, royal blue tunics, and brass helmets fitted with dashing chin straps worn English style, resting just below the lower lip.

It may have been at one of the parades of the Tabor militias that Baby Doe first saw the silver king himself. Tabor frequently donned the uniform of one or the other of his armies and marched proudly at its head, his powerfully built, six-foot frame swinging in precise time to the music of the band, his penetrating dark eyes darting from side to side as he acknowledged the applause of the crowd of onlookers with a quick smile showing tobacco-stained teeth beneath his long-handled mustache.

Leadville was full of diversions and Jake saw to it that Baby Doe sampled all of them. On Sunday afternoons they attended races staged by Bill Bush's Leadville Trotting and Running Association. Some evenings they strolled arm in arm down raucous State Street, a show in itself. In the glare of countless gas lights, crowds of flannel-clad miners, teamsters, smelter workers, neatly suited merchants, bonanza barons, and professional men milled to and fro. Occasionally, the crowd moved aside to allow a parlor house madam, lustrous in satin and paste pendants, to trot her span of matched grays down the street trailing a cloud of smoke from the long black cigar in her scarlet-painted mouth. Outside every variety hall brass bands rent the air with insistent percussive clangor as barkers clutched at passersby shouting promises of "cancans, female bathers, daring tumblers, and other dramatic attractions." Jake steered the fascinated Baby Doe past the one-girl cribs, the parlor houses, past the office of a practitioner of phrenology who for $2 would feel the bumps on his subject's head and come up with the description of a suitable mate. They made their way beyond the tattoo parlor, past the whooping, sour-smelling Bucket of Blood and the Pioneer, two of Leadville's more notorious saloons, to Mabel River's Athenaeum, an amphitheater lit by two huge bonfires blazing away in iron tanks suspended at roof level, where animal acts and lady wrestlers drew hundreds of spectators every night. Although the only other women Baby Doe saw in such places were those of the demimonde, she was not embarrassed. Baby Doe had no illusions about her status. As a divorced woman, she was considered by those in polite society to be scarcely a rung above a prostitute. Even actresses ranked above divorcées. One of Leadville's young bachelors was openly applauded in the *Chronicle's* society page for courting the charming leading lady at Wood's Theater. But none of Leadville's "eligible eight," as the press referred to a foot-loose but well-heeled group of young men in town, called on Baby Doe. Not that she wanted them to. For the moment, Jake filled the bill. He was attentive, generous, and kind, and what pleased her most was that he knew how to make her forget her abortive life with Harvey Doe. Jake knew she was game to go anywhere, anytime for a good time, but her preference was the theater.

After a sumptuous dinner of oysters and mountain quail at the posh Tontine or the Saddle Rock on Harrison Avenue, Jake often squired his inamorata to the Grand Central Theater to sit cozily in one of the secluded $5 boxes. Other nights they went to the Tabor Opera House to hear Charles Vivian, the touring English ballad singer and story teller all Leadville had taken to its heart.

Charles Algernon Sydney Vivian had come to New York in 1867 at the age of twenty-one with a shilling or two in his pocket and an urge to entertain in his soul. Wandering one Saturday night into John Ireland's Chop House, the animated hangout of amateur and semiprofessional actors, he gave an impromptu performance that was heard by the manager of the nearby American Theater who signed Vivian that night for a long and profitable run. His career launched, Vivian decided the next day to celebrate with his friends, but to his dismay he found that under New York's blue laws all the saloons were closed on Sundays. On Monday, he gathered his friends together and formed a club for the purpose of furthering good fellowship, especially on Sundays. Vivian and his companions called themselves the "Jolly Corkers," and within a short time the membership grew by giant steps. When one of the Jolly Corkers died, leaving his family destitute, the members decided the time had come to add good works to good fellowship, and out of the redirected and reorganized club grew the Benevolent and Protective Order of Elks.

When Vivian came to Leadville, his tickling patter, rollicking stories, and endless repertory of songs, some poignantly sentimental, others devastatingly comic, caused such a clamor for his performances that he extended his stay. It was a fatal extension. Within a few weeks, late in March, 1880, Vivian died suddenly of "quick pneumonia," a virulent disease that struck down many a newcomer to Leadville's high, cold altitude, caused some said by breathing the cold air that froze the top portions of a man's lungs. His funeral was held in the Tabor Opera House where only a few nights before the popular singer had made the rafters reverberate with thunderous applause. Afterward, hundreds of men and women lined Harrison Avenue as Vivian's casket was placed in a cortege of black-draped carriages for the trip to the cemetery. A band took its place at the head of the cortege and the procession started forward to the slow cadence of the "Dead March" from *Saul*. When the caravan reached the end of the avenue, the band suddenly struck up the stirring notes of "Ten Thousand Miles Away," and among the mourners—gruff miners, flighty matrons, stoic gamblers, bland-faced businessmen, and satiated harlots—unchecked tears welled from unblinking eyes as the strains of Vivian's favorite ballad drifted hauntingly downwind.

❧⁘☙

The outward signs of Leadville's sadness at the passing of its favorite entertainer soon disappeared. Life was too full and too fast for the ballooning population to tarry long in grief. Nearly forty thousand people were crowded into the camp by the time Baby Doe arrived in the spring of 1880. As the mining payroll rose to an astronomical $800,000 a month, Leadville took on the aspects of a full-fledged city. Entire blocks of substantial multistory brick buildings lined its major streets. Corner blocks sold for $10,000. The town's five banks were "over-run with deposits." Three newspapers hit the stands daily. The Clarendon Hotel, shortly to have in the Hotel Vendome an imposing competitor financed by the ubiquitous Tabor, was booked solidly for months ahead. The May Company and the Daniels and Fisher Company department stores offered to Leadville shoppers French plate hand mirrors, Balbriggan and silk hose, and the finest carpeting and yard goods available west of the Mississippi. From Charles Boettcher's hardware store, the largest in all of Colorado, miners could buy everything from rock drills to windlasses, from bellows to wheelbarrows, and no item of household hardware was too small to be stocked. Seven churches and seven schools, including a school of mines, channeled the minds of the young toward the rewards of learning, labor, and the love of God. Their efforts, however, were often mitigated by the influence of what the *Chronicle* counted as Leadville's 120 saloons, 110 beer gardens, 118 gambling hells, and 35 seraglios.

One of the more popular of the 118 gambling hells was Pap Wyman's, strategically located at the corner of Harrison and State streets, where he could draw simultaneously from the highest and the lowest elements in town. Wyman was the antithesis of the usual run-of-the-mill gambler. He gave orders that no drunk would be served a drink and no married man could play the tables. On the wall hung a sign stating, "PLEASE DON'T SWEAR." Over the band pit was a notice reading, "DON'T SHOOT THE PIANIST—HE'S DOING HIS DAMNDEST," which Oscar Wilde remarked was the most rational bit of art criticism he had ever read.

Wyman's was a favorite haunt of bachelor Jake Sands, and on the few evenings he did not spend with Baby Doe he could be found in deep concentration over a poker hand in the teeming room of gamesters. It was on one of these evenings, so the story goes, that Baby Doe went alone to the Saddle Rock Cafe and happened to meet the idol of her musing hours, the silver king himself, H. A. W. Tabor. How the introduction came about is not a matter of record. Perhaps, as some suggest, it was through the intercession of Bill Bush who had known

Baby Doe in Central City. In any case, meet they did, and Tabor fell hard for the earthy-voiced, winsome, and willing young divorcée, and within a matter of days Baby Doe became his mistress. She seemed not to be bothered that two of Tabor's earlier flames thought him "exceedingly vulgar" and could bear his presence only if they were blind drunk. As compensation for the loss of her affection Jake Sands accepted a diamond ring from Baby Doe, a ring paid for with Tabor silver, and with this gift the clothier's broken heart appeared to mend with no obvious scar.

Some time before he met Baby Doe, Tabor had shipped Augusta and their son down to Denver to inhabit the $60,000 Italian villa he built in a fashionable quarter of the capital to go with the office of lieutenant governor his new millions had secured for him. This left him free to fraternize with all the Leadville dancing girls, shady ladies, and actresses he cared to without having to face the steely cold eyes of Augusta peering reproachfully at him through blue-tinted pince nez. When he met Baby Doe, Tabor dumped his current plaything, Alice Morgan, an Indian club virtuoso at the Grand Central Theater, and moved his new love from the little boarding house room to the fanciest suite Billy Bush's Clarendon could provide. There, in the midst of the profligate excesses of the boom camp, their infatuation grew to a deep devotion as Tabor wove a rich fabric of reality and dreams to be ever more sustained by the erecting hordes of silver from his mines.

On the surface in the early summer of 1880, Leadville appeared to be rolling along in the harmonious meshing of capital and labor, supply and demand. But already, after only three years of boom times, there were forces at work that would shatter the happy-go-lucky prosperity of the Cloud City.

As production climbed, men flocked to Leadville seeking jobs, and in 1880 there was suddenly a surplus labor market in the camp. When the mine-owners realized they had a glut of workmen on their hands they took advantage of the situation and lowered the daily wage from $4 to $3. The gamble brought swift reaction among the miners. Led by union organizer Michael Mooney, the men of the night shift of the Chrysolite barred the way when the day shift reported for work at 7 AM on May 26. Their demands were straightforward, a return to a $4 a day wage, the right to choose their own shift bosses, and a maximum shift of eight hours. George Daly, manager of the Chrysolite, refused their demands but he promised to relay them to the mine-owners in New York. While the telegraph wires to Denver and the East Coast hummed with communiques between Daly and the owners of the Chrysolite group, Mooney and his men were joined by fifteen hundred other miners from Leadville's largest mines, the Little Chief, the Morning Star and Evening Star, and the Robert E. Lee. The strikers then marched from mine to mine, where Mooney gave impromptu and impassioned speeches encouraging others to join them. At the large mines his efforts were rewarded and new recruits joined the walkout, but at the smaller mines, where much of the work was done by a lessee himself and a small crew and where the wage system was an apportionment of the profit on the basis of shares, few were willing to support the daily-wage miners.

After Mooney's demands were presented to the owners of the struck mines, they categorically refused to discuss them and they ordered their managers to close down the mines altogether. When he heard about this move, Mooney threatened to wreck the pumps and let the mines flood. Daly and the other managers involved countered by asking for help from the Tabor Light Cavalry, the Carbonate Rifles, and the Wolfe Tone Guards. When the mines were fortified and garrisoned with the owners' private armies, the Chrysolite and Little Chief opened on May 31 with a skeleton crew of fifty miners who were not on strike, By now, Mooney's men were enraged and ready for battle. They massed at the mine portals, shouting oaths and taunts at the guards. They roamed through the streets of Leadville in defiance of sheriff's orders, hurling bottles and rocks through windows of the offices of the newspapers who sided with the mine-owners, and they marched up to Capitol Hill to the homes of mine-managers chanting slogans of vengeance. Leaving his post as president of the state senate, H. A. W. Tabor rushed up from Denver when he was told of the trouble. When they heard he was coming many of the strikers took hope. Up to now Tabor had been the miners' friend and his influence as the biggest of the carbonate kings was prodigious. But Tabor disappointed the hopeful. Finding Leadville on the brink of anarchy, he quickly helped organize a vigilance committee and with twelve hundred volunteers who were tired of the climate of ugliness and violence staged a show of force that made Mooney and his men soften their demands. They withdrew their request for the right to name their own shift bosses and they dropped their wage demand from $4 to $3.50 a day. But capital was not ready to negotiate its problems with labor, and on June 13, at Tabor's urging, Governor Pitkin ordered out the cavalry and telegraphed the beleaguered camp that "Lake County is declared under martial law." In the face of a town full of impassive, tough, trigger-ready troopers, the strikers beat a full retreat and the strike collapsed. With its collapse came a schism between large syndicated mine ownership and the mass of the mine laboring force that, like the schism between the white man and the Indian, need not have been created had there been less rigidity of thinking, less resort to force, and more honest discussion among the principals involved.

As a footnote to Leadville's decline, in 1896, the strike scene was repeated with even more ugly and tragic results. When the silver market crash came in 1893, miners working in the mines still in operation again took a cut in pay to $2.50 a day. But in three years some mines with higher-grade ore were again paying $3 a day. Union

agitators, grasping at the opportunity to strike a blow for labor, however fruitless and illogical, moved into Leadville and got a large group
of miners to strike to equalize the pay rates at all mines at $3 a day.
Whipped into a frenzy of righteous self-interest by shouting, fist-
pounding leaders, the miners staged a senseless raid, setting fire to
the hoist house and warehouses of the Coronado, a mine close to
downtown Leadville at the foot of Fryer Hill. In the shooting that
followed between mine guards and strikers, three miners were killed
and Fire Chief O'Keefe was fatally holed when he turned on the water
to douse the flames while his firemen, who allied themselves with the
strikers, stood by steadfastly refusing to fight the blaze. After weeks
of verbal and physical skirmishes that benefited neither side, the
strike dribbled away as mine after mine was abandoned when owners
found them flooded and lacked the capital to drain and reopen them.
Miners then turned on the agitators and summarily ran them out of
the dying camp. Wages stayed at $2.50, and hundreds of miners were
out of work.

But it was the strike of 1880 that marked the real beginning of
Leadville's glissade to hard times. In the aftermath of the upheaval
between mine-owners and miners, it came to light that a number
of mines with heavy production had been so poorly managed and
so badly exploited that they were close to exhaustion. In the face
of dwindling ore production, some owners had borrowed heavily to
keep dividends high, certain that new veins would shortly be opened
to vindicate the risk they took in mortgaging their properties. Caught
in the squeeze of rumor and revelation that swept Leadville were
Dave Moffat and Jerome Chaffee. A stockholder's investigation of
rumors that their Little Pittsburgh group was pinching out revealed
that the story was true. Moffat and Chaffee had been so busy taking
out ore they failed to conduct a concerted program of exploration and
development and consequently there were no new lodes waiting to be
mined. On the New York Exchange, Little Pittsburgh stock slipped
from $40 to $15 to $6 a share. Panic swept investors in Leadville
mines, and thousands of shares of silver stock flooded and broke the
market. In Leadville, one mine after another shut down, some never
to reopen. Five banks failed in four years. And when the Sherman
Silver Purchase Act was repealed in 1893, the once lusty Leadville
foundered on the very carbonate rock that had launched the camp on
its road to fame.

Long before the end came, however, some, like H. A. W. Tabor, had
already abandoned ship. Tabor had his course set for a farther shore
than the smoky, sulphurous little world of Leadville. From his Kansas
Free Soil days, Tabor had nourished a yen for politics, and with the silver lining he had steadily been giving the collection box of Colorado's
Republican party, he could think of no reason why greater offices than
the lieutenant governorship should not come his way. To pursue this

reasoning he moved his base of operations down to Denver where he could be in close touch with the machinations of government.

Other carbonate kings and merchant princes also forsook the dying town for Denver and elsewhere. James Grant built a bigger and better smelter in the capital in 1882. Peter Breene, who went from turning a windlass to wealth overnight, took up banking down on Cherry Creek. Sam Newhouse turned his riches into a copper empire and went pub-crawling annually in the company of an incognito Prince of Wales. Charles Boettcher added to his pile with a sound foundation in Portland cement on the outskirts of Denver, while John Campion stirred some of his millions into the first sugar beet industry in western Colorado.

One of those who stayed and whose presence embodied the rise and fall of the silver camp was Abe Lee. It was his lucky pan of gold flake that first opened California Gulch in 1861, and, after squandering two fortunes, one in gold and one in silver, Abe Lee became a derelict around town who made a few pennies a day selling water from a barrel to Leadville housewives.

In the late nineties Leadville shook loose from its somnolence when James J. Brown, managing the Little Jonny, turned up an abundance of sheet gold among the lead-silver ores that was so brittle it could be pried off the walls of the stope with a screwdriver. Out of this short-lived gold boom came to prominence two notable curiosities— Brown's wife, the legendary Unsinkable Molly, and the Ice Palace.

Margaret Tobin was fifteen when she rumbled into Leadville in a buckboard. She had hitched a ride with her brothers from their home in Hannibal, Missouri. Having their kid sister along didn't slow down the Tobin boys; Molly, as she called herself, could swing an ax and shoot a rifle as well as any of them, and what's more she could cook. But Molly had no intention of keeping house for a pack of brothers when she could do the same thing for a husband, with greater reward. When the saucy, titian-haired tomboy laid her Irish blue eyes on Leadville Johnny Brown, the name the town gave the happy-go-lucky Brown, his days as a bachelor were numbered. Molly became Mrs. J. J. Brown at sixteen, and quickly set about to plan a life of luxury and ease made possible with the gold pouring out of the Little Jonny mine.

She very nearly muffed her chance for the good life, however, when she accidentally set fire to $300,000 in greenbacks Johnny had stored for safekeeping in the potbellied stove of their cabin. But Johnny went back to work and duplicated the lost sum with much more to spare. It was Molly's deepest yearning to become a woman of culture. There had been neither the money nor the opportunity on the river bottom homestead in Hannibal to make a start, but now Molly made up for lost time. Johnny and she moved into a spacious and imposing mansion in Denver where Molly immediately began the study of art, music, and languages. Sandwiched between lessons were trips to Europe to practice her new polish. Johnny went along with her and watched indulgently as his impetuous, people-loving wife gathered a circle of foreign friends, all claiming titles, who were more than willing to come to Denver on extended visits to share home, table, and wine cellar of the J. J. Browns. Denver, however, was not so indulgent. To Molly's chagrin no one of the capital's new society invited the Browns to their dinner parties. In time, even Johnny tired of Molly's frantic cultural kick. He saw no reason to deny his humble Pennsylvania origins nor to pretend he liked sipping champagne in a damask-walled drawing room listening to the strains of a harp better than he liked downing a shot glass of whiskey in a Leadville saloon with the notes of a rinky-dink piano in his ears.

Shortly, Molly and Johnny separated, she to beat a well-trodden path between Denver and the playgrounds of Europe, and he to live in Leadville in the surroundings in which he was most at ease. Little or nothing else was heard from Johnny Brown, but not so his wife. Denver society savored as dinner table gossip the latest stories of Molly's flashy attempts to emulate her betters, stories picked up and passed over the back fences of Capitol Hill by grooms and sec-ond-story maids. Yet for all their derision, the Denver nabobs had to acknowledge that it was Molly Brown who was a mainstay in keeping the city's charities solvent, and they had to admire their loquacious neighbor when after the "Ludlow massacre," in which a number of striking coal miners in southern Colorado were shot and killed by the state militia, she rushed home from Cannes to storm the battleground bringing food, clothing, and money for the relief of the widows and orphans. Intercepted in Denver between trains en route from New York to the Colorado coalfields, Molly was asked by the press if, in view of the environment of hostility, it was prudent to risk going to the scene of the carnage.

"Of course it is!" cried Molly as she swung aboard her south-bound pullman. "To hell with the danger or the cost!"

Molly Brown would live to become accepted by Denver's blue bloods but it would take a human disaster of epic proportions to do it.

During the gold boom started by Leadville Johnny Brown, the merchants in camp, aiming to get the most out of the new period of prosperity, organized a Crystal Carnival in 1895. On five acres of flat-land they built a medieval castle all of blocks of ice with walls eight feet thick and fifty feet high. On the inside there were amusements to please every taste—dancing, skating, curio shops, restaurants, peep shows, ice statuary, and ore displays. The Ice Palace lasted for one year before it melted, and thousands of visitors rode the D&RG up to Leadville on a round-trip ticket that included admission to the mid-way of the world's largest icebox.

Again, declining ore prices, pinching out of veins, and the flood-ing of many mines brought an end to the camp's gold rush. In the

1930s, when large molybdenum deposits ten miles north of town at Climax were opened for mining, there was a moderate increase in population and prosperity, but never again did Leadville see the carefree, high-spending, fast-living days of its glamorous silver seventies.

11

The Carbonate Trail

FOR THOSE WHO TUMBLED INTO LEADVILLE LOOKING FOR THE HERALDED INSTANT riches only to find likely sites already overrun with prospectors, there was nothing to do but push on, following the trail of carbonate of lead that outcropped around Leadville in a wide, irregular semicircle.

Southwest of Leadville, over Cottonwood Pass, came a stampede of miners in 1878 to an abandoned gold camp Jim Taylor had named Tin Cup because he struck pay dirt in a nearby creek using his drinking cup as a gold pan. The reborn camp swiftly got a reputation for lawlessness. In its three headlong years of life Tin Cup had no less than five town marshals, necessitated by what appeared to be a year-round open season on lawmen.

In 1879, gophering silver-hunters trekked westward over the ever present mountains to a small meadow tucked among snowcapped peaks. Here in July came to life the camp of Gothic, flaunting a hundred tents and cabins by the time it was a week old. All around carbonate claims turned lucky finders to rich men. Jim Jennings led the list of the lucky, uncovering deposits of pure silver in the form of tightly tangled wire running $15,000 to the ton. Carefree, high-living Gothic whooped it up in its two dance halls on Saturday nights and looked with benign approval on the method adopted by rival candidates for determining which one was to be mayor: a roll of the dice, high man wins.

Over Elk Mountain another camp grew up to rival booming Gothic. Irwin was prospected in the height of winter, the deep snowpack no deterrent to a silver-hungry prospector. When the first pockets of ore were discovered on the mountainsides, the fortune-hunters moved down to the narrow little valley to build their cabins with the tidy townmaking instinct that ran through settlers of the West. The first thing they did was to clear the area by lopping off the trees at the level of the snow pack. Then they stripped the logs of branches and put together cabins. By the time the snow melted in the spring, so old-timers recalled, the

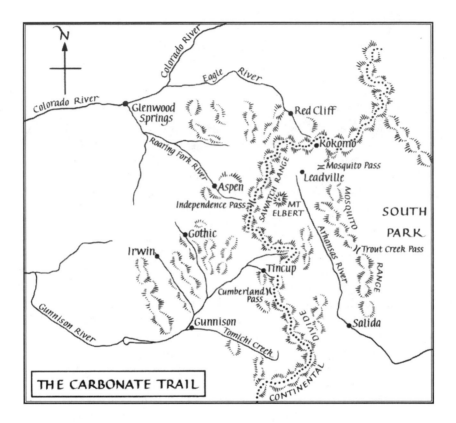

THE CARBONATE TRAIL

cabins had taken on a peculiarly topsy-turvy look as they gradually sank some ten feet through the melting snow to the ground. No transport, not even a surefooted burro, could reach these crow's nest camps until the thaw and it was a happy day when the first gray-furred, long-eared beast made its way up the trail in the late spring.

By 1881, Irwin consisted of three streets running side by side up the gulch, lined with false-fronted buildings, tents, and log cabins housing two hotels, the inevitable theater and gambling dens, twenty-three saloons, one bank, one newspaper, and a dozen or so mercantile houses. Only after the primary needs of the population were taken care of, did the builders turn their attention to the erection of a church and a school.

One spring day Irwin was thrown into a dither when no less a personage than General Grant rode the serpentine trail up to camp to see for himself the specimens of ruby horn silver coming out of the Forest Queen, Ruby Chief, and King Bullion mines. After a tour of the area, the miners conducted their illustrious visitor to a table laden with spitted venison and currant cake set in a clapboard building housing the Irwin Club. Exclusively a retreat for the males in town, the club was a source of infinite suspicion among the women in camp. They were convinced gross licentiousness held sway behind the windowless walls and doors of the building. The clothes-line gossip became so preposterous that the miners one day decided to purge the prurient thoughts from the minds of their womenfolk by staging a grand ball to show them that the Irwin Club was not a hurdy-gurdy house but a place to which a man could repair to seek the society of his own sex and speak of weighty matters of business without burdening or boring the ladies. Flattered by the invitation, the women in town searched through trunks for long forgotten gowns of grosgrain and mull and thumbed through well worn pages of *Godey's Ladies' Book* searching for pictures of the latest hairstyles. When the much-talked-of evening came a steady procession of miners, ungracefully clad in armorial suits of broadcloth and boiled shirtfronts and with stylishly dressed wives and sweethearts on their arms, made their way over the rutted streets to the clubhouse. When the crowd had all assembled the fiddler rapped his bow on the back of his fiddle, stamped his foot, and swung into the first quadrille. As feet flew over the dance floor first one and then another of the ladies, with stricken looks on their faces, begged to be excused and dashed into an anteroom to tear at bodices and petticoats to scratch skin seemingly set afire with an excruciating itch. On the dance floor soon the miners too felt a merciless stinging beneath their brushed serge and broadcloth and clawed at their clothes to scratch. In a matter of minutes the club was deserted as enraged females stalked off home with as much dignity as they could muster under the circumstances, followed by abject club members whose loudly vocal apologies were regularly interrupted by the need to stop and scratch. Smarting, in more ways than one, under what they considered gross insult, the ladies of Irwin were irrevocably convinced of the despicability of the club and its members. Had he been discovered, the town mysogynist who sprinkled itching powder on the dance floor would have lived to feel a prickling sensation he would have long remembered.

In the summer, miners treated themselves to a trip down-gulch in one of Sanderson's bumptious coaches to Gunnison on Tomichi Creek. Set at the junction of two watercourses in a narrow fertile valley, Gunnison was a natural crossroads and soon became the supply and transport center for the scores of silver camps nested around it. First surveyed in 1853 by Army captain John Gunnison, the valley some twenty years later became the apple of the eye of Professor Sylvester Richardson, geologist, physician, playwright, lawyer, carpenter, journalist, assayer, and a member in good standing in the band of intrepid townmakers who peopled the West. Drawn to this rich, grass-covered bottomland by the westward movement of prospectors to the surrounding Elk and Sawatch mountains, Richardson took one look at the valley and immediately saw in his mind's eye a vision of a town bustling with commerce to serve nearby mines and ranches. In 1874 he formed a town company with thirty eager takers buying up two shares, each at $100 apiece. Carried along by the contagious exhilaration of the mass of newcomers, the professor plunged into any enterprise that captured his fancy and in all of them he failed with a flair that would have soon elevated a less conscientious man to public office. He built a brick plant that limped along on kilns that temperamentally turned out bricks that crumbled to the touch. He sought in vain for the right pigments in Gunnison's soil for his paint factory. He experimented for years on soil compositions to develop a local concrete, but every heralded exhibition of his new product ended in total failure, and Richardson soon came to be companionably called "Old Cement" by Gunnisonites. He built a toll road for $2,000 and sold it to Otto Mears for $1,000, and last but not least he started a newspaper, the Gunnison *Sun,* that set firmly and finally after one spirited election season when the hard-luck professor vociferously backed for sheriff a man wanted by the law in another state.

Despite Richardson's failures, his town boomed. Mullin and Williard built La Veta Hotel, an elegant manse of "Queen Anne and modern architecture," rising four stories to tower above the rest of Gunnison's modest buildings, "a peacock among a lot of mudhens." The hostelry celebrated its opening with a banquet and ball in May, 1884. A reporter's review of the affair concluded that among the invitees, the graceful art of dancing "might well be called the 'lost art.'" But the disapproval by the critics of their terpsichorean technique by no means dampened the town's delight in the dance as a form of entertainment. To meet the challenge of demand, Fat Jack's Amusement

Palace advertised that it guaranteed to supply "sacred music to dance to on Sunday evenings."

One of Gunnison's favorite big spenders was (honorary) Colonel Jack Haverly, mining magnate and minstrel show king. With some of his theater receipts, he developed the Bullion King mine, financed Gunnison's first newspaper, and scattered the rest over the gaming tables in unstudied abandon. On his first evening in Gunnison, Haverly sauntered up to a faro dealer.

"What's the limit?" he asked.

Sizing up his sucker as a greenhorn, the dealer cavalierly answered, "The sky."

When the dealer saw Haverly pull out a roll of hundred-dollar bills and peal off a handful, he quickly grew respectful.

"S's's'sorry, sir," he stuttered, "being its kind of cloudy tonight, the sky over Gunnison is only twenty-five bucks high."

Over the mountains to the east, the frenzy to invest in the new silver region spread like wildfire and Gunnison was overrun with investors all eager to get their thumbs in the pie. Among them was ex-Governor John Evans, up to his favorite retirement pastime of railroading. As president of the Denver and South Park, his mouth watered at the thought of his narrow gauge being the first to tap the lucrative ore trade of the new district. With his line already at Buena Vista, Evans confidently assured a committee of mining men that only the South Park would be able to buck its way over the continental divide to satisfy their needs. He was patently wrong.

The D&RG, after winning a doughty battle fought against the AT&SF in the courts and behind granite boulders of the Royal Gorge of the Arkansas for the right-of-way through that scenic channel and engineer's nightmare, set rails posthaste for Gunnison. Meanwhile, the South Park resolutely surveyed its route and included in its plans an 1,800-foot tunnel from the headwaters of Chalk Creek through the divide to the headwaters of Quartz Creek. Work began on the tunnel, aptly named Alpine, at an altitude of 11,500 feet in January, 1880. Engineering estimates gave the following June as a completion date. But nature, with its prodigious Rocky Mountain winters, dictated the actual finish date—one year and seven months after the first hunk of rock was blasted from the curving entrances of the tunnel portals. Nevertheless, despite blizzards and snowslides, not a day passed that work did not go forward, and on July 26, 1881, the east-side crew broke through to meet the crew working from the west. So precise was the engineering that the alignment of the two headings was within 11/100ths of a foot. Some 460,000 board feet of California redwood went into the timbering of the sixteen-foot-wide tunnel, and when it was all finished John Evans had to face the unpalatable fact that the D&RG had beat him into Gunnison by four profitable months.

Gunnison did not depend for its survival on mining alone. The town became an important trading and supply center for the scores of ranches that grew up around it during the bonanza days of mining. And when the money-metal bust flung the mountain towns into a decline from which they never recovered, Gunnison rode out the storm to settle into a quiet and long-lived prosperity.

More than anything else in the world, Henry Gillespie wanted to strike it rich. He had been bitten by the gold bug back in the 1860s when as a boy in Manhattan, Kansas, he ogled with envy the passing parade of emigrants streaming west under the cry of "Pikes Peak or Bust!" Every time he saw the Leavenworth and Pikes Peak stage thunder through town his ache to follow it to the shining mountains grew worse. But Henry was not a reckless individual. He stayed put until he graduated from Kansas Agricultural College and then he went to work for the Kansas Pacific railroad building westward to Colorado as fast as its borrowed money and the guiding hand of General William J. Palmer could get there. By the time Gillespie got to Colorado, the gold rush was over and silver was the commodity that lit a new fire in the eyes of prospectors all over the territory. Gold or silver, Gillespie didn't care—to live in a bonanza environment and perchance to turn up one for himself was all he sought. It was to A. E. Breed's Caribou up in Boulder that he went first to take a mundane but well-paying job as bookkeeper. The profits he recorded in the ledgers were not his, but he was satisfied just to write down the symbols for amounts of money running to six and seven figures and on his day off to poke around the gulches looking for color. When Breed sold the Caribou in 1873, Gillespie was out of a job and for four dismal years he went back to railroading, working as a time-keeper and as a yardmaster. Then the great Leadville strikes came and Henry hurried to the new camp. Opening a mining exchange, the conservative Kansan waited for the claim that would make him rich to come across his desk. But for two years none came his way, and for a time it looked as if Gillespie was destined to be no more than a bystander in the bonanza West. Then in October of 1879, Bill Hopkins, a prospector whose scraggy beard and tattered clothes marked him as having been away for a time from Leadville's sartorial amenities, stumped into Gillespie's office with a tale of discovery that brought the tall young broker swiftly to his feet.

Pointing out the window to the snowcapped Sawatch range to the west, Hopkins told how he and two other men had fought their way up and over the twelve-thousand-foot pass above the tiny gold camp of Independence and down a tumbling river they named the Roaring Fork to a broad basin where silver outcropped in regular intervals around a spring at the foot of one of the mountains girdling the basin. As soon as they had staked out claims, a party of hunters happened by with the word of Meeker's massacre and the Ute uprising. After the hunters moved on, the trio of prospectors decided that Hopkins should race back to Leadville to record their finds and have their rock samples assayed, while the remaining two men stayed to defend their claims and themselves in case the Ute discovered them. When Hopkins found they had struck high-grade ore he came to Gillespie so the broker could spread the word and encourage miners to follow the Roaring Fork to Hopkins' diggings. The more white men there were around the new site the more protection there was against the rampaging Ute. During the time that Hopkins spun out his story, Gillispie's mind was busy. Here was his own opportunity and he not only spread the word but led the rush. With the money his brokerage brought in, Gillespie bought tools and a burro and headed back over Independence Pass with Hopkins.

On the way, the two men heard to their relief that through Chief Ouray's efforts the northern Ute were pacified and the danger of Indian attack was over. When they reached their destination, Gillespie took one look at the ore in place in the two mines Hopkins' partners were working and bought both claims, the Spar and the Galena, for $25,000. When the transaction was completed it took the young broker but a short time to see that the key to his fortune lay in extensive development of his mines and that, he knew, took cash, lots of it. It then occurred to him that not only did his mines require development, but the surrounding beauty of the basin and the mineral riches held in its mountains was a combination that urged the development of a town.

While Gillespie let his visions multiply on a grand scale, scores of other hopefuls crowded along the narrow trail from Leadville to the new diggings. They were handsomely rewarded for their trouble. On Aspen and Smuggler mountains, mines soon abounded: the Mollie Gibson coughed up ore assaying at 3,300 ounces per ton, and from the Smuggler mine came a nugget that weighed 2,060 pounds and was 93 percent pure silver. So many miners were imported to operate the mines that at the day's end the sides of the mountains glowed with myriad trails of light as miners lit their way home with flickering mine candles.

In addition to running his mines, Henry Gillespie took on the role of promoter and toured the eastern United States seeking investment money for the new silver camp that was now called Aspen. The camp had been born in the usual way, with tents and clapboard buildings set on neatly platted, spacious streets, but it wasn't until the mid-1880s when Gillespie latched onto Jerome B. Wheeler, the then president of R. H. Macy Company, that Aspen felt the touch of a master builder. The younger man heard Wheeler was in Colorado Springs and dashed over the mountains to persuade the stocky, dark-bearded New Yorker to spend a few days in the new camp. Impressed by Gillespie's articulate presentation, Wheeler agreed to take a look. And one look was all it took. Wheeler immediately succumbed to the spell of the mounting riches and scenic splendor of Aspen, and with the help of his financial wizardry the area soon sported a smelter, mills, quarries, a bank, an opera house, and a three-story hotel, the Jerome, featuring a genuine elevator operated by water power.

As in most of the other out-of-the-way camps, freight rates for ore carted out of Aspen by wagon were exorbitant, and it was clear to Wheeler and the rest of the mining men in town that, like all the rest of Colorado's remote treasure troves, the town needed rail service if the full potential of its mines were to be realized. This idea occurred to Wheeler and his friends at almost the same time that it occurred to the directors of the D&RG. Under the circumstances, it would have seemed that the situation was ripe for a highly gratifying meeting of minds. Not so. While the narrow-gauge D&RG geared itself for a track-laying orgy from Leadville—north to the Colorado River, downriver to Glenwood Springs, already a spa for bone-weary, muscle-aching Coloradoans, and south along the Roaring Fork to Aspen—Wheeler and two men from Colorado Springs, lumberman H. D. Fisher and Michigan ironmonger James J. Hagerman, gave birth to the rival Colorado Midland Railroad. Cannily insisting on broad gauge, which would make it possible to connect their line with mainline transcontinental trains, Wheeler et al began their line at Colorado Springs, brought it over Ute Pass to the west to Fisher's pineries, across South Park, over Trout Creek Pass and north, paralleling the D&RG tracks into Leadville. From there, the Midland pierced the divide to the west, threaded its way along the Frying Pan River to join the Roaring Fork halfway between Glenwood Springs and Aspen, curved left, and headed up the valley to the waiting silver mines.

Both lines reached the outskirts of Aspen within two months of each other and what had been a race became a war as each sought an exclusive right-of-way to the mines located in the mountains south of town. The shortest way to the mines was to bring the rails directly through the middle of town, and to the delight of spectators, in late September of 1888, the two roads squared off for possession of Ute Avenue. From dawn to dark the pace of grading and track-laying accelerated as each day brought the rivals closer to their goal. After dark on the night of September 18, the Midland track-layers in a spurt of industry reached the coveted avenue and started to lay their rails across it at the very spot the D&RG tracks, coming up fast from half a block away, would intersect it.

Suddenly there appeared a gang of D&RG men armed with Winchester rifles moving toward the intersection. With each clang of a Midland hammer on a spike the riflemen moved a step forward to protect their crossing. The Midland men pushed ahead, the swinging iron rails catching the light of the rising moon in a sparkle of silver. At the moment when the contending gangs stood sledgehammer to rifle barrel, "grim furrows of resolute determination" marking their brows, Aspen's mayor hurried into the moonlit street and called a halt. After a conference, the two roads submitted their dispute to the courts where it was duly, and dully, settled in favor of the Midland. The D&RG, after having come so far, decided to stay, and laid track around the other side of town. It took some consolation from the fact that there was more than enough freight business in Aspen for both lines to prosper.

With two railroads to get them there, silver kings, merchants, lawyers, actors, and capitalists flocked to Aspen, H. A. W. Tabor among them. With his usual fiscal abandon he plunged nearly $3 million into the development of the Tarn O'Shanter series of mines located above Aspen on Castle Creek. At the peak of their production, he took out $20,000 a month in pure profit. Occasionally, he hooked his private car onto the Colorado Midland daily passenger train and brought Baby Doe to the thriving new silver center. When that happened, the expansive Tabor gave his miners a twenty-four-hour holiday and free drinks.

Also to Aspen came Jake Sands. Never a man to overlook beckoning opportunity, Jake put aside whatever disappointment he felt at Baby Doe's alliance with Tabor to maintain a cordial, and as it turned out, profitable relationship with the glamorous pair. In 1885, he sold his haberdashery store in Leadville, leased two of Tabor's Tarn O'Shanter mines, and opened what became Aspen's toniest men's clothing store. Caught up in the whirl of high society's sudden appearance in the booming silver town, the courtly, good-looking Sands, now that he had lost Baby Doe, did not remain long a bachelor. He married, rented a stylish two-story brick residence on Main Street, and became the guiding light in civic and charitable enterprises of Aspen. One of his triumphs was the staging of a benefit baseball game between the fat and lean men in town that drew as much money as it did laughter.

Playground for Aspen's nabobs was Glenwood Springs, fifty miles downstream on the Roaring Fork. Here, Walter Devereux, one of Wheeler's lieutenants, and his two brothers introduced to a horse-happy public the gentleman's game of polo. Here, too, for those wishing to remove the dust of the racetrack and the polo field from their well-fattened skins was a five-hundred-foot natatorium that took advantage of Glenwood's natural hot springs to provide a bath of refreshing and relaxing temperature. Capitalizing on the fashionable habit of "taking the waters," the Colorado Midland obliged Aspenites with a daily excursion train down to the Springs and return: the fare, $2 including bath. As their train paused at the Springs in midwinter,

mainline passengers of the D&RG were treated to the tongue-cluck-
ing sight of mining magnates and their ladies, all fetchingly attired in
the latest word in bathing costumes, merrily cavorting in the thera-
peutic waters in the middle of a raging snowstorm.

By the end of 1892, Aspen was the largest silver camp in the
world, with a population of eleven thousand people, ten passenger
trains a day, electric lights, six newspapers, and a passion for horse-
racing. Annual silver production ran $10 million. But in a few hours
on that fateful day in November, 1893, when the last "aye" echoed
through the chambers of the Congress marking the repeal of the
Sherman Silver Purchase Act, the whole kit and kaboodle became
worthless. Eighteen hundred miners were laid off, banks slammed
closed their doors in the agonized faces of hundreds of small depositors, and millionaires who yesterday wallowed in the downy comfort of their wealth saw it torn away, leaving them destitute. Crestfallen Henry Gillespie saw all of his possessions—his fine home, his
ranch, stables, conservatories, hatcheries, and horses—all go under
the auctioneer's gavel. Aging Jerome Wheeler, staring in disbelief at
his accountant's ledgers which showed he had $500,000 in assets and
$1.5 million in debts, declared himself bankrupt and wept as the last
and the most cherished of his holdings, the Jerome Hotel and the
Wheeler Opera House, were liquidated. Even Jake Sands felt the sting
of silver's demise as a monetary commodity. He lost his mine leases,
his store, and worst of all to the handsome haberdasher, his prestige.
As a last resort he pawned the $5,000 diamond ring Baby Doe had
given him when they parted and left Colorado to make a new start.
Eventually, he turned up in Globe, Arizona, as the owner of a small
pressing and tailoring shop.

And so Aspen went the way of all silver camps into a somnolence
that by all signs should have been permanent. But when the frontier civilization gave way to an urban society and skiing in America
became a sport instead of a way of winter transport, old-timers remembered the slopes of Aspen and Smuggler mountains that were made for
downhill and crosscountry runs, and Aspen boomed to life again to
become one of the country's foremost shrines of winter sports.

❦

Next in the carbonate circle was Red Cliff. Some twenty miles north
of Leadville, miners dug up good silver deposits on Battle and Horn
mountains and the site became a roaring if short-lived silver camp.
The town's busy newspaper, the Red Cliff *Comet* chronicled it all. It
gave full and detailed coverage to the camp's first St. Patrick's Day
in 1880, which happened to coincide with the day that the first log
was sawed in the camp's new saw mill. In celebration of these two

momentous events, the miners staged a grand parade. They strapped the freshly sawed plank to the back of a disinterested burro, and then they festooned both plank and burro with evergreens in honor of the Irish, and all the males in town formed up behind the beast to march down the main street with the wavery notes of a cornet providing the cadence for their swinging feet. The parade, however, took an inordinately long time to get through town because the marchers broke ranks to stop at every saloon they came to, on both sides of the street. In a short while, reported the *Comet,* it became clear that the participants, including the burro, were in some danger of being unable to finish the route as the notes of the cornet grew shakier and the burro staggered forward in lurches and stumbles followed by the reeling line of miners caterwauling various renditions of "The Wearing of the Green."

The *Comet's* columns covered every newsworthy event and spared no one, not even the "soiled doves." Big Hat, one of Red Cliff's more popular parlor-house girls, made the front page when in a moment of inebriate despair she cast herself into Turkey Creek to end her life. The attempt was totally unsuccessful, chiefly because the creek had only four inches of water in it. Despite her dramatic gesture she was pulled unceremoniously onto the bank by some passing miners, soaking wet, slobberingly maudlin, and very much alive.

Passengers arriving in Red Cliff on the narrow-gauge D&RG in the fall of 1881 might have been impressed by the imposing structures of the town—its five hotels, including the Star with its French chef, its three business blocks, and the ever-present fixture of a boom camp, an opera house. But for all of its signs of permanence, Red Cliff's fall was as abrupt as its rise, and it slipped finally and precipitously into anonymity after the silver crash of 1893.

So strong was the bonanza fever that few men in Colorado could resist prospecting in their spare time, and if they had no time they at least found a way to grubstake those who could go gophering for claims. In 1878, George Robinson, a merchant outfitter in Leadville, staked two friends to a prospecting trip in exchange for one-half of their profits if they found pay dirt. By June, 1879, the two pick and pan envoys had staked out ten good claims, roughly eighteen miles northeast of Leadville. With their behind-the-counter partner they organized a company capitalized at $10 million furnished by a group of New York investors. Meanwhile, others arrived to sink their picks into the rich pocket of earth. The Wheel of Fortune, the White Quail, and the Rattler were all mines to rival the Robinson group's high-grade production.

The town of Robinson sprouted a smelter and a hotel, and became the hub of a series of lesser camps, including Kokomo and Carbonateville. Both Kokomo and Carbonateville straddled Ten Mile Creek upstream from Robinson and were but brief stars in the galaxy of silver camps, lasting for only about four years. Their transience could not however be blamed on the apathy of their prospectors. No occasion was so sacrosanct that a man failed to keep his eyes open for a lucky find. When a miner named Scotty died in the middle of the first winter that saw the town's birth, his friends wanted him to have a decent burial in the newly laid out cemetery. So they hired a gravedigger, promising him $20 if he would dig through the ten feet of snow piled on the ground and through six feet of hard frozen earth under it to make a proper grave. While the grave-digger dug, Scotty's remains rested chillily in a snow bank. When at the end of the day the grave-digger failed to appear and collect his money, Scotty's friends went out to investigate. When they found the site of the grave, the grave-digger was missing, and there was a note pinned to Scotty's body: "Struck it rich at 4 feet below grass roots. Gone to town to record location. Will be up to plant old pard in the morning."

George Robinson's personal fortunes flowered as well as his town and he was on the brink of fame as lieutenant governor-elect when his life was accidentally ended by one of his own mine guards. In 1880, one of the Robinson group of mines, the Smuggler, was the subject of disputed ownership, and as the contest wore on relations among the claimants became touchy as tinder. On an icy November night, when Robinson was alerted that his opposition was planning to rush the mine that very night, he hurriedly got together some men, armed them with rifles, and stationed them around the portal of the Smuggler. Robinson gave orders that the guards were to shoot anyone not authorized to approach the mine. As the hours passed, Robinson grew anxious; about 1 AM he decided to check to see if all was well at the mine. As he approached the barricaded portal, he was immediately challenged.

"It's all right," called Robinson without giving his name. As he turned to leave, there was the crack of a rifle and the mine-owner fell with a bullet in his side. He died as first light came over Mt. Lincoln.

Robinson did not live to see his town's demise; his period of good fortune was as short as the camp's. Unlike most of the other silver diggings, Robinson was not turned into a ghost town overnight by the 1893 debacle. Its decline came much earlier. As soon as the high-grade ore gave out in the late eighties, Robinson slumped into inactivity. Today the remnants of the camp are entirely buried beneath the settling ponds of the voracious Climax molybdenum mine.

12

Statehood and Status

IN 1864, AS SOON AS CONGRESS PASSED THE ENABLING ACT PAVING THE WAY FOR Colorado's entry into the Union as a state, the territory's statehood advocates wasted no time in getting the issue before the people. At the head of the movement was Governor Evans, who would be asked to resign over the Sand Creek affair, *News* editor William Byers, David Moffat, Jerome Chaffee, and a score of others all known as the "Denver Crowd." For months, Byers' paper, as spokesman for the group, harped on the virtues of statehood: federal appropriations would be forthcoming, also railroad land grants, and eastern investment capital. But it was an uphill fight against two important factions. Dead set against the Denver Crowd was the "Golden Crowd," headed by brooding W. A. II. Loveland, flanked by Henry Teller, ex-US marshal Alex Hunt, and the much respected Judge Moses Hallett. Their objection was not to statehood but to the possibility that the Denver Crowd would pull it off, and in so doing snatch for themselves all the prize political positions, leaving the Goldenites powerless. Loveland and company need not have worried, at least not right away. The independent-minded, ragtag electorate emphatically turned thumbs down on two successive statehood referendums. They refused to believe Byers' claim that the cost of running a state would be only 116,000 more than the cost of operating the territory. And when he suggested that the operating funds could quite easily come from a $2 poll tax, the populace seethed. Prices were already sky high, with the Indians throttling commerce across the plains. Incomes were down and business was poor, mainly as a result of the reduced gold output from refractory ores. To the Colorado householder the 50 cent poll tax already in use was about 50 cents too high.

To play safe, the Golden Crowd gave full support to the replacement of Evans in the governor's chair with crafty A. A. Cummings. The new governor, to the satisfaction of Loveland et al., painted dire word pictures of the danger of prematurely achieving statehood to a receptive constituency, and labeled the entire matter a scheme for personal aggrandizement fostered by the Denver Crowd.

However, on a third go-round, with spellbinding William Gilpin resurrected from political limbo to run for governor, the statehood issue carried, if only by a bare 155 votes. But the glee among the Denver Crowd was shortlived when Democratic President Johnson, harried to distraction by hardline radical Republicans who kept the Congress in a prolonged posture as an arena of viperish retaliation against the South for its secession, vetoed the Republican-sponsored Colorado bill on the basis that too few people resided in the territory to warrant statehood. The fact that Henry Teller had gone to Washington to lobby vociferously against the bill accomplished exactly what the Golden Crowd had in mind. An attempt to override Johnson's veto in the Senate failed by three votes.

The Denver Crowd were down but not out. Their chief strategist was outspoken, intrigue-loving Jerome Chaffee, whose stamp-mill millions allowed him to buy Clark and Gruber's mint and to create out of it the First National Bank of Denver and at the same time left him free to manipulate men and monies. Chaffee realized that no statehood measure could pass the Congress if Colorado Republicans were not unified. As a consequence, in a risky move that astonished his rivals, he pressured his Washington Republican friends to get Johnson to remove A. A. Cummings and appoint Alex Hunt, Golden's man, to the Colorado governorship. But the sop to the Golden Crowd did not work. Chaffee had not counted on the tenacity of Henry Teller, and when in 1868 the Denver Crowd marshaled their forces for another try to get Congress to grant statehood to Colorado, Teller demanded that the issue be put to the people once more. Chaffee, Evans, and Byers, sufficiently uncertain of their hold on the voters, declined. By their action, statehood for Colorado was tabled for five years while the party leaders in Denver and Golden got on with "the politics of business."

In the early seventies, thanks to Nathaniel Hill's breakthrough in the treatment of refractory ores, Colorado climbed out of its depression. And immediately the foremost piece of business to engage the minds of both the Denver and Golden crowds, and one that was partially responsible for the enmity between them, was the connection of the new Eldorado to the outside world. It was John Evans who declared more than once in public that Colorado without a railroad was worthless. And both factions were still smarting at the rebuff delivered to Colorado by the Union Pacific's selection of a transcontinental route through Wyoming. It was inconceivable that the gold ingots of Colorado had to accept transport in unwieldy, slow-moving ox-drawn wagons, when a scant one hundred miles to the north a first-class, cross-country railroad served an economy based on the lowly cow. There was, in all this turmoil, one flickery light on the horizon for the Denver bunch. Congress had ordered the Kansas

Pacific in 1866 to become the eastern division of the Union Pacific and to head west to join the mainline at some point not more than fifty miles westward of the longitude of Denver.

The KP was 310 miles east of Denver and, when asked, the line's officials hastened to assure Coloradoans that their railroad would build to Denver immediately. The officials failed to mention that the KP was short of funds and land.

Over in Golden, however, William A. H. Loveland could not have cared less about the Kansas Pacific. His Colorado Central, originally organized to build west over Berthoud Pass in the hope that the Union Pacific would buy the route *and* the railroad, quickly announced that it would reverse direction to head east and north to connect with the UP at Cheyenne. To benefit by this change in plans all Denver had to do was to put up $200,000. Despite some reservations among the Denver Crowd, the voters jumped at the chance to signify their willingness to appropriate the money and then were stunned to find that Loveland had no intention of putting Denver on his direct line. The capital would be allowed a branch track.

This humiliation was capped by the not very surprising visit to Denver of Colonel James Archer of the KP who shot nervous smiles at his listeners as he explained that his road would now require $2 million and two years to reach the city. Denver was crushed. She had been fleeced and duped at every turn, and the bitter frustration made itself felt in acrimonious infighting among the Denver Crowd.

But at the bleakest moment there rode into town a man of talent and foresight with a particularly apt surname to set Denver free from her trackless cage. The bubbling George Francis Train, promoter and financial wizard, who had organized the Crédit Mobilier that made possible the UP and the Crédit Fonder that made possible the towns along its route, told a rapt audience at the Denver Board of Trade that they had only one realistic course open to them. Forget the devious Loveland. Reject the reneging Archer. Get up $2 million and build their own line to Cheyenne. The first on their feet with enthusiastic cries of approval were John Evans, lawyer Bela Hughes, and Dave Moffat.

In two days the Denver Pacific was capitalized and its stock on the market. Byers, in his new brick offices on Larimer Street, appealed in the *News* for every man in Denver to invest every penny he could scrape together in the new railroad to insure the value growth of his property. With $300,000 in public subscription, a $500,000 bond issue, and more to come, the Denver Pacific broke ground and headed for Cheyenne at a remarkable grading pace of two miles a day. On June 17, 1870, all Denver went out of its collective mind with joy at the sight of the first locomotive, the *David H. Moffat,* chugging its way into town. The crowning ceremony was to be John Evans swinging a hammer on a silver spike to nail down the last tie. The spike, fashioned from Pelican-Dives ore, came down to Denver the night before in the custody of Billy Barton, owner of Georgetown's Barton House. While he waited for dawn, Billy tapped a barrel in every saloon on Larimer Street, got roaring drunk, ran out of money, pawned the spike for $10, drank that up, and passed out. When he failed to show at the noontime ceremony, the territorial attorney general produced an iron spike hastily wrapped in silver tea paper and none of the delirious onlookers noticed the discrepancy.

In August, late but not least, there came rolling into town the laggard Kansas Pacific. In charge of its last $2 million lurch across the plains was energetic General William J. Palmer, who found the money the railroad needed to revise its timetable and accelerate its arrival in Denver.

All of a sudden Denver had two railroads, and the rush of newcomers to her city limits was nearly as overwhelming as the 1859 immigration. In its first month of operation the Denver Pacific deposited 1,067 passengers in the capital and hauled thirteen million pounds of mixed freight, much of it money-metal ore headed for eastern refineries. For the first time Denver was now nearer in terms of travel time to New York than she was to the remote mining camps strung out among Colorado's mountain bastions.

The face of Denver changed. Urbanity replaced rusticity. Mud-chinked log cabins and clapboard false fronts were torn down to make room for whole blocks of imposing two- and three-story brick buildings. Ada Lamont's rough-hewn bordello down on Indian Row gave way to stylish, brick-faced, lace-curtained parlor houses on McGaa Street. Later, when William McGaa dropped from favor his street's name was changed to Holladay in honor of Ben Holladay, the well-liked stageline operator. But when the brides of the multitude took over the street, Holladay's friends complained that it was a slight to his memory. The town agreed and once more the name was changed, this time to an innocuous Market Street. East of the business district fine stone residences made their appearance on newly graded streets planted here and there with fast growing elms. Gaslights cast their soft yellow glow over the town. And in the markets, local meats, poultry, and produce were available all year long, at new low prices. Beef sold for 8 to 10 cents a pound, potatoes for 2 cents, cabbage for 3 cents.

Along with the change in its physical appearance came a change in Denver's population. Mountain men like Uncle Dick Wootton, who felt hemmed in by the encroachment of civilization, fled to pastoral haunts.

As the Indians were withdrawn, the squaw men also removed themselves. Some, like William McGaa who found his earlier acceptance by Cherry Creek residents repudiated by the new society, drank themselves to death.

To take their place there arrived "Broadway dandies in yellow kid gloves," arthritics, rheumatics, and consumptives hopeful of cures or at least relief from their suffering in the high, dry, invigorating air,

"English sporting tourists, supercilious-looking," and coveys of English, Dutch, and German investors, Italian musicians, and Chinese section gangs.

Young Stephanie di Gallotti, possessed of an appealing contralto voice, arrived in Denver on a concert tour accompanied by her husband, Baron di Gallotti. The Gallottis were not in America out of choice but sought refuge in this country as a result of an abortive plot on the life of the king of Sardinia in which the baron was involved. Other than manipulating court intrigues, the baron's talents seemed to have been few, and it was left to Stephanie to support the two of them by her singing until a hoped-for pardon came from King Vittorio Emmanuele, and they could return to the palatial life they had once known in Naples. Stephanie, thirty-five years younger than her husband, impressed her Denver audience with her full, clear voice and charm onstage. So successful was her reception and so well did the aging baron feel in Denver's climate that they stayed on. But to Denverites, who never did fancy opera much, the novelty of an Italian baroness making the walls reverberate with the stirring notes of *Tannhauser* soon wore off, and as the winter came on Stephanie had fewer and fewer engagements. With their money nearly gone, Stephanie pawned her jewels and was reduced to singing in saloons and afterward passing the hat to make enough to pay their hotel bills. All the while, the baron hounded the post office, certain each day that the red wax sealed pardon carrying the royal stamp would be waiting for him. One summer day it was, and the baron fell dead of the shock.

The baron's widow left Denver but in a few years she came back and married a local bartender and went to live in Leadville. Now fortyish and tending toward stoutness, Stephanie tried performing again, but her first concert showed her voice was too far past its prime to lull even a drink-deafened miner on a Saturday night. In time, her second husband died and Stephanie lived out her days in a small log cabin, part of Leadville's drab background. On the day after Queen Victoria died a reporter for the *Herald-Democrat* suddenly remembered some wild story he had heard to the effect that the former saloon singer was related to the dead queen. In need of a lead feature, he took a chance and went out to see Stephanie. When he saw her face he knew that there was some truth to what he had heard. Stephanie bore a startling likeness to Victoria. In answer to the reporter's questions, Stephanie explained her family tree, and showed the newsman a photograph of Victoria inscribed to "Cousin Stephanie." The reporter then went away to write a poignant story about royalty shut away among the mine tailings of a rough and ready western boom town.

If Denver was wide-eyed at the arrival of Italian and English nobility, it was wowed by the appearance of Russian royalty.

Alexis, the blond, wavy-haired, twenty-one-year-old fifth son of Czar Alexander II, arrived in New York on a chilly windy November day after forty wretchedly seasick days in His Imperial Majesty's ship *Svetlana*. Wined and dined in New York, Philadelphia, and Boston, the grand duke was invited by President Grant to Washington before going on to Chicago and Milwaukee. At the White House, General Phil Sheridan, one of the guests at the lavish state dinner for the royal visitor, happened to mention that he was about to go on a buffalo hunt to the great plains. Alexis' blue eyes brightened.

"Oh, how *I* should like to shoot a buffalo!" exclaimed the Russian, to whom the hunt was the sport of sports.

Sheridan looked sidelong at Grant, who nodded almost imperceptibly. "Then you must come along with us," invited the ruddy-faced general.

Alexis beamed and ordered his aides to skim off a few thousand rubles from the royal treasury to pay for the five-car train loaded with five kinds of spirits, in addition to champagne, that would take him to buffalo country. Grant protested that the party was on him, but Alexis insisted on paying for at least the rental of the Pennsylvania Railroad's plush palace cars.

While the president and the grand duke argued politely over who would pick up the tab, Phil Sheridan wired ahead to Bill Cody to gather up a thousand Sioux for local color, sent Colonel George Forsythe to lay out a camp on the North Platte fitted with good beds, stoves, and privies, and reserved rooms at Denver's American House for the royal party to relax in after the hunt.

The train carrying the buffalo-hunters stopped at Omaha long enough to pick up Colonel Custer and then it raced across the miles to the fanciest bivouac the US Army had ever seen. On January 14, the third day of hunting, Alexis bagged his first buffalo, leading his companions on a whirlwind chase over the hardpacked ground. He got two more trophies in the next two days and was perfectly content when a raging snowstorm enveloped the party to put away his Springfield and let John Evans, now president of the Denver Pacific, send a locomotive to pull the ducal train down to Colorado's capital. The entourage was met by an appropriate sampling of city fathers done up in statesmanlike black broadcloth and stovepipe hats. The reporter who covered the event was impressed with the duke's "elephantine hands" encased in "pearl-colored gloves."

At the white-tie reception given him the next night by the Pioneer Club, whose membership consisted chiefly of the Denver Crowd, Harriet Randolph and the other ladies who were bidden to take a turn around the dance floor with His Excellency were impressed less by his elephantine hands than by his elephantine feet. Alexis, it turned out, was a crack shot but an egregious dancer. No mind. It was a gala affair and Denver's new social elite were in hog heaven. They had outdone themselves in decorating the dining room of the American House to a level that was fit for royalty. Floor to ceiling

French-brocaded drapes hung at the windows, and gas-lit chandeliers cast a warm glow on bunched American and hastily stitched together Russian flags. The music, the food, and the champagne were the best that could be found in the frontier capital. Some of the women even risked the latest Paris fashion for respectable women, a daringly cut bodice, showing considerable pink flesh below the collarbone. The handsome duke enthusiastically took it all in, flirting openly with the fan-waving, fluttering misses North and Fluery, and quite gallantly catching in his ample arms the governor's wife, who fainted at the end of the grand march. When the last oyster was consumed and the rustle of skirts and tinkle of crystal stemware died out, it was left to William Byers, the public voice of Denver, to sum up the occasion. With classic simplicity, befitting the momentous event, the *News* declared on the following day, "we have had a ducal ball, we have had a live duke."

The duke was also elated. At the end of his trip he threw his huge arms around Phil Sheridan and General Custer in a hug that brought home to his hosts the appropriateness of the bear as the symbol for Russia.

Ducal balls were all very well for Denver's new society, but most of its five thousand population still sat somewhere below the salt and found outlets for their social needs in diversions of a more universal nature. Depending on their inclinations, they had two streets to choose from: Holladay and Blake. Conveniently, the two streets were a block apart. On Holladay were the fancy houses operated by a battalion of efficient, chic, and sophisticated madams. On Blake Street were collected the variety theaters, music halls, and gambling joints. In between the two streets, for those whose tastes tended toward the exotic, was Hop Alley where opium-flavored fantasies and games of fan tan and pi-gow waited.

Mattie Silks, the acknowledged dowager of Holladay Street, took up residence on the infamous avenue in 1875, aged twenty-nine, plump, petulantly alluring, and proud to claim that she had never been a prostitute but had immediately started out her red-lit career as a madam. That was when she was nineteen in post-Civil War Springfield, Illinois, where the streets swarmed with mustered-out blue coats. But soon after the war was over, Mattie closed up her Springfield house as the boys in blue entrained one by one for home and business dropped off. Taking a cue from her departed clientele she too took a train, the Kansas Pacific, and headed west, stopping to set up shop at the "end of the trail" towns that lived for the annual visit of hundreds of drovers who caterwauled their way into town after a celibate six weeks driving their cattle to the railhead from the Texas plains. Eventually, Mattie's westward progress brought her to Denver, and there on the site of the old Leavenworth and Pikes Peak Express stable she bought the first of several famous Denver houses whose numbers were memorized by most of the town's male population. Mattie kept twelve "boarders," served the finest food, the choicest liquors, and spent her handsome income keeping twenty-two thoroughbred racehorses and swaggering Cortez D. Thomson.

Cort Thomson was a Texan of medium height, blue of eye, virile, handsome, with sandy hair and a copper-tinctured mustache, and too proud to work. An ex-Quantrell Raider, he had perfected footracing as an adjunct to his quasi-military role in that notorious outfit. Unhorsed, he could streak to the safety of a clump of beeches before gunsights could be trained on him. After the war, Thomson turned his racing prowess from an escape mechanism to a sometime profitable way of life. In Georgetown at the height of its silver boom he was hired to train the newly organized volunteer fire company to be winners in the intercamp tournaments. With the money he got he bet on himself in a few well-chosen racing challenges and made enough to deck himself out in a flashy wardrobe and roll down to Denver in the style to which he fancied himself accustomed.

When and where Mattie and Cort met is nebulous, but all agree that their mutual attraction was immediate, deep, and nearly indestructible. Mattie seemed not to care at all that Cort developed into a spendthrift, a drunkard, and a compulsive gambler. He was her "fancy man." That his highstepping, piston-action legs and his efforts to promote sports events of various kinds brought more debts than credits seemed not to bother her. Only one of Cort's shortcomings could provoke her to anger. When she found that Kate Fulton, a rival madam on Holladay Street, had found Cort's jaunty attractiveness too much to resist, Mattie drew herself up to her full height in her Renaissance silk and lace gown, patterned after one worn by a de' Medici, and challenged Kate to a duel.

On the appointed day, while William Byers' *Rocky Mountain News* loudly deplored the entire affair as demoralizing and disgusting, all of Denver's sporting fraternity gathered just outside the jurisdiction of the law beyond the city limits at the Olympic Gardens to cheer their favorite. Cort seconded Mattie. Her adversary chose Sam Thatcher, a humorless card shark, as her second. The weapons were pistols. After the ladies paced off their distances, there was a sudden hush as the count began, "one...two...three!" Eardrums oscillated, eyes blinked, and noses twitched at the acrid smell of black powder as the shots rang out. When the swirling smoke drifted away, Mattie stood, and Katie stood, but Cort Thomson lay on the damp ground writhing with a bullet in his neck. Which of the two women shot the debonair sprinter could never be proved. But since Cort survived and Mattie ultimately married him, the question became academic.

If a gentleman caller at Mattie Silks's house was guaranteed a good time in the pleasantest of surroundings he was absolutely bedazzled by a visit to 1942 Holladay Street. Statuesque and stunning Jennie

Rogers, alias Leah Fritz and a few other names, was barely literate, but she communicated well enough to collect $17,000 in blackmail and build Denver's most lavish parlor house. Hiring the capital's foremost architect, William Quayle, whose latest triumph was the First Congregational Church, Jennie watched with some satisfaction the deepening envy of other madams on the line as her three-storied stone eminence took form. In classic style gargoyles graced the facade. A stone face of Jennie herself crowned the apex of the roof pediment, and four other faces, two men and two women, said to be principals in the blackmail scheme, were artistically spaced below Jennie's. Inside, the parlor walls and ceilings were covered with the finest French plate glass framed in bird's-eye maple, and sliding glass doors opened onto a spacious ballroom featuring parquet floors and gilt side chairs.

The girls of the parlor houses were gay and fulsome but were careful not to be too forward with gentlemen visitors. They saved their hijinks to embarrass city officials in public. When Oscar Wilde's visit was advertised the poet was pictured in the press against a background of sunflowers and lilies. Immediately, a troupe of girls from Holladay Street houses staged a street corner parade clothed in sunflowers and lilies and little else. They were arrested and charged with "meretricious display" by an exasperated police sergeant. Three of the frail sisters followed up with the "Dance of the Houris" in the nude at Nineteenth and Larimer streets and were hastily carted off to face police fines for "naughty capers." But the caper that set Denver on its ear was the reaction on Holladay Street to one of the city's periodic and lame efforts to regulate the demimonde. The town council in a moment of passionate righteousness decreed that all girls on the line seen in public had to wear a yellow armband to signify their profession to the world. In a matter of hours there was a run on yellow articles at every couturier's and clothing store in town, and the ladies of the evening trooped through Denver's exclusive districts, restaurants, and shops dressed head to toe in bright blinding yellow. The city council quickly revoked its decree.

Below the lighthearted veneer of the frail sisterhood however lay a world of grim despair and savage violence. The girls were often brutalized by their customers, beaten by pimps, and cheated by their madams. Some were driven to self-destruction, especially at Christmas, a hard time for the lonely, ostracized women in the cribs and parlor houses. Some took overdoses of laudanum, their passport to forgetfulness available by the bottle without prescription at the local pharmacy. Others sought a desperate escape by swallowing bichloride of mercury or another of their powerful antiseptics. Scarcely a month went by without a notice in the papers of an attempted suicide, sometimes successful, on Holladay Street.

One night of violence Victorian Denver never forgot was Allhallows Eve, 1880, when Hop Alley was ravaged by riot. Two thousand

Chinese lived in the neighborhood. Most had come as coolies working on the railroads, but some were imported to work the mines of the
Little Kingdom, South Park, and Leadville. As Colorado slipped into
the depression of 1873, the presence of the Chinese became a fester-
ing canker. Cheap labor was in great demand. White miners seethed
with resentment at the Chinese who worked for next to nothing and
filled the jobs they once held. In 1874, Central City's white miners
burst upon the Chinese community, cut off the queues of the men
they could lay their hands on, and chased them out of town. When
H. A. W. Tabor brought in a gang of coolies to work the Matchless in
1878 he was openly censured. As the state election of 1880 drew near,
expulsion of the Chinese became an explosive issue. On the night of
October 31, 1880, a few days before election, the Democrats, the party
of the working man, staged a torchlight parade through downtown
Denver carrying banners calling for the expulsion of the Chinese,
who were "taking the food from the mouths of whites." The crowds of
onlookers on the streets worked themselves into a frenzy as marchers
chanted, "the Chinese must go!" A crazed voice rose above the tumult,
"Let's run 'em out!" With the force and speed of a torrent, gangs broke
away and headed for Hop Alley, streaming down the streets breaking
windows and kicking in doors. The mob burst into a Chinese saloon,
breaking the furniture, kicking the patrons to the floor, and smash-
ing the bottles and mirror behind the bar. Then they systematically
moved through the alley wrecking stores, shooting open doors, and
dragging the hapless residents out on the street to beat them senseless. One of the mob lassoed a fleeing Chinese, and a hundred hands
grabbed the rope to drag the struggling bundle of humanity along the
debris-covered street until it struggled no more. Then with whoops of
drunken lust they hung the body to an arm of a lamppost and surged
on their bloody way with no one to stop them. Denver's finest were
without a police chief. He had been suspended on graft charges a few
days previously. It was left to Mayor Richard Sopris to put down the
riot himself. Awakened and told of the trouble, Sopris immediately
ordered the saloons closed and called out two civilian militia companies, the Chaffee Light Artillery and the Governor's Guard. A number
of Chinese were put in protective custody, and gradually five hundred
civilian deputies brought order back to Denver. But the wound was
closed only on the surface and race riots in Colorado and elsewhere
were regular and ugly occurrences until Chinese immigration was
curtailed by Congress in 1882.

The second of Denver's notorious streets in the seventies and
eighties went in less for violence and more for sleight-of-hand. The
gambling dens on Blake Street offered every game the West had
known. Some but not many were fairly run.

Queen of the gamblers was a dark-eyed, petite beauty who called
herself Madame Vestal and who ran an establishment housed in an

enormous tent on Blake Street. Renowned as a twenty-one dealer, Madame Vestal's table was always crowded by those who cheerfully lost their money for the privilege of listening to her cultured, southern-accented voice. What led an attractive and obviously well-educated woman to the life of a gambler never failed to fascinate players in Madame Vestal's tent. But no one in Denver guessed she was the infamous Confederate spy, Belle Siddons.

When the war erupted Belle had just returned to her home in St. Louis after graduating from the Female University in Lexington, Kentucky. Missouri had allied itself with the North but southern sympathizers abounded in the state, including young Belle. St. Louis in wartime was a military town. It was the headquarters of the Union Department of the Mississippi and swarmed with Army and Navy officers. For most girls just out of college, to be swept up in the whirl of the pomp and circumstance of military social functions was a heaven-sent opportunity to land husbands. But not for Belle. All of her feminine wiles were aimed at eliciting information on troop movements and rumors of strategy from her blue-uniformed escorts. She was so successful that Union general Curtis had her arrested in 1862. When she got out of prison she moved to Jefferson City, Missouri, and married a surgeon. At his premature death of yellow fever, Belle taught for a time at an Indian agency and then found life as a country school marm too tame. She next turned up in Kansas exhibiting adroitness at dealing cards that showed she had done her homework. From Kansas she moved her operations west to Denver. Ultimately, Belle moved on, had a tragic affair with gang leader Archie McLaughlin, who terrorized the Black Hills, and finally dropped out of sight in San Francisco in 1881, a drawn, dissipation-ravaged skeleton. But in her heyday in Denver, Madame Vestal's tent was a favorite hangout—her games were honest, her whiskey unwatered, and her ropers-in all male.

A few doors down Blake Street there held forth the king of Denver's sporting tables, Ed Chase. Like his consort in the gaming business, Chase also began life on the right side of the tracks. Born in Saratoga, New York, the tall, blue-eyed, prematurely graying gambler attended select Zenobia Seminary sharing classes with the likes of Leland Stanford. Unlike his classmates, Chase's interest in the law was restricted to getting around it, his enthusiasm for engineering was geared to the operation of a roulette wheel, and his yen for banking was concentrated on that in baccarat. He learned the gambling game during the summer when he worked at his parents' exclusive hotel in Saratoga. When the Pikes Peak craze reached New York, young Ed Chase saw his future. He arrived on Cherry Creek in June, 1860, and when it got around that he knew how to run a pool hall and card games, some local boys, including Jerome Chaffee, grubstaked him to his own establishment. Chase prospered, if his customers did not, and soon opened a gaming resort called the Progressive Club. After a stint with the Third Cavalry that included attendance at the infamous Sand Creek affair, Chase got back into civvies and treated Denver in the late sixties to its plushest card room and variety hall, the Palace.

If the possibility of a winning faro hand failed to appeal, the patron of the Palace could indulge his thirst for thrills watching the supple-limbed and golden-voiced Barbour sisters on the Palace stage. Chase himself was so captivated by the talented sisters that he made one of them his third wife. Or, if the Palace brought a man no luck, he could push open the doors of Chase's Inter-Ocean Club a few doors down the street from the Palace and try again. It was from this establishment, according to Forbes Parkhill, that a young man rushed early one morning to dash across the street to the bank. Brandishing a sealed envelope he stopped in front of the cashier's window.

"Here, look inside and lend me a stake," he cried.

The cashier gingerly opened the envelope to see four kings and an ace, one of the two unbeatable hands, according to the Hoyle of the day. Unimpressed, the nongambling cashier declined the loan. In despair the young man gathered up his cards and tore out into the street to seek a loan elsewhere and ran smack into the president of the bank coming to work. Presenting his case on the street corner, the young man quickly excited the interest of the mutton-chop-whiskered banker who moved rapidly but sedately into the bank, counted out the necessary cash, and accompanied the young man back to his game. When the bank president returned in about ten minutes with his money plus $500 interest, he sought out his strait-laced cashier.

"I know you don't gamble, my good man, and I commend you for it," said the president. "However, I think you should know that four kings and an ace will always command the entire assets of this bank—do I make myself clear, sir! The entire assets!"

In 1870 Chase opened the Cricket Club, and it quickly became the town's most notorious variety hall. Loud, inebriate crowds packed every performance to whistle and shout and stamp their feet at tawdry can-can girls, hootch dances, and stand-up comedians whose humor was as broad and coarse as the bottom of the South Platte. There were eight shows a day, around the clock, with no rehearsal. Rehearsals were unnecessary, Chase decided, since the audiences were generally in no condition to discern art from exhibition. To drum up their availability, every afternoon Chase piled a few of his barely clothed, painted girls and his zingy brass band into a garishly painted wagon to wheel around town giving free but brief samples of what the Cricket Club had to offer.

There were protests among the gentry against Chase's activities, and from time to time Denver's bluecoats dutifully staged raids on the gambler's establishments. But the charges against him were conveniently thrust into first one pigeonhole and then another. When they finally came before the courts some two or three years later the well-liked Chase got off with fines totaling about $10 for each count. So adept was the gambler at keeping friends at City Hall that he remained alive and in town to a venerable ninety-one years old, a Sphinxian storehouse of stories about Denver's famous and infamous who patronized his various establishments.

When Jack Langrishe found that Denver audiences lapped up the sensational shows on Blake Street, he risked producing a few of his own in his usually respectable Denver Theatre. Importing fulsome Marietta Ravel he put her in the lead of *The French Spy* and *Tartine, the Pride of the 14th,* both salty tales of soldiers and wily women calculated to create total recall and relish in the minds of the ex-Colorado Firsters. Bawdiness down on Blake Street, where it belonged, was one thing, but when the disease threatened to invade the sacred precincts of Denver's new society, that was untenable.

"Denver during the past week," carped Byers in the *News,* "has been treated to the most sensational...drama...the kind that depends for its success on clap-trap...red fire, legs, and anatomy in general."

Langrishe bowed to his public and tempered his fare and steadily lost money to the live-wire Blake Street spas. His declining income resulted not only from unmet competition but also from the effects of another national depression that was steadily eating its way across the country.

Like the national Panic of '57, the Panic of '73 was a result of reckless speculation, graft, and shady financial manipulations among bankers and businessmen interested in quick profits. For Denver it was a time of recession. The climbing prosperity wavered, leveled off, and began to slip backward. Isabella Bird, who found herself stranded in the capital, wrote to her sister that "The Denver banks have all suspended business. They refuse to cash their own checks or to allow their customers to draw a dollar, and would not give greenbacks for my English gold!...[E]very-one, however rich, is for the time being poor."

Part and parcel of the panic and its aftermath were two scandals that had considerable effect on Colorado. One was the blowing wide open of the monetary shenanigans of Denver's hero, George Francis Train's Crédit Mobilier. Formed to construct the federally financed Union Pacific Railroad across the country, the Crédit charged the government double for everything it used and gave the stockholders of the Crédit the difference as dividends. The system worked smoothly until the directors of the Crédit began squabbling over the loot and someone in Congress sniffed a rat. To quiet governmental fears, the directors of the Crédit bribed Grant's vice president, Schyler Colfax, and several members of Congress with free stock in their shady company. The bribe, however, failed to work and the names of the government leaders implicated in the deal came to light. Congress ultimately whitewashed the whole affair, but not before the people of the country got a good hard look at the effects of unbridled avarice. In Denver, the respectable and Republican new society was appalled. Colfax, whose sister lived on Capitol Hill, was a frequent and much admired visitor.

But the Denver Crowd was in for another surprise that struck even closer to home.

When Grant took office for his second term in 1869, he replaced Colorado's Governor A. C. Hunt at the expiration of his term with General Edward McCook, a close friend of Secretary of War William W. Belknap. No sooner had McCook taken office than it became apparent that he was there to line his own pockets with the profits to be gained from contracts to supply Indian annuities. With thousands of dollars worth of beef, blankets, and other goods annually going to the Ute, Arapaho, and Cheyenne, such contracts were nearly as money-making as a gold mine. With judicious gifts and promotions McCook was well on his way to acquiring the contracts he wanted. Not only did he succeed in getting the contracts, but he pulled a handsomely rewarding double-cross by buying cattle at $7.50 a head and selling them to the government for $35 a head. When at last after some slick detective work, Jerome Chaffee, now a territorial delegate, and Dave Moffat got the evidence they needed, Grant removed the accused McCook from the governorship. The repercussions traveled all the way to the War Office where McCook's old friend Secretary Belknap was impeached and tried for getting rich by selling Indian annuity contracts to his friends.

One of the effects of these events in Colorado was the erosion of the Republican majority and the election of Democrats to office for the first time since the Civil War. One of these was Thomas MacDonald Patterson, who dumped from his delegate seat Republican Jerome Chaffee, the very man who laid open the corruption of fellow Republican Governor Edward McCook.

Tom Patterson was a Black Irishman, reared in a humorless Presbyterian family who emigrated from County Carlow to America when he was nine. Long Island and then Indiana were Patterson's boyhood homes. He trained to be a printer and a jeweler, served in the Union Army in the Eleventh Indiana Volunteers, and after the Civil War, studied at Wabash College in Crawfordsville, Indiana, for two years before he turned to the study of law. When he was admitted to the bar he hung up his shingle in Crawfordsville only to take it down after five years to try out his courtroom ability in the opening West where a man with a deep concern for the welfare of the common man might have a freer field in which to work. He swung off the Denver Pacific at Union Depot in 1872 and within two years he was the city attorney. When the Democrats met in the spring of 1874 in Colorado Springs to nominate a territorial delegate, hard-hitting but dignified Tom Patterson was their man. And when election time came he was also the man of the people who were fed up with the fraud and vulgarity apparently rife among the Republicans. They gave Patterson the nod over Chaffee by 2,163 votes, the largest plurality given an office-seeker in Colorado Territory to that time.

Patterson was a blessing in disguise for the statemakers. In 1873 when President Grant got back to Washington after tiptoeing over the bricks of silver bullion on a Central City sidewalk, he recommended that Colorado be once again considered for full-fledged membership in the Union since "It possesses all the elements of a prosperous state." On the strength of Grant's statement, a statement that the Denver Crowd had calculated would come after the lavish lengths to which they had gone to impress the visiting chief executive, Chaffee worked up another statehood bill for Colorado before he gave up his seat to Patterson. It was a last-ditch, desperate move by the Colorado Republicans still divided and disgraced by the shadow of Crédit Mobilier and McCook scandals, and it had little likelihood of passing the Democratic-dominated Congress. But Chaffee and the rest had forgotten one thing. Tom Patterson thought statehood was a good idea, too, and although it was not yet time for him to take his seat in the House he rushed back to Washington and worked side by side with the surprised Chaffee to get the bill through. Just when the Congress seemed ready to table Colorado's petition, Patterson aimed his most eloquent oratory at his Democratic colleagues, convincing them that the people of Colorado would without delay, when statehood came, put the Democrats in charge of running the new political entity. The enabling act passed, and Chaffee and Patterson went home to Denver to observe the convention called to write a state constitution for Colorado. Unhappily for Patterson he had to sit by and observe what would be the trend in the new state—of the delegates elected to the convention, twenty-four were Republicans and fifteen were Democrats. The Republicans were fast mending their fences.

In general, the delegates wrote a document that followed a middle course, firmly separating Church and State, encouraging economic expansion but providing checks on deceit and fleecing, and allowing mining, the life-blood of the region, to go untaxed for a specified number of years. On the issue of woman's suffrage, for which Augusta Tabor and James Belford had worked so hard, the convention allowed that women could vote in school but in no other elections. In July, 1876, miners, merchants, farmers, doctors, gamblers, lawyers, tradesmen, actors, and bankers accepted the constitution by an overwhelming majority, and on August 24, Denver danced in the streets as the telegraph tapped out the news, the new Eldorado was admitted to the Union. On the one-hundredth anniversary of the country's founding, Colorado became the Centennial State.

In the election that followed, Tom Patterson for one had reason to stop dancing and start crying. After promising a Democratic victory in Colorado, Patterson saw the entire first team for the new state fall into the Republican column. John Routt, a squarely built, square-shooting veteran of the Union investiture of Vicksburg, who had been named to fill the deposed McCook's term, was elected governor. Chaffee and Teller went to the Senate, and the Red Rooster of the Rockies, James Belford, went to the House defeating none other than Tom Patterson himself.

Statehood gave Colorado a great push forward. Investors appeared from all directions to ride the shiny new iron rails to the new Eldorado. Extensive development of the older mining areas in Gilpin and Clear Creek perpetuated the stream of gold ore pouring into newly built smelters and refineries. Out of the granite peaks of the San Juans came silver, a new money metal, and from the headwaters of the Arkansas the richest of all the silver lodes vomited forth a continuous flood of wealth. Broad-gauge and narrow-gauge track webbed its way over mountain passes and across verdant valleys to link together every major mining camp with at least one railroad. Farmers, after surviving two disastrous locust plagues, loaded bumper crops for transport to distribution centers. Flour mills, pickle factories, cracker works, candy manufacturers, and pottery workshops went into production. With the discovery of coal deposits in the southern counties, iron and steel manufacturing gave the Pueblo area the look of a little Pittsburgh. With Colorado-made steel available, a new industry in machine tools, especially mining machinery, came to life in the brand-new state.

In Denver the new wave of prosperity was reflected by a phenomenal jump in population, from 5,000 to 30,000 between 1870 and 1880 and from 30,000 to 100,000 between 1880 and 1890. New residential areas leapfrogged each other along wide boulevards. Stone and brick mansions with glittering Tiffany doorknobs dotted Capitol Hill on the east side of town. Downtown teemed with traffic—six-passenger shiny black hacks, tasteful broughams with small monograms of their affluent owners gilded on the doors, and open victorias fought for passage between dray wagons, horse-drawn streetcars, and fashionably turned-out pedestrians hurrying about their business. Beginning its gradual march eastward, the shopping center moved one block, from Larimer to Lawrence Street. H. A. W. Tabor placed his stamp on downtown Denver by donating the land for a new post office, creating an office building at 16th and Larimer, leasing the elegant five-story Windsor Hotel, and constructing the opulent Tabor Grand Opera House.

Although only two blocks away from these grandiose structures flourished Sodoms to match any mankind had yet put together, and morphine and cocaine could be bought at the corner drug store for 10 cents a box, and bunko steerers and pickpockets stood on every street corner, Denver's new elite, as true Victorians, closed their eyes to the reality of Holladay and Blake streets and acted out their new role with studied earnestness. Ten miles away, mountain lions stalked the rugged slopes of the Rockies and an occasional renegade Ute slit open a settler's gullet, but in Denver, the moguls of mining and their elegantly gowned wives spent evenings at the theater followed by a late supper at Charpiot's, just as New York's sophisticates capped their evenings with a midnight snack at Rector's. Respectability was *de rigueur.*

Lord Roseberry, stopping in Denver at the time, watched with fascination as an audience sedately sat on its hands at a play so poorly acted that English audiences would have catcalled it off the stage. "Denver," commented his lordship, "conscious of a shady record in the past really likes to be bored in this way under the impression that respectable people are always bored, and that, being bored, a Denver audience is respectable." Some of the newly rich even had elaborate coats of arms constructed to use on stationery, carriages, and china. Even their vices took on an air of respectability. Cheeky madams were sure enough of their ground to send engraved invitations to the opening of their houses to the most influential men in town. It was rare that anyone declined. An evening meeting of the city council, reported the *News* on August 21, 1880, had to be canceled for lack of a quorum because most of the councilmen were at the opening of "a newer and fashionable den of prostitution on Holladay Street." Wives pretended not to notice, and amused themselves by giving private musicals in their mausoleumic new homes and by attending receptions at the Denver Sketch Club. Here they sipped weak tea and viewed the newest work of a local Titian, trying not to spill their cups as they moved among the life-sized alabaster figures of Venus and Apollo used by students to paint the human body. Only the depraved modeled from life.

Occasionally someone pricked the taut self-consciousness of the newly urbanized populace. One who dared was Eugene Field, an irreverent man who ironically is renowned today as a writer of sentimentally sweet children's verses. Field came to Denver from the editor's desk of the Kansas City *Times* to become pilot of the *News* after W. A. H. Loveland, angling for a platform on which to air his political aspirations, bought the oldest paper in Denver and changed the flag at its masthead from Republican to Democrat. Very quickly the readers of the *News* were treated to the stiletto pen of the master puncturer. To Field, prominence was fair game. Reporting on printing company executive Kent Cooper's visit to Colorado Springs Field wrote, "Colonel K. G. Cooper went swimming in the hot water pool at Manitou last Sunday afternoon and the place was used as a skating rink in the evening." Nor was well-meant effort sacrosanct. On the morning after a production of Shakespeare by a local gentlemen's club when those who were in the audience opened their morning *News* to see if what they had seen was good theater they read that "The Reverend George W. Miln played Hamlet at the Opera House last night. He played it until 11 o'clock."

Denver's new society could withstand the barbs of a newcomer. But when one of their own stood before them revealed as a transgressor their cloak of respectability suffered a major rent.

William Byers, one of the first to open a business in the embryo settlement on Cherry Creek, had been a model of astuteness and decorum in the busy years since 1860. A regular churchgoer and a dedicated family man, it was he who had warned the starry-eyed gold-seekers at the beginning of the Pikes Peak rush to "leave not your character at home, nor your Bible, you will need them both...in a community whose God is mammon, who are wild with excitement and free of family restraints." And throughout Denver's early years he practiced what he preached. Byers' *Rocky Mountain News* was constantly carrying on vigorous crusades for the maintenance of morality in the town he helped to build. Through his fairness, his farsighted-ness, and his untiring efforts to bring status to the new Eldorado, he had earned the respect of all who knew him, friends and enemies, from gambler Charley Harrison to Governor John Evans. Byers' *News* and his sideline dabblings in land and mining brought him to afflu-ence, although because of his generosity he was seldom in the black. His influence was wide, especially among Republicans whose precepts he staunchly espoused. With statehood a near certainty after Grant signed the enabling act in March 1875, the estimable editor let it be known among his political affiliates that when the Republicans met in June of the following year he would not be at all adverse to accepting the nomination for governor. His friends were delighted, and he was fairly sure that he would be offered the nomination.

Unlike the previous winter, the winter of 1875–76 was excep-tionally mild, and Denver was treated to a warm spring that uplifted spirits already on the rise at the momentous prospect of Colorado taking her place among the states. On a sunny and mild April 5, two months before the Republican convention, William Byers left his office for lunch in a buoyant frame of mind to match the exhilarat-ing weather. At 12:20 PM he caught the horsecar that would take him the few blocks to his modest home on Denver's fashionable Capitol Hill. The moment he took his seat a stylishly dressed young woman with "eyes of coal gray tint, regular features, dazzling teeth, penciled eyebrows, small poised head, and wavy auburn hair" threw herself at the surprised editor, with a stream of abuse pouring from her pale lips. Astonished passengers caught only a few of the woman's shrilled cries. "Oh infernal villain," she yelled, "you have always lied like hell to me...damn you, damn you, damn you... As soon as Byers could fight off his attacker he leaped up and made his way to the front of the car. The conductor stopped the car and Byers jumped off. Still screeching her venom the woman followed him onto the street, where she suddenly reached into her handbag and pulled out a pearl-handled pistol. Byers grabbed her arms and held her fast. A moment later Lib Byers, who had seen the performance from the window of their home as the streetcar passed by, wheeled up in the family two-seater buggy. The frantic editor let go his tormentor and swung aboard the buggy as the woman's pistol went off. The bullet whizzed behind the seat tearing the cushion but missing its intended mark. Unscathed, Byers and his wife streaked for home as the distracted woman stormed after them. A few minutes later Byers' son Frank intercepted her and she was safely conducted to the custody of Denver police.

The Golden papers, delighted at being able to point a righteous finger at one of the august members of the Denver Crowd, broke the story first.

It had all begun some four years earlier, reported the Golden *Transcript*, when the woman in question, Mrs. Hattie Sancomb, a milliner of Lawrence, Kansas, wrote to Denver's city fathers inquiring about business opportunities in the new Eldorado. Byers, as head of a committee to foster new commerce, answered that unlimited opportunities were available. The attractive Mrs. Sancomb, recently divorced, hightailed it for Colorado. Choosing Golden as her place of residence, the pretty milliner settled into a small house and then made a trip into Denver to thank the editor of the *News* for his advice. Byers, who found less and less warm companionship with his prim wife Lib the busier she became in the affairs suitable to her station as a society matron, found himself mightily attracted by the chic and personable Mrs. Sancomb. The attraction, it turned out, was mutual. Their correspondence grew from warm to torrid and their meetings followed suit. And then in 1874, the conscience-stricken Byers girded his resolve and broke off the affair. At this point the electric Mrs. Sancomb searched her heart and found that the only balm that could restore it to health was $5,000. Byers, however, refused to knuckle under her threat of blackmail, tore up unanswered her cajoling, pleading, and finally threatening missives, and for his hard stand got shot at on a public street in broad daylight by his enraged paramour.

Denver papers reluctantly picked up the story, their editors manfully giving their colleague the benefit of every doubt. Professor Goldrick, Byers' friend from the first wild and woolly days of Denver, pretty much summed up the consensus: "Mr. Byers was a big fool to allow himself to become 'intrigued' with any divorced woman..." but, he went on, "the whole thing was a well laid plan, to...'rule or ruin' one of the most prominent pioneer citizens of Colorado."

Goldrick's accusation that the scandal was a plot hatched by the Golden Crowd, perhaps to undermine Byers' political popularity, failed to stand up, and even when it was revealed that Hattie Sancomb's divorce had resulted from her flagrant adulterous conduct with a cavalier Army colonel at Fort Leavenworth, dishonor was too deeply etched on the armor of Byers' respectability to be so easily wiped off. When the Republican convention met on June 14, Byers' name was conspicuously absent on the rolls of the nominating committee. Impeccable John L. Routt got the nomination, and when it was clear that he had been dumped by his long-time associates, William Byers sold his newspaper, and at forty-seven retired to his ranch where he had twenty-five years to ruminate on the frailty of man.

If the majority of Denver's nouveau riche were painfully self-conscious, there was one of their number whose artlessness, gauche as it may have been, was at least refreshing. Sporting a dazzling solitaire diamond on one finger of his bony hand and diamond cuff links the size of silver dollars, H. A. W. Tabor wheeled and dealed his way from one fortune to another. Commuting between Leadville and Denver, with money-making side trips to Chicago, New York, and Washington and inspection tours of his mining properties in Aspen, Rico, and Ouray, the silver king gave Denver's new society plenty to talk about. He leased the stately new Windsor Hotel, decorated it in turn-of-the-century opulence, and brought Billy Bush down from Leadville to manage it. To make the Tabor Grand Opera the spiffiest theater west of the Mississippi he sent to Japan for cherrywood, and to Europe for the most expensive carpets, plush seats, and crystal chandeliers he could find. When he saw workmen raising a picture of Shakespeare to place at the top of the proscenium at the decorator's direction, Tabor stopped them.

"Hell!" he said, "what did Shakespeare ever do for Denver? Put a picture of me up there." They did.

The Tabor Grand Opera opened September 5, 1881, to a glittering audience with Emma Abbott in the title role of *Lucia*. But for those who had come to see the woman who would occupy the Tabor box, it was a wasted evening. The Tabor box was empty. With a wife who would not, and a mistress who could not, be seen in public with him, the silver king's esthetic triumph was somewhat diminished. But if he could not have public acclaim he could at least enjoy the private attentions of his beloved Baby Doe. To make their trysts more convenient, now that Tabor's far-flung investments and political aspirations drew him more and more often to Denver, he established Baby Doe in the most elegant of the Windsor suites, and whenever evening found him in town he donned his pure silk nightshirt trimmed with point lace to share with her the fleeting hours of the night.

Life with Augusta had become too trying, and Tabor stopped visiting the house on Lincoln Street on Denver's Capitol Hill altogether. Instead, he repeatedly sent his emissary Bill Bush to ask Augusta to agree to a divorce. Each time Bush carried an emphatic "No!" back to his boss. Countering Tabor's request in March of 1882, Augusta filed suit for separate maintenance, claiming her millionaire husband had failed to provide for her support since January, 1881. Irked to distraction, Tabor slipped down to Durango where he owned the livery stable and local stage line and where he had friends among the judiciary to obtain a sub rosa divorce. The plea was desertion. The decree was granted in record time and in secret. Throughout the summer Tabor, smugly in possession of his clandestine divorce, watched Augusta's separate maintenance action languish in the courts as he and Baby Doe further cemented their already close relationship. Some six months after the Durango divorce, the lovebirds impetuously ran off to St. Louis and were married, again in secrecy, by a justice of the peace.

Tabor's desire for political fame was almost as deep as it was for the beauteous Baby Doe, and he was determined to have both. All through the summer while he dallied with his ladylove, he kept his eye and his checkbook trained on the upcoming senatorial election. Two seats were in the balance. One was for a full six-year term. The other, owing to the elevation of Henry Teller to Secretary of the Interior, was for thirty days, the remainder of Teller's term. The Colorado legislature was scheduled to meet in January, 1883, to elect men to fill these posts, and on Denver streets it was commonly conceded that Lieutenant Governor Tabor as Republican state chairman and holder of the key to the party's treasure box would get the six-year term. However, although Tabor's dalliance with Baby Doe was the everyday grist for Denver's rumor mill, the man in the street did not know about Tabor's secret divorce and secret remarriage. More seriously for Tabor neither did Augusta. But it was soon clear to Tabor's attorneys that their lovelorn client's political aspirations could well come a cropper if the truth leaked out. To accede to Augusta's suit for separate maintenance would acknowledge her as Tabor's wife. And that, thanks to his St. Louis fling, would leave him with two wives. If Augusta discovered the Durango divorce she could have it declared invalid since she was not informed of the action. Again, Tabor would end up with two wives. And no Victorian, however leavened by the permissiveness of the frontier, would send a bigamist to Congress.

Solution to Tabor's seemingly catastrophic predicament came from anunexpected quarter. A nosy county clerk down in Durango came across the record of Tabor's divorce and sent a copy to Augusta. Shocked, Augusta ran to her lawyer Judge Amos Steck who, by some coincidence, had represented Baby Doe in her action to divorce the bumbling Harvey. Judge Steck in turn contacted Tabor's attorneys and laid before them an ultimatum. Let the separate maintenance suit go through or he would have to release the news of the invalid Durango divorce to the papers. Tabor countered by offering Augusta $280,000 if she would drop the separate maintenance suit and start divorce proceedings. Judge Steck apparently realized that Tabor would stop at nothing to win his freedom and suddenly did an about-face and put pressure on Augusta to give in. Otherwise, he warned, Tabor would use his vast influence and vaster fortune to ruin her in the eyes of the public with trumped-up charges, nasty newspaper coverage, and get the divorce himself, leaving Augusta penniless. Whether or not the judge was right or whether there was collusion no one knows. Nevertheless, on January 2, 1883, Augusta filed for divorce, claiming desertion. The court quickly granted her suit. In exchange for some cash, the Lincoln Avenue home, and an apartment house, Tabor got his freedom.

But when the divorce came, it was too noisy and too late. The legislature met in the middle of January and in a seesaw battle between ex-Governor Pitkin and Tabor that reached a hopeless stalemate, a third candidate got the six-year Senate term on the ninety-seventh ballot. Tabor's vote-getting dollars spread among the law-givers were not enough to overcome the sensational and damning news accounts of the divorce proceedings. Editors printed in boldface letters Augusta's anguished plea before the court: "I wish the record to show that this divorce was not willingly asked for...Oh God! Not willingly! Not willingly." The silver king, whose money, if not his person, was still vital to Colorado Republicans, was given the thirty-day unexpired term of Henry Teller.

Disappointed but not dejected, Tabor left immediately for Washington promising Baby Doe that he would send for her near the end of his term and marry her in public in a wedding ceremony attended by the biggest names in the nation's capital. When Tabor took his seat as senator he drew somewhat less than rave notices. "Vulgar," "ruffianly," "uncouth," and "shambling" were some of the adjectives his colleagues used to describe him in their midst. Some complained that he put a crimp in the Senate's operations by constantly inviting key men out for a drink. But for the silver king, the thirty-day stint was a lark, and he enjoyed every minute of it. And, true to his word, when his term drew to a close he sent for Baby Doe.

On March 1 at 9 PM, there gathered in one of the parlors of Washington's stately Willard Hotel Colorado's congressional delegation, including Henry Teller, Nathaniel Hill, Jerome Chaffee, and Judge James Belford, circumspectly solemn in boiled shirts and frock coats. Also in attendance were the McCourts from Oshkosh, who came to see their winsome daughter, gowned in $7,000 worth of brocaded white satin, wed to the Carbonate Croesus. None other than the president of the United States, Chester Arthur, also presented his silver-edged invitation to be admitted to the ceremony. Conspicuously absent, however, were the First Lady and the wives of the august Coloradoans. As tenders of society's public mores, they had flatly refused to attend the marriage of a femme perdu to a debauche.

The Reverend P. L. Chapelle of Washington, D.C., performed the Roman Catholic rites. Afterward the happy couple and their guests proceeded to the "collation chamber where, after the bride had cut her cake, the viands were partaken of" and gaiety was the order of the evening.

The next day, however, the astonished whispers of the wives of the Colorado congressional delegation, who knew Tabor and Baby Doe were both divorced and who could scarcely believe the pair had gotten a priest to marry them, reached the ears of Father Chapelle. Hurriedly, the priest returned the wedding fee and indignantly asked Tabor why he did not reveal he had been divorced.

"You didn't ask me," answered Tabor guilelessly.

Chapelle then publicly denounced the deception to clear his name, and very quickly Washington's gossip columnists tracked down for their avid readers all the details of Baby Doe's divorce, Tabor's divorce, and their secret St. Louis marriage. At the news the respectables exchanged smug smiles and "I told you so's," and even President Arthur, it was duly noted, was aggrieved at the revelation.

Tabor and his bride denied any intent to deceive and retreated happily to Denver where they set about buying a $50,000 house and fitting it out with five gardeners, two coachmen, two footmen, three carriages, six horses, and several alabaster statues of nudes on the rolling lawn that caused gasps and shudders among Denver's new society. Once in their lavishly furnished domicile, Baby Doe waited for the ladies to call. It was a wait without end. No one came. But there were other compensations— baby Elizabeth Pearl was born in 1884 and baby Rose Mary Echo Silver Dollar arrived in 1889. There were the gala dinner parties given for Tabor's mining cronies and visiting investors and exciting trips over Colorado's miles of rails in palatial private cars. Perhaps best of all were the hours Tabor and she had together when the calumny and scandal of their earlier years were forgotten and they amused themselves with myriad plans for their silver-lined future.

With Denver's transformation from a raw frontier town to metropolitan civility came a new form of criminal—the swindler, con man, and bunko steerer.

In the summer of 1872 a well-dressed man who went by the name of Arnold quietly showed a handful of rough diamonds around town to a select group of highly respected businessmen. When questioned about the source of the gems, Arnold smiled benignly and mentioned "the San Juans." The word passed quickly and a new fever swept over Denver. Diamonds in the San Juans! The *Rocky Mountain News* fed the rising excitement of the inhabitants with tantalizing daily stories reporting all the latest speculations and rumors. Ex-Governor Gilpin, who was ready to believe anything about his beloved Rockies, gave an illustrated lecture to a jammed house on the new diamond regions of Colorado and New Mexico. But the whole business blew up in a gossamer cloud of deception when an investigation turned up the word that Arnold was a well-known San Francisco swindler with a long history of phony mining investment schemes.

Denver's blood pressure was slowly seeking its normal level after the Arnold affair when there appeared another con man of considerable skill. Erlanger's ploy was a fictitious savings bank. Dangling the lure of high interest rates before the eyes of scores of small depositors, Erlanger unctuously persuaded them to withdraw their hard-come-by cash from the commercial banks and place it in his savings bank. Erlanger, however, failed to mention to his rapt listeners that his bank was his pocket, and when it was full he skipped town leaving behind him a passel of sadder, but regrettably no wiser, citizens.

But the master of all the gamesmen was Jefferson Randolph Smith, whom Denver with some misplaced affection soon learned to call the "King of the Thimbleriggers."

A Georgian, like many another who came to Cherry Creek looking for quick riches, Smith had spent his early life bumming around the Southwest, finally putting on the spurs and chaps of a cowboy. The long, lonesome hours riding the range and the hard cattle drives suited the growing young drifter. At the end of one trail he took the first step toward a new life when he idled some time at a circus and lost a month's wages on the lightning sleight-of-hand of the shell game. As his dollars disappeared they opened his eyes to a new and captivating world. Without delay, the slight, wiry cowboy joined the circus and apprenticed himself to the bunko artist. He was an apt pupil, and in a year or so the twenty-eight-year-old Smith was ready to start out on his own. He chose first to grace booming Leadville with his presence. Here he perfected his specialty, a variation on the shell game, involving bars of soap. Setting up his box on a street corner, Smith drew a curious crowd as he picked up a bar of soap, appeared to wrap around it a $20, $50, or $100 bill, and then deftly enclosed the currency-vrapped soap in plain blue paper. While his fingers flew in the wrapping operation his voice sang out an invitation: "Come gentlemen, cleanliness is next to godliness; buy a cake of soap for the richest bath you ever had. Five dollars can get you one hundred! Step up and clean up!" In a few minutes one onlooker would push through the crowd and lay down his $5. Carefully selecting his cake of soap from the stack of wrapped bars, the "sucker" ripped off the blue paper and triumphantly held aloft a crisp hundred-dollar bill. Others quickly bit and, as Smith planned, found nothing but a 5 cent bar of soap beneath the blue wrapper.

"Soapy" Smith's game was a steady money-maker, and before long he expanded his business to include nearly every bunko scheme known to the West and some that weren't. His corporate structure would have made the dean of a business school proud. Soapy was board chairman, president, and chief executive officer. Joe Simmons, his sidekick from cowpoking days, came aboard as vice president in charge of operations. Doc Baggs took over the marketing of watered mining stocks. "Judge" Van Horn, whose talents ran to jury-fixing and ballot-box-manipulation, headed up the legal department. Contracts were in the capable hands of Charlie Bowers, a solemn basset-eyed swindler, whose repertory of secret fraternal signs and handgrips was unlimited. A pack of light-fingered, glib-talking salesmen completed the force as Soapy diversified from the shell game to lotteries, rigged prize fights, hijacking, and taking tribute money from civil authorities and crooked purveyors of services.

The Soap Gang wormed their way to prominence with patient endeavor. Police looked the other way when an irate victim of the soap game squawked about his treatment. Nor did they appear to notice the string of "bandit barbers" who, once they got a man in the chair, gave him everything from a haircut to a manicure and slapped him with a bill for $20. If he complained, the brawny barbers took him into the back room, worked him over until his freshly shaved face was a pulpy mass of bruises, and then threw him into the alley.

As long as Soapy's attentions were centered on strangers and not residents he was left strictly alone, and sometimes even commended. When Glasson's Detective Agency moved into town and abrasively set about, uninvited, to uncover Denver's underworld and its connections with business and government, the Soap Gang moved quickly. Breaking into the agency's office, they pistol-whipped Glasson and put the torch to his incriminating files. For that bit of heroism Soapy Smith received the warm kudos of both the law and the people of Denver.

But a little later when another grifter with the penny-dreadful name of "Rincon Kid" Kelly arrived to challenge Soapy's supremacy and brought open gang fighting to Denver's streets, the town reacted in alarm and beat the drum for a city-wide clean-up. The cry was loud enough and the sweep of the broom vigorous enough that both the Rincon Kid and Soapy Smith left town to let the furor die down and let Denver settle again into its old habit of peaceful coexistence between the moral and the immoral.

Robust, high-living Denver, playground of Colorado's nouveau riche, mecca of Europe's ennui-ridden society, gave no outward sign in the eighties that she was aware that her life-blood was in jeopardy. Almost unnoticed in Colorado had been the demonetization of silver in 1873. When its effects began to be felt, the Bland-Allison Act, making silver the supplementary coinage of the nation and authorizing mints to buy silver at a fixed price, stabilized the market. But the western silver bloc got a scare when Cleveland came to the presidency in 1884. He supported the eastern capitalists who were ranged against the westerners, whose seemingly unlimited supply of money metal was a threat to the eastern power structure. Despite support from Cleveland, however, the gold bugs lost out, for the time being, when the silver bloc pushed through Congress the Sherman Act, requiring that the US Treasury increase its silver purchases to 414 million ounces of silver each month. For the moment the crisis was over and Denver, oblivious to the portents, continued to have the time of her life.

13

"Holy Moses!"

COLORADO WOULD HAVE ONE MORE SILVER BONANZA BEFORE THE WHITE METAL
became next to worthless on the open market.

Down on the eastern edge of the San Juan Mountains, near the headwaters of the Rio Grande, a handful of ranchers settled a compact little valley watered by a willow-banked creek and isolated from the rest of the main trails and toll roads of Colorado by sheer rising peaks. The only ready access to the valley was through Wagon Wheel Gap, a narrow canyon on the southern edge carved by the creek as it flowed down to meet the Rio Grande. The pastoral tranquility of the picturesque valley had gone undisturbed throughout the boom days of Silverton, Ouray, and Telluride and the Leadville bonanza. But in 1889, a cowpuncher named Nicholas C. Creede changed all that.

Creede was a drifter. Born William Harvey in Fort Wayne, Indiana, in 1842, Creede changed his name and headed west when his best girl up and married his brother. Taking odd jobs he worked his way across Illinois to Iowa where the Civil War caught up with him and he joined the Union Army. He spent most of the war as a scout, his eyes learning to catch the smallest detail, his powers of deduction disciplined and sharpened, and his body honed to iron toughness on the long forays into the wilderness in all kinds of weather with the barest ration of hardtack and sowbelly for subsistence. It was ideal training for a man whose consuming occupation would be scouting the countryside for riches in the earth.

In 1889, Creede was no spring chicken. He was forty-seven years old and he had spent the last nineteen of those years in Colorado looking for his big bonanza. So far, his search had been less than moderately successful. In 1878, he had turned up a little silver float west of Salida, and a flash in the pan strike a few months later netted him a quickly spent $13,000. But like all those who ferret the ground for fortunes, Creede was not discouraged. He moseyed over southward from Salida, poking up and

---

down the gulches and creeks until he came to Wagon Wheel Gap. On a whim he made his way into the peaceful little valley and took a job punching cattle. In his spare time, he scrambled over the sides of the sharp-peaked mountains that lipped the valley on the north, examining every outcrop of rock with his scout-trained eyes. If there was pay dirt in the gulches around the valley, he was determined that it would not elude Nicholas Creede.

In May, Creede was working his way up the east fork of Willow Creek when he found a few specks of silver float, enough to cause him to follow its trail a mile or so upstream into a canyon so steep and narrow that by three o'clock in the afternoon the sun was shut out, casting the rocks in gray shadow. Here on the west side of the creek the particles of silver appeared to concentrate. Creede looked up at the steep canyon wall and picked out a crevice that would give him some footholds. Then he started climbing slowly, casting his gaze over the rock underfoot, reaching for small shrubs to pull himself upward, half the time on his knees. About halfway up, Creede hauled himself onto the top of a small ledge jutting out beside a tall pine tree. The exertion of the climb in the high altitude left him gasping, so he decided to rest on the ledge before tackling the rest of the slope. While the rest of his body relaxed, Creede's ever-restless eyes wasted no idle gaze on the tawny beauty of the valley below framed by the canyon walls and bathed by the late afternoon sun but darted over the rocks around his perch in the sky. Suddenly, above the ledge he caught sight of a yellow-green outcropping of porphyry. Prying a chunk loose Creede turned it over carefully in his hand and then let go with a yelp that reverberated all the way down the gulch.

"Holy Moses!" Creede cried, "chloride of silver!"

His elation was justified. The Holy Moses lode turned out to be a vein of silver ore five feet wide assaying at $80 a ton. The mine was a respectable producer, and Creede, with the help of three men, worked it until the winter cold drove them from their wind-driven slope. Ruminating over his discovery in front of the potbellied stove in the ranch bunkhouse, Creede realized that to develop the Holy Moses properly he needed capital and a railroad to get the ore to the smelter as cheaply as possible.

There was one man in Colorado who could provide both. Without further hesitation, Creede tucked some samples of the Holy Moses lode in his saddlebag, got on his horse, trotted the ten miles to Wagon Wheel Gap, swung aboard a D&RG coach, and rode to Denver to see its president, David Moffat.

In the meantime, Creede's discovery was heralded over the state and a steady stream of men wound their way to the once peaceful valley, their eyes glazed with visions of mounds of silver awaiting their shovels. Here and there along Willow Creek sprang up a log cabin next to which was a crude shaft and a happy miner surrounded by sacks of ore waiting transport down the mountain to the railhead at Wagon Wheel Gap. The fall of 1890 saw a mass meeting of miners and the formation of the King Solomon Mining District. Along the east fork of the creek, the few tent cabins strung along the narrow creek bank called themselves the camp of Willow.

In Denver, Nicholas Creede made his point with considerable effect on Dave Moffat. The upshot was that the boy wonder of Colorado railroading sent the manager of his silver-rich Leadville mine, the Maid of Erin, to verify Creede's find. The ex-scout's story checked, and Moffat promptly bought the Holy Moses for $70,000 and put the eagle-eyed Creede on his payroll as a prospector, guaranteeing him a salary and one-third interest in any discoveries he made.

Creede wasted no time going back to work. Gophering his way over Campbell Mountain and down to west Willow Creek, he started up the side of Bachelor Hill when he ran across three Germans busily digging a shaft to get at a vein of purple quartz that assayed high in silver. Ted Rennica, the leader of the group, elatedly told Creede how he and his friends had started out with a $25 grubstake and had just about reached the end of their food supply without finding any promising leads when their independent-minded burro broke loose from its tether and strayed off into the rocks and brush of the hillside. The Germans went after it, and to their disbelieving eyes the trail led directly to the rich vein they now were feverishly working. Rennica and his partners named their claim the Last Chance. They told the attentive Creede that they already had enough ore sacked to bring them close to $50,000.

Creede noted well all that the German told him, and with his practiced eye he carefully sized up how the vein of the Last Chance lay. Then, acting on little more than a hunch, Creede skipped back to the tent town at the base of Campbell Mountain and sent off a letter of resignation to Dave Moffat. With that item of business taken care of, he packed some supplies and headed back to the Last Chance. After surveying the surrounding area he finally stepped off a claim adjacent to the mine of the Germans in a direction his hunch told him the vein of silver-bearing purple quartz led. His hunch paid off. A few feet below the surface he uncovered the continuation of the Last Chance vein. An assay showed that the ore was as rich or richer than that of the Germans', and it was all his. Quickly, he hired some miners, dug two shafts, one hundred feet apart, to a depth of eighty feet and started production. In December, when Moffat's D&RG spur from Wagon Wheel Gap reached the junction of the west and east forks of Willow Creek, wagonload after wagonload of bright purple ore waited to be shipped to the smelter. When Moffat himself heard about Creede's new discovery he wired the cowpuncher, offering to buy him out. But Creede refused. He knew he had a good thing and he was not about to share it with anyone. Between December of 1891 and December of 1892, Creede took $2 million out of the Amethyst, as he named his purple-ored bonanza.

With the announcement of the richness of the Amethyst, and the coincidental impetus of the arrival of the railroad, the latest and last of Colorado's silver camps burst into new life. The tent town deep in the narrow gulch on the east fork of Willow Creek was renamed Creede in honor of the cowpuncher's discovery. Overnight, clapboarded, false-fronted buildings were hammered together, squeezed in between the rock wall of the canyon and the tumbling creek. The town spilled out downstream into the north end of the valley where there were a few more feet of real estate to build on. Here in Jimtown, a name that quickly was corrupted to Gintown in recognition of the fact that every other building housed a saloon, lots that were free for the having in the morning sold for $200 by nightfall. No townmaker had a chance to get his manipulative hands on Creede. It grew too fast and like Topsy. Ten thousand people crowded into camp scrabbling for pieces of land. Lots were claimed by the simple expedient of driving four stakes into the projected property, one at each corner of the lot, and tacking a piece of paper with the claimant's name on it to one of the stakes. The rule of the camp was that construction had to begin on a piece of real estate before nightfall of the day it was claimed or it would be forfeited. As the race to build swept the frenzied camp, prices soared. The demand for lumber outstripped the supply provided by the seventeen hastily erected sawmills and lumberyards that filled the air with the whine of sawing twenty-four hours a day. Telegraph wires quickly strung from Creede to Wagon Wheel Gap and Alamosa to connect with lines to the East hummed with the news of another Leadville. Hundreds of adventurers stamped with the impression, as Leadville editor Carlyle Channing Davis put it, "that the streets are paved with silver, and that all one has to do is get a wheel-barrow and a shovel, go out into the suburbs and bring in a load of standard silver dollars," clapped on their derbies and headed for Creede.

In Denver, serving, as always, as a gateway to the mineral riches of Colorado, the silver-seekers found more fuel to feed their excitement over the new boom camp. The name of Creede, said one traveler passing through the capital, "faced you everywhere from billboards, flaunted at you from canvas awnings stretched across the street, and stared at you from daily papers in type an inch high." The D&RG ran double sections to Creede every day of the week. The cars were so crowded that men sat on each other's laps, on the arms of seats, stood in the aisles, and clung to handholds in the vestibules. In four months the cost of construction of the line was paid for, so heavy was the traffic.

The thousands of people who stormed the new diggings created a monumental housing shortage. Rough board buildings calling themselves "hotels" dotted Creede's crooked single street. A new arrival could rent one of the sixty cots jammed into a hotel's only room for

$1.50 a night. If he cared to risk catching pneumonia in Creede's chill mountain air, he could forego the blanket that went with the bunk and pay only 50 cents. A better deal was to wangle a berth for $1 in one of the ten Pullman sleepers that Moffat obligingly kept on a side-track to ease the shortage of beds in town.

Richard Harding Davis, sent by the august *Harper's Monthly* to cover the phenomenal rise of Creede, found "not a brick, a painted front nor an awning in the whole town. It is like a city of fresh card-board. In the street are ox teams, mules, men, and donkeys loaded with ore crowding each other familiarly, and sinking knee deep in mud. Furniture and kegs of beer, bedding and canned provisions, clothing and half-open packing cases, and piles of raw lumber heaped up in front of the new stores...It is more like a circus tent which has sprung up and may be removed on the morrow, than a town." In the stores, some of them still simply wooden frames with can-vas stretched over the boards, shoppers found that flour, bread, and meat cost them 25 percent more than in Denver. In the hastily rigged saloons, a miner could slake his thirst with beer at 25 cents a glass or whiskey, its volume generally increased with the addition of plugs of tobacco, alcohol, and a little distilled water, for $1 a glass.

On a brittle, freezing day in January, 1892, Lute Johnson rode into town and immediately set up his handpress to issue the first edi-tion of the Creede *Candle*. The paper was immediately popular and it succeeded in quickly putting out of business two other weeklies that dared to compete for the eyes of Creede's readers. A few months later, Johnson gave over his editorship to Cy Warman, a colorful maverick known around Colorado as the Poet of Cochetopa.

Warman started out his working life as a locomotive engineer, but he soon found that he could make as much or more money from the verses and ditties that sprang into his mind as he highballed his twenty-car-freights down the track, so he compromised. He went to Denver to edit a railroad magazine. In July, 1892, Warman was hired by the owner of one of the *Candle's* rivals, the Creede *Chronicle*, who hoped Warman could save his fast failing sheet. But it was too late, and when the *Chronicle* shut down for the last time, Lute Johnson rubbed salt into the wounds of his defeated opponent by immediately hiring Warman to be the editor of the *Candle*. Warman's allegiance to the *Chronicle* had hardly been established when the paper folded and what with finding himself out on the street and short of funds for his next meal he found no difficulty in taking up his new banner. Under his felicitous hand the *Candle* burned brighter than ever.

The town teemed with humanity. Miners, bankers, gamblers, prostitutes, grocers, speculators, drifters, and Ute filled the narrow, mud-running streets at all hours. At night jostling crowds moved up and down Creede's main street under the garish light of giant flares that burned all night to illuminate the town, prompting Cy Warman to pen a line that became famous throughout the West. "It's day all day in the daytime and there is no night in Creede."

The flares that lit the town soon gave way to electricity. The enterprising general manager of the Denver Consolidated Electric Light Company organized a light company for Creede on February 1, 1892, and cavalierly promised lights on by midnight of February 6. In every gambling den and saloon in town, odds were immediately posted and Creede, in the midst of its headlong rush to riches, paused briefly to drop a few thousand dollars on the latest popular game of chance. On February 2, it started to snow and for four days carpenters and linemen worked frantically in a raging blizzard from 7 AM to midnight in temperatures reaching as low as -18^0 F to erect buildings, boilers, and dynamos, and poles and wiring. At 10 PM on the sixth, two hours before the deadline, the fires were lit beneath the boilers and crowds of people began to gather at the powerhouse. At 11:15, steam was valved to the engine. Slowly the flywheel turned, then it gained speed to spin in revolutions faster than the eye could discern, setting the dynamos whirring. At 11:45 the switch was thrown, and from twenty-three arc lamps hung along the street, light flooded Creede to be greeted by a cheering mass of mittened and mackinawed onlookers, both winners and losers.

Shortly after the advent of electric lights, a few people in town decided Creede needed a town government. They elected Evan Morton to the post of mayor and named six councilmen to help him run the camp. No attempt was made, however, to give the new government a police force to back up its jurisdiction. At their first meeting the mayor and the council took on the weighty business of deciding on an emblem for the camp. After some deliberation a crossed pick and shovel was appropriately selected. The next item on the agenda was the important matter of how to pay for the luxury of self-government. There was only one way, and in recognition of the inevitable, the council dutifully created a source of revenue through that mixed blessing of civilization, the levying of taxes. In setting up their revenue structure the council clearly recognized that although the prosperity underlying the booming town was brought about by the production of its silver mines, the economic base of Creede itself rested solidly on two other concomitant forms of commerce. The action of the council was reported, with a rare absence of editorial comment on the front page of the *Candle*, "...assessments: saloons $5.00; fancy houses, each girl $2.00 a month and landlady $10.00; other branches of business, $2.50 a month..." Reading their papers over a dollop of rye, the gamblers, saloon keepers, and the frail sisters on the row rocked with mirth at the announcement. With no means of enforcement, Morton and company had about as much chance of collecting the revenues they needed as drawing a royal flush, and the camp went merrily on its governmentless way.

Trading his silver star for a green eyeshade, Bat Masterson arrived in Creede in the first days of its beginning to manage the Denver Exchange, a gambling and drinking house of comparatively high caliber. Masterson was just the man for the job. Dressed invariably and impeccably in a lavender corduroy suit, white shirt, and black string tie, the dapper ex-lawman exhibited the calm competence of a man long accustomed to being in command. A man might lose, and heavily, at the tables in the Denver Exchange, but it would be an honest loss. And a man might linger over his drink in quiet conversation secure in the knowledge that he would not suddenly have to dive for cover to avoid being caught in the crossfire of a shoot-out. Masterson tolerated no gunplay and quickly rid his saloon of any incipient troublemakers. Not that Creede had many. For the first few months of its existence the town was uncommonly free of the killings and violence that marked the birth of most Colorado boom camps. To Masterson, who had been around the West long enough to know whereof he spoke, this was but a foreboding calm before the storm. As he remarked to one of his dealers, "I don't like this quiet. It augurs ill…I have been in several places that started out this way and there were generally wild scenes of carnage before many weeks passed… It seems as if there must be a little bloodletting to get things into proper working order." Masterson would soon see the accuracy of his observation.

Creede was wide open. The town government was a laughing-stock. What authority there was rested in the hands of a tall, thin figure whose skin bore the pallor of dissipation, and whose right hand jerked impulsively toward his gun whenever his darting eyes picked up the sight of a stranger in town. Bob Ford had reason to be jittery. Ten years before he and his brother, Charlie, had connived to kill their friend Jesse James and collect the $10,000 offered by Governor Crittenden of Missouri for the capture of the outlaw dead or alive. The Fords traced James to a small cottage outside St. Joseph, Missouri, where the outlaw was hiding out peaceably enough under the name of Howard. James let the Ford boys stay with him, and stay they did, watching and waiting for their chance to take the notorious gunman. On a warm April day James decided to open the front door of his house to let the air circulate through it. But before he did so he removed his holster and guns, in case a passerby might think it suspicious that he was wearing arms in the house. Tossing his guns on the bed, James opened the door, and turned back into the room. As he did, his eye caught sight of a crooked picture on the wall, and with his usual meticulousness James stepped up on a chair to straighten it. This was the moment Bob and Charlie Ford were waiting for. They both drew their guns. At the sound of clicking hammers James started to turn but he did not live to face his antagonists. A slug from Bob Ford's gun sliced into his back and he arched onto the floor at the feet of his killer, his open eyes and mouth frozen in an expression of incredulity at the moment of death.

Bob and Charlie turned themselves in, were convicted of murder, and sentenced to be hanged. Two hours after sentence was pronounced, Governor Crittenden pardoned them. But instead of the $10,000 they had been promised, they were rewarded for their deed with only stares of abhorrence and vicious epithets. After a year or so of ostracism, Charlie committed suicide and Bob left Missouri, turning up in New Mexico where he turned a few bucks relating to sideshow audiences how he cut short the life of the infamous Jesse James, and then he ambled into eastern Colorado where he ran a small saloon near Walsenburg. When the Creede boom erupted, Ford packed up his stock of rotgut, his cases of plug tobacco, and his decks of marked cards, and slipped into the silver camp. On Creede's main street he put together a barn of a building and opened Ford's Exchange. The saloon, not unexpectedly, prospered. It prospered so well, in fact, that on the side Ford ran a battalion of pimps who worked Creede's main street, alleys, and dives for customers to sample the wares of such inelegantly named tarts as Slanting Annie, Lulu Slain, and the Mormon Queen, and operated a troupe of bunco artists who sold greenhorn arrivals everything from salted silver claims to phony lottery tickets. Smug and taciturn when sober, garrulous and touchy when drunk, Bob Ford bossed Creede with taut insolence. But prosperity and prestige brought no peace of mind to the killer of Jesse James. Gnawing at his guts was the knowledge that someday one of the outlaw's friends would find him.

Not long after Ford became the acknowledged boss of the bustling camp, there came into his domain another threat not to his life but to his prestige. Jefferson Randolph Smith, observing the latest wild escalation of silver fever in Creede from his faraway headquarters in Denver, had no intention of sitting out the Creede boom. Gathering up a few of his lieutenants, Soapy caught the D&RG downtrain and comfortably sat back in the green plush seat to map his strategy as the mountain-studded countryside rolled by the windows of the coach. When he arrived in the new camp, Soapy and his friends sauntered around the town taking stock of the saloons and gambling dens. As they edged around the knots of men standing on the street corners and through the crowds milling in the narrow street, long looks followed the slight, dark-eyed Smith whose lint-free, neat black suit and wide black sombrero set him apart from the rumpled and haphazardly dressed populace. When the brightness of Creede's arc lights caught the two-carat diamond stickpin in Soapy's black cravat, observers nudged one another, exchanging looks that Bob Ford might have read, and rightly so, to mean that he was about to be challenged as camp boss.

Smith started his takeover on a very small scale. First it was the shell game, then a couple of rounds of his favorite soap gimmick, and

soon the king of the thimbleriggers was well on his way to corona-
tion in Creede. Over in Ford's Exchange, the killer of Jesse James
grew more jittery than ever. After watching Soapy Smith pull away
a good chunk of his racketeering profit, Ford acted. Through one of
his men he sent Smith an ultimatum telling the con man to leave
Creede voluntarily standing up or he would see to it that he left lying
down in a white pine box. Smith was delighted. This was the crack
in the wall he had been waiting for. He quick-stepped over to Ford's
saloon and closeted himself with its owner. Behind the firmly closed
doors of Ford's stuffy little office, and with all the finesse and unction
of a long-perfected technique, Smith soft-soaped Ford into relenting
his hard line. The smooth-talking Soapy assured Ford that there was
enough swag to be had in Creede to support both of their operations
and that he had no designs on Ford's role as camp boss. To prove he
meant what he said, Soapy cut Ford in on what was about to be the
bunco artist's most audacious scheme.

A few weeks earlier in a back corner of a junkyard in Denver,
one of Smith's friends, gambler Bob Fitzsimmons, found a full-sized
figure of a man made of cement and plaster of Paris. So realistic was
the manikin, despite the few spots of erosion, that Fitzsimmons pondered aloud to Soapy that there must be some way to exploit his find
for profit. Soapy saw immediately the virtue of what Fitzsimmons
was saying. But he also knew that Denver was by now too sophisticated to swallow a hoax of a kind that came to his mind. Raw young
Creede, on the other hand, was precisely ripe for just such a play on
its powers of credulity. Under great secrecy, Smith had the figure,
which he dubbed "Colonel Stone," wrapped and crated securely and
shipped to Creede, where in the early morning hours after its arrival
he had it clandestinely hauled up one of the gulches forking Willow
Creek canyon and partially buried. Then Smith and Fitzsimmons sat
back to wait.

Sure enough, on April 9, a prospector stumbled onto the partially
exposed figure of Colonel Stone and raced down the rocky incline
to town where he babbled the news of his find to any and all who
would listen. Within two days the word had percolated all the way to
Denver. "J. J. Dore finds the pretrified body of a man near Creede," headlined
the *Rocky Mountain News* on April 11. As soon as Dore blurted out
his story, Smith gave his men the signal to retrieve Colonel Stone and
bring him down to Creede. Dutifully, Fitzsimmons and several others
let Dore lead them to the site of his discovery and, with poker faces
that would have been a tribute to their profession, they unearthed the
statue and carted it in a hired wagon down to Soapy's temporary headquarters in the Vaughn Hotel. Barroom speculation on the creature's
identity favored the notion that it was an unfortunate of Frémont's
first expedition that passed up the valley in 1842. But Soapy squashed
those rumors by hinting that the figure had a far more significant

origin which he would shortly announce as soon as he had finished his scientific investigation of the case.

In a day or so the stage was set. Colonel Stone was moved to lie in state in a room in Bob Ford's Exchange, his lifelike features eerily lit by flickering kerosene lamps. The "petrified man" lay on a dais that was slightly inclined so that the audience could have a full view of his realistic face. In front of the dais were benches and beside it was a hastily built lectern. Soapy sent his men out on the muddy springtime main street of Creede to hawk handbills proclaiming "The Prehistoric Man! The Missing Link! See him in the flesh (petrified) and hear his anatomy and life story described in all its particulars by Professor Jefferson Randolph Smith! (Ladies will not be admitted.)"

The combination of Colonel Stone and Soapy Smith was a sellout. For weeks, legions of men crowded into Ford's saloon, the chime of their 25 cent admissions in Bob Ford's pocket dissolving any doubts that remained in his mind about Soapy Smith.

But what Ford failed to realize was that by making a deal with Smith he had taken the first step down from his pinnacle of power. The rest of the way he would snowball. With Ford securely hooked, Soapy continued his career as lecturer in anthropology just long enough to clear sufficient money to set up his own headquarters. Down the street from Ford's place, Soapy opened a dazzling watering and gaming house he called the Orleans Club and he let it be known that he, not Ford, was the boss of Creede. Before Ford could mount any opposition, Soapy, in a series of moves that left his opponent blinking, imported his hard-riding brother-in-law, John Light, to take over as town marshal, persuaded the townspeople to hold new elections for a municipal government, selected candidates from among the town's respectable businessmen for the jobs of councilmen, justice of the peace, and coroner, and rigged the election so that his favorite sons won. With the forces of law and order at his command, Soapy ran Creede with a style all his own. No grifters were allowed into town to ply their trade without Soapy's okay. No tramps were tolerated. Word went out that visitors to Creede were not to be swindled unless they were obviously well-heeled and showed the easy come, easy go characteristics of spendthrifts. New, legitimate businesses were encouraged to set up shop in town. Bordellos, bars, dance halls, and casinos were forced to contribute to the cost of operating the city "government" by forking over to Soapy a percentage of their take. Suspicious newcomers to Creede were treated to having the heels of their boots shot off by crack shot Marshal Light as a warning that no hit-and-run crimes would be tolerated. Generally, after slamming home his message, the marshal bought his victim a new pair of boots. Soapy freely indulged in charity. Every "bride of the multitude" who died while Soapy reigned received a decent burial. And the con man had a special fondness for religious causes. When he caught sight of a gang of rowdies bedeviling a street corner preacher, he ordered the taunting group to appear at the parson's church for services on the following Sunday.

"But Mr. Smith," apologized the preacher, "I have no church."

Soapy cogitated a moment and then grabbed the parson by the arm. "You will by next Sunday!"

Hauling the bewildered preacher from one saloon to the next Soapy collected $600, and in three days the clapboard house of worship was ready. On Sunday, Soapy and his gang led an overflow congregation in the hymn singing, their mixed baritones and tenors fervently heard above the rest on the line "Free from the law. O blessed condition!"

When Parson Tom Uzzell came down from Leadville to breathe a little fire and brimstone into the tobacco smoke-filled air of Creede's saloons, he collected a tidy $75 on his first night's rounds. But during the night while he slept the untroubled sleep of the innocent, some of Soapy's henchmen in a prankish mood stole into the parson's room and lifted his trousers with the money in the pockets. When Soapy heard about it, he was livid. He made his men return the trousers and the $75 with $25 of their own money as interest.

When Soapy's high-toned friends from Denver dropped in to see him, it was a cue for an elaborate champagne supper at which Soapy played the genial host and raconteur. Yet he was just as cordial to renegade outlaws as he was to men within the law. When Dave Rudabaugh, a cattle rustler fleeing the law for the killing of the sheriff of Las Vegas, New Mexico, slunk into town, Soapy took him under his wing, fitted him out with a new haircut, some new clothes, suggested he grow a mustache, and put him to work in one of the mines in which Soapy had an interest. Soapy's credo was simple. As long as he behaved himself, according to the rules set down by Soapy Smith, any man was welcome in Creede.

Life went rosily forward for Soapy until the following spring when one of the dealers at the Orleans Club, Joe Simmons, came down with pneumonia. Had it been any other of his men, Soapy would have seen to it that the man was properly cared for and have put the matter out of his mind. But not with Joe Simmons. Joe was special. The two men had grown up together, and when the name of Soapy Smith became synonymous with power in Denver, Soapy sent for Joe to join him. Together they had built and kept running smoothly the highly efficient underworld organization that was the bane of Denver. Now, with his friend lying grievously ill Soapy was distracted. For days he never left the dealer's bedside, bathing his hot dry skin with fresh cool water from Willow Creek and patiently spooning small mouthfuls of broth between Joe's fever-dry lips. But for all of Soapy's ministrations, Joe's condition grew worse and finally he slipped into a coma and died on March 19, the feast day of his namesake, St. Joseph. Grimfaced with the strain of his vigil showing on his gray, hollowed cheeks,

Soapy announced Joe would be buried the next day and according to his last wishes there would be no funeral. Instead, as Joe had wished, his friends would gather around his grave and drink to his health.

On the morning of March 20, Creede awoke to a sky of blowing snow, but the plan to bury Joe Simmons went forward despite the inclement weather. The cortege was like no other Creede had seen. In the black-draped ore wagon carrying Joe's casket rode a case of Pommery champagne, the bottles clinking merrily as the wagon made its jarring, creaking way up the rutted, rock-strewn road to the cemetery located atop a rise west of Gintown. His grief masked by an irreverent gaiety, Soapy walked beside the wagon, cracking jokes with his cohorts whose flashy brocaded vests lent a touch of garish color to the gray air filled with swirling snowflakes that laid a soft mantle of white over the hard features of the landscape. Up the steep, slippery road they climbed, boots sliding and wheels spinning. Every time one of the pallbearers slipped to his knees in the mixture of mud and snow on the road, the air turned purple as he let go a string of oaths. Midway up, Soapy called a halt for a moment's rest. He looked around at the glum faces of his grunting, panting friends. "Damn thoughtful of old Joe," he quipped, "providing a snowstorm. No need to ice the champagne." His words brought quick bursts of laughter from the men, and they started up the road again with new determination. Just when they reached the crest of the hill, there was a sudden rumbling and clinking as the coffin and the basket of wine began to slip backward in the bed of the wagon. As the men started toward it, the tailgate abruptly banged down and the casket shot out of the wagon and down the hill. "Save the champagne!" roared Soapy, as a dozen hands grabbed the basket of wine just before it tumbled after the tobogganing coffin. In the midst of shrieks of hysterical derision and heartfelt curses, Soapy and his friends dragged Joe's battered casket back up the hill at the end of a rope, and when their friend was safely deposited in his grave, Soapy popped open the Pommery and passed the bottles around. Soapy led the toast, and Joe's friends drank to his health as requested, rendered one shaky chorus of "Auld Lang Syne," drained the bottles of champagne, and weaved their tentative way down the snow-sodden road to town.

Despite his outward indifference to life and death, the passing of Joe Simmons seemed to work a change in Soapy Smith. He became more and more preoccupied and now and then appeared to have lost his self-confidence. His usual gregariousness gave way to brooding solitude. And no more did he talk of the great works he would accomplish in Creede.

The change in Soapy was reflected by a change in Creede, and Bat Masterson's earlier prophecy came to fruition as Creede slipped headlong into a savage period of carnage.

On March 31, Reddy McCann, a faro dealer, got himself a bellyful of whiskey and proceeded to shoot up Creede's main street at four in the morning, shattering windows and narrowly missing townspeople asleep in bedrooms fronting the street. Marshal John Light, awakened by the noise, followed the raging McCann to the Branch Saloon, where he found the dealer at the bar. Light reached for McCann's gun. McCann lunged at the marshal, shoving him backward. Light whipped up his hand and slapped McCann's face, flipping the surprised dealer's cigar out of his mouth and sending it flying over the bar. McCann, his cheek reddening from the blow, pulled his gun, but before he could cock it, Light's revolver flashed fire and the dealer slumped to the floor, mortally wounded. The coroner's jury exonerated Light and called it self-defense, but not everyone in town agreed with the verdict and after several weeks of threats and near misses by snipers, Light resigned his post and fled Creede.

Under Soapy's manipulative guidance, the town elected a new marshal, a tough German saloon owner from Upper Creede. Peter Karg had worn his badge for only three weeks when Jack Pugh, one of Bob Ford's confederates who ran a livery stable, set out on a vendetta against Mayor Osgood. Their dispute was over a lot that Osgood claimed and Pugh had jumped only to have Osgood throw him off. Pugh's grudge magnified with each passing day and on May 4 he stalked into Osgood's Holy Moses Saloon determined to settle his account with the mayor. He sat in a far corner all afternoon downing one shot of booze after another and disinterestedly taking a hand of whist and seven-up as he watched Osgood move around the saloon. By 10 PM Pugh had begun to vocalize his opinion of Osgood. With his speech clogged with liquor and epithets he branded the whiskey rotten, the air foul, the girls surly, and the rest of the patrons scum. When the marshal dropped in for a nightcap, Osgood quickly warned him that Pugh was working up to an explosion and he was afraid he would be the victim. Karg went over to Pugh's table and quietly suggested that the stableman drink up and go home. The marshal's words lit the fuse and Pugh exploded, jumping to his feet, upending the table and sending cards, glasses, and poker chips cascading to the floor. Once on his feet the belligerent Pugh waved his fists in front of Karg's glowering face.

"If you don't go home on your own," roared the marshal, "I'll drag you there myself!"

Pugh grabbed Karg by the shoulders and spun him around. "You try, you damn Dutch son of a bitch, and I'll kill you!" shouted Pugh.

Karg broke loose and turned to face Pugh. The liveryman reached for his gun. Karg pulled his and fired. Pugh fell near the stove, and he lay there for nearly an hour before some of his friends arrived and carried him to his house where he died the next morning. Again, the coroner's jury returned a verdict of justifiable homicide. But by June, Creede needed a new marshal. Like John Light before him, Karg decided that by the number of threats against his own life there were enough people in town who were ready to avenge Pugh's death and he, too, resigned and rode out of Creede.

Two other deaths made headlines in Warman's *Candle*. They did not involve lawmen, but they were symptomatic of the degenerating orderliness of Creede. As the word of the lawlessness and violence of the silver camp spread outside the valley, consternation swept through the offices of companies who wrote insurance for residents of the boomtown. Some canceled their policies outright, others refused to write any more insurance. Particularly reluctant, understandably, to take any more risks in the face of Creede's high homicide rate were the accident insurance companies.

The climax to Creede's bloodletting came in June, 1892, three days after a disastrous fire leveled half the camp. Bob Ford's Exchange had gone up in smoke and he was just putting the finishing touches on a two-by-four and canvas temporary structure to take its place when there walked into the makeshift barroom Ed O'Kelly, a peppery young tough who had arrived in Creede a few days previously and had made Ford's saloon his headquarters. Unknown to Ford was the fact that O'Kelly's wife was the sister of Cole and Bud Younger, two members of Jesse James's old gang. To Ford, O'Kelly appeared to be merely another one of countless swell-headed, dude-dressed drifters that bellied up to his bar. As O'Kelly entered Ford was signing a subscription paper pledging a share of money to help bury one of the town's prostitutes who had died the night before. O'Kelly and Ford exchanged the guarded ritualistic greetings strangers gave each other on the frontier, and then Ford spotted the iridescent stone mounted in the gold stickpin in O'Kelly's yellow silk tie.

"An opal, ain't it?" said Ford. "Shouldn't think you'd wear one of them stones. It's bad luck."

O'Kelly laughed. "Maybe for you, but not for me," he said as he downed his drink and strutted out of the saloon. Within a few minutes he was back, his easy laugh of a few minutes ago replaced by a visage hardened with hate. In his hand he carried a double-barreled shotgun. Just inside the door he waved at the few people standing at the bar to move away, and then he called to Ford whose back was to him. When Ford turned, O'Kelly snarled an unintelligible oath and fired both barrels at the saloonkeeper. The blast raked Ford's head and neck and he pitched backward to the floor, dying in the blood from his gaping wounds that quickly pooled under his head.

As O'Kelly marched toward the door, the new marshal, who had heard the blast as he sat across the street in his office, came running toward Ford's place. Meanwhile, several patrons at the bar grabbed O'Kelly crying, "String him up! Murderer, let's lynch him!" The marshal called for his deputies and demanded the men release the struggling O'Kelly. Not that Ford was particularly popular. He was not. His reputation as a craven coward had followed him to Creede. But even a coward by the frontier code was traditionally allowed the opportunity to meet his antagonist face to face, gun to gun. Grudgingly, with muttered threats and hooded warnings, the crowd of men released O'Kelly to the custody of the marshal. But the word of Ford's killing leaped from mouth to mouth through the town and soon knots of citizens collected ominously on the street corners. From the window of the Orleans Club, Soapy Smith watched the rising lynch fever spread over his town. At one point he sent two of his men to help Marshal Rossen spirit O'Kelly away to an abandoned prospector's cabin up on Willow Creek, a couple of miles from town. As nightfall came the groups of men coalesced into a mob, and with blue steel gun barrels gleaming in the glare of Creede's all-night lights they marched on the marshal's office. Soapy Smith was there to meet them. With flashes of his old spellbinding oratory he cajoled, humiliated, threatened, and uplifted the mass of agitated would-be hangmen until at last they put away their weapons and wandered off, to let the work of the law be done.

Ed O'Kelly was convicted of second-degree murder and sentenced to life imprisonment. After serving ten years, he was pardoned in 1902. He would have lived longer had he stayed in prison, but his fiery temper earned him a fatal bullet on a summer's day in 1904 when he manhandled a policeman on a street in Oklahoma City. The remains of Bob Ford were first buried in the Creede cemetery, not far from those of Joe Simmons. Later, however, his common-law wife, Dot, had them exhumed and taken to Missouri for reburial in the Ford plot, writing an end to the legendary saga of Jesse James.

After his brief return to his old self, Smith once more lapsed into the apathy that was his habit since the death of Joe Simmons. How deep his withdrawal went was shown some days after the shooting of Bob Ford when Soapy was visited by a solemn delegation of townspeople. The spokesman of the group declared that it had come to many men in Creede with revealing irony after the night that Soapy dispelled the mob that would have lynched Ed O'Kelly that the only man in town who was able to preserve law and order was a man of the underworld. So, said the spokesman for Creede's citizens, the people of Creede had banded together and resolved to rid the town of its one-man rule and establish a genuinely popular government. Without further embellishment, the leader of the group then asked Soapy Smith to leave Creede. As if he, too, sensed his time was up in the boom camp, the dethroned king of the thimbleriggers pursed his lips in a rueful smile and after a few minutes' contemplation agreed to go.

❧

While the personal fortunes of Creede's citizens rose and fell, the vast silver fortunes realized from the metamorphosed rock of Campbell,

Bachelor, and Mammoth mountains continued on the upswing. The Champion mine, with two thousand ounces of silver to the ton; the Mammoth, aptly named for its twelve-foot-wide lode; the Phoenix; the Pipe Dream; the Happy Thought; and the Commodore were but six of the steady producers in addition to Nicholas Creede's incredibly rich Amethyst. Mine tunnels and shafts lay one and two thousand feet above Willow Creek. To get the ore out, brawny miners loaded it into sacks that weighed out at two hundred pounds or more when filled, and loaded them onto the backs of surefooted burros who picked their way down the narrow, precipitous trails to the wagon road along the creek bed. Here the sacks were transferred to wagons, or in winter to iron-bound wooden sleds, drawn by spans of draft horses, for a breakneck trip to the depot and the waiting ore cars. In the first eight months of rail service, Creede mines sent forth ten cars of ore a day at an average value of $1,000 a car. In 1892 alone, ore containing five million ounces of silver was trundled out of Creede to the smelters.

But there were signs on the horizon that would mean the end for the boom camp. Out of the surfeit of its mines would come the seeds of disaster. When, after a long and bitter fight in the US Senate, repeal of the Sherman Silver Purchase Act came at the end of October, 1893, in a matter of minutes the price of silver plummeted to 50 cents an ounce and with it plummeted the fortunes of Creede. Where there had been plenty, there was suddenly nothing—no money, no jobs, no hope. Out-of-work miners, bankrupt merchants, and ruined bankers packed up and straggled out of Creede with what means they could find. The depression took its toll of everyone in town, even among those in the most enduring of professions. "Lulu Slain," reported the Creede *Candle,* "a frail daughter laid aside the camellia for the poppy and passed into the beyond early Wednesday morning. She and the Mormon Queen had been living in a small cabin in upper Creede but the times grew hard and the means of life came not. They sought relief from life with morphine, the inevitable end of their unfortunate kind, a well-trodden path from Creede. Lulu's dead; the Queen lives."

On Nicholas Creede, the man who brought the colorful camp into being, the silver crash had little or no effect. By the time the Sherman Silver Purchase Act was repealed Creede had already made his fortune. But the cowpuncher millionaire would have only five years to enjoy his wealth. In the late fall of 1892, Creede was forced to leave the high altitude of his adopted home on the advice of his doctors. Moving to Los Angeles, he retired to the easy life in sun-baked California. However, life turned out to be not easy but filled with the strife of domestic problems. After several years of wretched bickering and recriminations, Creede and his wife agreed in late 1896 to separate. The silver king gave his wife $20,000 and saw her off to her family home in Alabama for what he hoped would be the last sight of her. But in six months she was back, hounding him for money and attention. Under her barrage of demands, Creede grew increasingly despondent, and on August 2, 1897, a servant found him in the garden of his palatial home, like Lulu Slain, dead of an overdose of morphine.

A few hardy souls stayed on in Creede, some to make a new start offering accommodations to hunters and fishermen and the few tourists who now and then ventured up the scenic valley. But never again was there "day all day and no night" in the once jubilant silver camp.

14

Populism and Panic

THE CHAIN OF EVENTS THAT WOULD SPELL RUIN FOR THOUSANDS OF COLORADO'S silver miners and near ruin for the state itself began several years before the new Eldorado's major silver deposits were found. Almost from the time of its founding, the United States had conformed to a bimetallic money standard, coining gold and silver under an established ratio of sixteen ounces of silver to one of gold. But, by the latter half of the nineteenth century, silver had become worth more than one-sixteenth as much as gold, and, instead of selling silver to the mints for coinage, silver producers more and more sought the higher prices paid by industrial and commercial users. Partially in response to this economic fact and also in response to a general European trend to adopt an exclusively gold standard, the US government suspended the coining of silver in 1873. For the first time silver was without a firm base and was forced to seek its own price level in the market place. The "crime of 1873," as it was later called, was scarcely noticed in Colorado at first, and then when the maws of the mines of Leadville, Aspen, Creede, and the San Juans spewed forth their tons of white metal, the price of silver sagged and the market grew sluggish. Mine-owners and miners alike watching waning profits, dwindling wages, and rising debts clamored for a return to bimetallism.

Their cries for the reinstatement of a cheaper money were loudly seconded by farmers and other low income groups still smarting under the penalty of having to pay debts incurred in Civil War-issue greenbacks worth 35 cents each with gold dollars worth 100 cents each.

In Congress, "Silver Dick" Bland of Missouri, a long-time advocate of bimetallism and leader of the unsuccessful fight to kill the demonetization bill of 1873, was quick to respond to the public pressure. He introduced a bill calling for a return to the unlimited and free coinage of silver at the old 16 to 1 ratio. The House passed his bill and sent it on to the Senate where despite the skillful maneuvering of Henry Teller and the other "silver senators" it ran into a temporary roadblock of eastern "gold bugs." But under the amendations of the chairman of the Senate Finance Committee, William B. Allison, Bland and his silverite friends won a partial victory. The Bland-Allison Act, providing that the Treasury buy and convert to coin not less than $2 million and no more than $4 million worth of silver a month, passed both yea-minded houses by a large majority in February, 1878. But when it got to the desk of the president, "Sound Money" Rutherford Hayes vetoed it on the grounds that it would be dishonest for the government to coin and use dollars lower in value than those pledged to pay the public debt. Congress, however, with an eye toward the fall elections, saw nothing dishonest about providing the voting public with cheap dollars to pay off their debts and overrode Hayes's veto.

Mollified, Coloradoans took some considerable satisfaction in watching the price of silver climb back toward the old mint figure of $1.29 an ounce. For six years, the silver state rolled merrily along in its prosperity.

But despite the subsidy, after reaching $1.21 an ounce the price of silver started to slip downward again as the market was glutted with the white metal. Consternation over the "silver question" set in all over again. Some mine-owners could stand a slip to $1 an ounce, while others would have to close down for lack of a profit. A slide to below 80 cents an ounce would bring mass shutdowns of the mines and catastrophe to all. To make matters worse, President Grover Cleveland, who came to office in 1885, made it abundantly clear that he was staunchly in the camp of the eastern antisilver bankers and capitalists who fervently believed their power was being eroded by the free-coining, cart-wheeling western silver magnates. Throughout the mining West resentment against antisilver forces flared. California, Colorado, and Nevada together produced nearly 75 percent of the nation's output of silver and to them the silver question was fast becoming an all-consuming issue since their economics rested so heavily on the money metal. In 1885 the Colorado Silver Alliance was formed, and four years later the National Silver Convention with delegates from all the western states and territories met in St. Louis. To the reporters covering the event it was clear that free coinage of silver took precedence over all other political questions and even transcended party loyalty.

Nevertheless, when Democrat Grover Cleveland was turned out of office in 1888 and Republican Benjamin Harrison assumed the presidency, the free-silver advocates took heart. In Congress when a protective tariff bill fostered by Representative William S. McKinley came up, the western mining states seized the opportunity to lay down a simple ultimatum: do something for silver or nothing would be done for McKinley's tariff bill. After the initial surprise at the audacity of the representatives of the Union's newest states wore off, Senator John

Sherman of Ohio set his jaw and did what had to be done to get the necessary votes for the tariff measure but to forestall unrestricted free coinage of silver. The Sherman Silver Purchase Act of 1890 guaranteed that the US Treasury would buy 4 1/2 million ounces of silver a month, double the amount guaranteed by the Bland-Allison statute. At the news of the enactment of the Sherman Silver bill, there was much crowing and dancing in the streets of Denver. But it was a short-lived gaiety. Again, after a spirited rise, the price of silver inched downward. And again the old bugaboo of silver, Grover Cleveland, entered the picture. Even after his defeat by Harrison, Cleveland was still the Democrats' standard-bearer and it was clear that he was going to be their nominee for the presidency in 1892. All Coloradoans could think during this period was Cleveland's slashing statement that the Sherman Act was a "dangerous and reckless experiment in free, unlimited, and independent coinage." As a delegate to the Democratic convention that year Tom Patterson, who was now co-owner with John Arkins of the *Rocky Mountain News*, bolted the party when the platform decried the Sherman Act as "a cowardly makeshift frought with...danger..." and carried a statement that called for bimetallism only if parity were achieved between gold and silver.

When the Republicans met that summer to nominate their man and write a platform, Colorado's delegates fought hard for a free-silver plank along with their allies from eight other mining states. But their efforts were useless. Party leaders were in no mood to make an issue of coinage and the platform was, to the silverites, wholly disheartening.

But there was abroad in the country a grass-roots movement that would provide a haven for silverites discouraged by the unyielding stand taken by advocates of the gold standard.

Western homestead farmers, disgruntled that Congress failed to do anything about falling prices caused by huge farm surpluses or about exorbitant railroad freight rates, looked for new representation. Laborers were equally disturbed that few members of either the Democratic or Republican parties were interested in an eight-hour day, arbitration of labor-management disputes, and better working conditions. In addition to these two groups there were assorted malcontents, some of whom pointed the accusing finger of favoritism at the government for making it easy for bankers and capitalists to collect double interest on their investment by allowing banks to buy government bonds on which they got 6 percent interest and then borrow money from the US Treasury using the bonds as collateral and loan that money at another 6 percent interest. No such partnership, grumbled the dissidents, existed for the farmer or the laborer. Others insisted that the refusal of the government to continue to make available cheap money in the form of greenbacks was a conspiracy backed by capitalists to keep the blue-collar worker up to his eyes in debt. Still others carped about the evils of the sweeping McKinley tariffs that gave unwarranted protection to manufacturers. Despite the diverse complaints of the disenchanted, they were all agreed on one point. There was too little money in circulation, and what there was was too dear.

In this motley collection of complainers the free silverites of Colorado and the other mining states found a natural home, and their representatives were vociferously present among the 1,400 delegates who met in Cincinnati in May, 1891, to form the People's party. When the Populists, as the new party came to be labeled in the press, met in convention in July, 1892, to field a slate of candidates and write a platform, they selected General James B. Weaver of Iowa, a greenback party stalwart, as their presidential nominee. The convention then turned out a platform that in the political arena of the day reeked of radicalism. Shot through the document were reflections of the European movements aimed at social reform, combined with promotions of sectional self-interest. It called for free coinage of silver at 16 to 1, the abolition of national banks, a graduated income tax to soak the rich, government ownership of railroads and communication, restriction of immigration, and an eight-hour day for labor.

Colorado seethed with excitement once it found a rallying point. Even before the train carrying the livid Patterson back to Denver from the Democratic convention arrived, co-owner of the *News* John Arkins had an edition on the streets proclaiming the paper's support of Weaver and the Populists.

All across the state from Boulder to Telluride, Creede, and Aspen, and every mining camp in between, the Populist doctrine hummed. Denver, the hub of the state's political activity, quickened. In the onyx-paneled lobby of the Brown Palace, the capital's newest and grandest hotel, knots of the faithful free-silverites exchanged epithets leveled against the ogre Cleveland and the eastern gold bugs. Joining the *Rocky Mountain News* in coming out strongly for the Populist ticket was the Leadville *Herald-Democrat*. The regular Democrats tried to buck the tide by founding their own paper, the *Denver Post*, but it was to no avail. Free silver and the Populists were epidemic. Richard Harding Davis, fresh from his distasteful journalistic sojourn in unexpurgated Creede, listened appreciatively to the sophisticated arguments for free silver coinage put forth in the ultra-exclusive Denver Club and fell all over himself touting the civility of this "thoroughly eastern city—a smaller New York in an encircling range of white capped mountains." The last thing the unsophisticated man in the street wanted to hear was that Denver in any way resembled the hated East, whose big business and banking establishments he believed were somehow to blame for the economic straits in which Colorado now found itself after nearly ten years of unremitting boom times. The headlong prosperity of the eighties caused speculation and high borrowing. Now debt was nearly universal. Colorado's dry farmers on its eastern borders saw drought eat up the profits of good years. Miners who mortgaged their homes, merchants who borrowed on their businesses, and farmers who took out loans for machinery and buildings bleakly faced ruin as wages and profits dwindled.

When election day came Colorado went whole hog for the Populists, right down to the level of county clerk. Old party loyalties dissolved, and for the first time the state left the Republican column in a presidential year. Catapulted into the governor's chair was sixty-seven-year-old Davis H. Waite, a white-bearded and fiery reformer from Aspen. Spared in the Populist sweep were senators Henry Teller and Ed Wolcott, but only because their terms of office were not yet up.

Nationally, however, the story was different. Weaver, the Populist candidate, carried only four states; Harrison and the Republicans suffered a last-minute loss of face owing to a bloody steel strike, after having run on a platform of domestic peace and security; and Grover Cleveland, the enemy of silver, swept into the White House.

Anguished as Colorado was at Cleveland's election, its despair grew to distraction when India, one of the world's major silver buyers, suddenly ceased coining silver and the price of the money metal plummeted to 62 cents an ounce. Major mine- and smelter-owners called a moratorium on production until the price of silver rallied. Denver rocked with panic. People streamed into the streets, eyes glazed, mouthing deprecations as they stormed the banks and held impromptu meetings to denounce the conspiracy against silver. Ten banks failed in three days. Thousands of miners and their families drifted into town as mines shut down. Relief camps supplying beds and food were set up by the Army. Men of wealth were suddenly beggars. A contingent of destitute built a raft and sailed it down the Platte to join Coxey's army for its march on Washington. Telegrams to New York papers reported a million people thrown into the streets and rampant threats of secession and the formation of a "western empire." Tom Patterson published a stirring "Appeal to the Country" in the *Rocky Mountain News* hinting at a permanent political coalition of westerners and southerners against the eastern power bloc.

On July 11 at a mass meeting in Denver Coliseum, Governor Waite stepped to the podium and declared war on the capitalists, European and eastern, on the "monarchy and monopoly against the right of people to self-government." Tumultuous applause greeted his battle cry and when it died down he delivered the words that made him the most quoted man in the country for that week. "And if," shouted Waite as he pointed an accusing finger at the air above the heads of his listeners, "the money power shall attempt to sustain its usurpations by 'the strong hand,' we shall meet that issue when it is forced upon us, for it is better, infinitely better, that blood should flow to the horses' bridles than our national liberties should be destroyed."

Overnight "Bloody Bridles" Waite became a national celebrity, but his campaign never got out of the corral. Except for their stand on silver the Populists found few friends among Colorado's influential Republicans, who rebelled at the seeming irresponsibility and radicalism of the social reforms championed by the party. And even when they found Wake's evangelical approach levied at the silver question the Silver Republicans recoiled. Henry Teller saw fit to issue a statement denouncing the governor's words as "rabid frothings," and the Denver *Republican* charged that Waite's importunate outburst was "criminal folly" that could only hold up to ridicule Colorado's serious purpose in its fight to reestablish bimetallism.

In Washington, the declining price of silver was having a decisive effect. Cleveland watched in dismay as the US Treasury accumulated depreciated silver at nearly $50 million a year, while the country's gold reserves were steadily being depleted due to a deficit balance of payments and a run on government bond redemption in gold by European interests fearful of a radical Administration that would insist on payment in depreciated silver.

Combined with his monetary problems, the president saw the nation slowly being paralyzed in a depression whose causes were much the same as those that sent Colorado on a downward economic spiral. The lot of whipping boy fell to the Sherman Silver Purchase Act, which Cleveland's financial advisers assured him was the cause of the economic crisis. Therefore, on a hot, humid August 7, Cleveland called Congress into special session instructing the legislators that he wanted the government once and for all cut loose from the millstone of the silver subsidy. In Washington the silver senators of Congress, representing 2 percent of the nation's population, girded themselves for the fight over the repeal of the Sherman Act.

In the meantime, in Denver, old "Bloody Bridles," who was trying to carry out some of his campaign-advertised reforms and at the same time find a local solution to the silver crisis, found himself in a war, but not the one he expected. For the moment the silver question had to be abandoned as civil insurrection hit the capital.

❧❧❧

Shortly after he took office the earnest, headstrong governor ordered Denver's Police and Fire Commissions to clean up their graft-ridden ranks. The commissioners, however, were perfectly content with the system of kickback and payoff under which they operated with Denver's underworld, so they ignored Waite's order. Infuriated, Waite demanded that the commissioners resign. They blandly refused, claiming that as legislative appointees they were excepted from executive order. To emphasize their immunity the commissioners occupied their offices in City Hall day and night and gave notice that they intended to stay put. In the governor's mansion, the white-bearded Waite fumed, and in a moment of recklessness he called out the state militia. As three hundred uniformed troops prepared to surround City Hall the beleaguered commissioners sent for help. And to whom did they appeal for aid in this matter of principle? To the legislature? To the citizenry at large? Not at all. They turned to their friend and tactical wizard, Jefferson Randolph Smith. Soapy, who had shaken off the gloom of his Creede days and returned to Denver, was equal to the occasion.

Harking to the Clausewitz line that "war is nothing but a continuation of political intercourse with the admixture of different means," Soapy quickly set about gathering up the admixture. He gave orders to his lieutenants to round up platoons of Denver's blacklegs, con men, and triggermen. While the muster of banditti went forward, Charlie Bowers with a squad of safe crackers and lock artists led a midnight raid on hardware stores for arms and ammunition. Soapy stationed his armed forces along the streets approaching City Hall, on the front steps, at every second-story window, and on the roof of the building. While the commander in chief of the rabble army gave instructions to his forces, punctuated by short bursts from the shiny new stolen repeating rifle in his hands, Doc Boggs commandeered five hundred sticks of dynamite and collared a gang of ex-miners to rig the explosives for detonation at various strategic locations in and around the City Hall. In the morning Denver awoke to the sound of measured footsteps crunching on the brick pavement as the militia, dressed in wrinkled uniforms hastily snatched from stored trunks, moved into place in front of City Hall. Forgotten was the specter of unemployment and debt. Here was diversion that had all the trappings of comic relief in a classical tragedy. Men scrambled into their clothes, clapped their derbies onto uncombed hair, left their wives to light the household fires, and scampered after the troops to watch the fun. Hundreds crowded the streets in front of City Hall as Adjutant General Brooks ordered cannons hauled up and trained on the rebel stronghold. They cheered as the troops dragged up a battering ram ready to force the front door of the hall. Messengers on skittery mounts dashed to and from the governor's mansion. By 9 AM the town strummed with excitement.

Seven companies of Colorado Firsters were called to standby duty to put down the rioting and looting that was certain to follow if a shooting war developed. In a series of conferences with his scouts, militia General Brooks assessed the situation and dispatched a message to the frenzied governor. "If a single shot is fired," scribbled Brooks, "they will kill me instantly and they will kill you in 15 minutes. But if you say fire, we'll fire!"

Waite strode back and forth in front of his desk, sputtering in rage at the rebel's defiance of his authority. The minutes ticked by as gun captains and squad leaders stoically waited for the word to attack. At last, to the governor's mansion rode a delegation of the Denver Crowd to prevail upon Waite to spare the capital a blood bath. Their cool persuasion brought the governor down to earth, and at their suggestion he ordered the militia to disband and agreed to let the courts settle his dispute with the commissioners. The news of Waite's capitulation was greeted at City Hall and on the streets by thunderous groans and cheers. When Soapy Smith, rifle still in hand, appeared at the front door of City Hall onlookers rushed forward to shake his hand and thump him on the back. Denver was still Soapy's town.

For weeks, the issue serpentined through the judicial process and the decision when it came was in Waite's favor; the offending commissioners and scores of their employees in police and firemen's ranks lost their jobs. But the decision came too late for Waite. In the public's eye, the quixotic governor was a laughingstock, good for barroom jokes and clubroom innuendoes.

Having come a cropper on civil reform, "Bloody Bridles" now turned his attention to the persistent crisis of sagging silver prices, mine shutdowns, and unemployment. The imminence of congressional action to scrap the Sherman Silver Purchase Act sent shudders through Waite and the rest of Colorado's free-silverites. They railed against eastern "money changers," and painted vivid pictures of Coloradoans as "serfs of the men who hold the pursestrings of the world." To alleviate the silver sickness scores of proposals were put forth by leading state officials and mine- and smelter-owners. Relief for debtors in the form of modification of attachment laws, and the creation of a state depository of silver bullion and issuance of certificates of deposit which would be accepted as currency in the state were two popular schemes. But the one that the audacious Waite pounced on in exultation was an ingenious if fantastic scheme, promoted by A. C. Fish, president of the Pan-American Bimetallic League, calling for Colorado silver to be sent to Mexico to be minted into Mexican dollars which would in turn be made legal tender in Colorado. Here was the answer to the need for more and cheaper currency to relieve the debtor class. Two birds with one stone: free-silverites and Populists satisfied with one sweeping design. Waite wasted no time sounding out Mexico's President Diaz on the idea. When he received an encouraging reply, the governor called a special session of the state legislature to set in motion the adoption of the plan. In the meantime, the nation's newspapers picked up the story and had a field day ridiculing what they termed "Colorado's fandango dollars."

In Washington, Colorado's two senators, Teller and Wolcott, submerged their mortification over the notoriety Colorado was receiving owing to Waite's championship of fandango dollars, and allied themselves with twelve other solons from the silver states to try to stop repeal of the Sherman Act. The House quickly voted for repeal but in the Senate the silver bloc saw to it that the scrap dragged on for a tense eighty days in the longest filibuster then recorded. Finally, however, on October 30, 1893, Cleveland got his way. The Sherman Act was repealed by a vote of 48 to 37. Two-thirds of the Senate Republicans, Teller and Wolcott's own party, voted for repeal.

Telegraphers tapped out the grim word. In a matter of minutes the price of silver skidded to 50 cents an ounce. Denver papers hastily got extras on the streets. Small groups of solemn-faced men gathered on street corners and in the bars and saloons to read the news. There were no violent outbursts of protest. For most people, already caught in the grip of a depression, there was only a quiet and bitter despair. In a way, the repeal of the Sherman Act was an anticlimax. The steady decline of the price of silver had already ruined small operators. Only the handful of silver barons were left to experience the last ignominy. Of these, the improvident were wiped out.

Improvident beyond imagination was the biggest of the barons, Horace A. W. Tabor. With an estimated annual income of $4 million from his vast silver holdings it was easy for Tabor to believe that the well would never run dry. He and his Baby Doe lived in opulence. No detail of their possessions was not touched with the brush of wealth. Tabor was shaved from a Limoges mug with a brush of sable hair. Three special-order carriages carried them about town—one of dark blue enamel upholstered in light blue satin, one of brown enamel with red upholstery, and one painted black and cushioned in white satin. Livery for the footmen matched. Baby Doe's opera coat of hand-picked ermine skins was the envy of the wives of the Denver Crowd. An elaborate Tiffany tea set stood unused on the crotched mahogany sideboard in the dining room of the Tabors' plushly appointed house. Baby Lillie was christened in a dress of the finest lace trimmed with natural seed pearls. Around her tiny neck was a golden chain from which hung a diamond locket. Wherever they went the Tabors traveled by private palace car with an entourage of uniformed valets, handmaids, and governesses.

As the money poured in Tabor succumbed to the delusion of infallibility, investing in everything that struck his fancy—mahogany forests in Honduras, a life insurance company in Chicago, railroads to everywhere, diamond mines, streetcar systems, grain futures, gambling spas, and, perhaps the least rewarding of his investments, the Republican party. He laughed off the caution of an ardent young congressman named William Jennings Bryan that "economists were beginning to say that silver was as common as dirt, that the rest of the world would not accept it in exchange." In the velvet-draped library of the ornate house on Denver's fashionable Thirteenth Street, Tabor poured over one prospectus after another to find new sumps in which to plunge his treasure. Never once did his mind contemplate the need for a rainy day poke. But in time Tabor's harebrained investments needed bolstering. Instead of liquidating his weak holdings he shored them up with money borrowed on the Matchless and his other heavy silver producers.

Then came the silver debacle of 1893. Saddled with mortgage-ridden mines whose ore was now worthless, with a portfolio of diversified investments not worth the paper on which the stock certificates were printed, and a bin full of outstanding personal debts, unpaid more out of inattention than intention, the silver king caught the full force of the storm. Standing by in bewilderment as his accountant totaled up his debts and assets, Tabor stared for a long time at the columns of figures as the truth seeped into his consciousness. Tabor, the Carbonate Croesus, the Bonanza Baron, the open-hearted, open-handed friend of all of Colorado, was bankrupt.

All he had left after a spending spree of a dozen years in which $42 million slipped through his fingers was Baby Doe, two young daughters, and the boarded-up, worthless Matchless.

Tabor was not alone in his debacle. Blustery Bill Hamill of Georgetown, Jerome Wheeler of Aspen, Jack Morrissey, and a score of other noteworthies were all toppled from their silvery thrones. But if Colorado's man-in-the-street miner gloated over the fall of the tycoons, he did not gloat for long as he realized that his coat tails were tied to those who had owned the mines and smelters. After fifteen rollicking years of plenty there now faced both bankrupt owners and out-of-work miners the prospect of a long exile in poverty.

Part III
Gold Again, 1890—1910

15

Cripple Creek—
"The World's Greatest Gold Camp"

BY 1890, IT WAS GENERALLY SUPPOSED THAT ALL OF THE GOLD WORTH MINING IN Colorado had long since been located. Stream after stream and gulch after gulch had been prospected time and time again until all of the high-grade outcroppings were found. There was plenty of silver left in the state, but it was clearly only a matter of a short time before the nation's antisilver forces would succeed in removing its price support to let it flood the market to worthlessness. On the basis of the overall picture, an economist of the day would have said that Colorado's thirty-year money-metal boom was over. But he would have been dead wrong.

Ironically, only some twenty miles west of Pikes Peak, completely overlooked by the fifty-niners in their rush to find riches at the base of that famous landmark, there still lay waiting to be discovered a treasure box that would make the Clear Creek bonanzas and the placers of South Park look like hors d'oeuvres before a feast of gold. The site of what was to become "the world's greatest gold camp" gave no outward indication of the wealth it held. It was not unlike dozens of other high, gulch-crossed basins of the Rockies. Some of it was above timberline, barren except for summer's tenacious wildflowers; the rest was sere in winter with an adequate but not abundant glass cover in spring and summer. For a number of years the region had been used as a summer pasturage, and a few homesteaders' cabins dotted the sparsely grassed flats lying between the deep gulches. Southward through the basin among occasional clumps of aspen and stunted spruce meandered Cripple Creek, so named by drovers after a frightened calf leaped a fence and plunged into a gully breaking its leg. To the west of the eleven-thousand-foot-high valley rose a sharply angled prominence with the absurd name of Mount Pisgah, and to the north and east were a series of granitic peaks of lesser height to completely rim the basin on three sides.

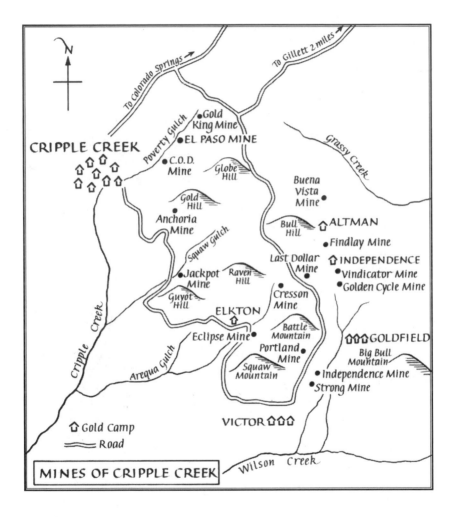

MINES OF CRIPPLE CREEK

One of the homesteaders' cabins belonged to Bob Womack, a happy-go-lucky cowpoke chiefly renowned between Cripple Creek and Colorado Springs for his never-miss ability to lean precariously from the saddle of his galloping cayuse and snatch a bottle of bourbon from the ground with his teeth. With his father and his brother, Will, bland-faced Bob knocked around the various gold diggings of Colorado for nearly ten years before he built a little cabin and settled on one end of the abandoned acreage Will picked up on Cripple Creek. Here, between lackluster performances as a cowboy, Bob prospected. He wasn't sure what he was looking for but he went on looking, concentrating his efforts on a draw he named Poverty Gulch to match his financial status. One day in 1874, he picked up a piece of gray-colored rock, lighter than it looked, and on the strength of its heft alone he sent it on to Denver to be assayed. The verdict was gold at $200 a ton. Bob was elated. At thirty years of age he had found his fortune. He went back to prospecting, following every rivulet that trickled into Poverty Gulch looking for the mother lode, but for sixteen years nary a piece more of float did he find.

But that did not stop him from talking about that one lucky piece of pay dirt. No one in teetotaling Colorado Springs or its wet neighbor, Colorado City, where Bob went on Saturday nights to drain as many shot glasses of bourbon as he could pay for, took his find seriously. His drinking buddies constantly poked fun at him. When a newcomer bellied up to the bar someone was sure to lean around him and call down the row of grinning faces to Womack.

"Tell him about your gold mine up on Cripple, Bob!"

And Bob, grinning with the rest, would do it, his eyes growing brighter as he described his gulch and how the water in it came off the slope of the hill and how gold by the ton could be locked somewhere in the hillside.

One day a pair of bunko artists who may have heard one of Bob's spiels did something about it. They salted with gold flake an eighteen-foot hole on loamy ground on one side of the basin, and then when they had sufficiently titillated the curious populace with rumors of mysterious conferences, night rides, and armed guards around the edge of their hole, they let it leak out that they had discovered ore worth $2,000 a ton. After hoodwinking a hundred or so buyers, the tricksters were exposed and the event went down in local history as the "Mount Pisgah Hoax." Now people laughed even harder at Bob Womack.

But the amiable cowboy did not give up. He had nothing to lose by poking away at the earth and talking up his idea to any who would listen. Eventually he caught the interested ear of Dr. John P. Grannis, a dentist who like many another had come to die or be cured of tuberculosis in the high dry air of Colorado Springs. Luckily for Bob he was cured. Grannis borrowed some money and staked Bob for one-half interest in anything of value the cowman found.

On a bracing fall day when small patches of Cripple Creek's hills radiated color from aspen leaves turned to gold in the frosted night air, Bob Womack came across a one-half-inch streak of pale yellow-bronze rock that shone with a brilliant luster. With his rudimentary rock sense, Bob guessed that his pick had turned up a promising lead. He quickly stepped off his claim and with hands shaking in excitement he scratched his name and the date, October 20, 1890, on a board and nailed it to one of the claim stakes. Then he packed some specimens of the lustrous rock in his saddlebags to take down to the Springs for assay. As he swung to and fro in the saddle to the rhythm of his horse as it loped down the cow trails toward town, Bob let his mind race away in fanciful visions. He would call his find the El Paso lode, and its profits would make him rich and respected.

The man to whom Bob went with his rock samples was Professor Henry Lamb, a metallurgist new to the Springs' perky little seat of higher learning, Colorado College. Lamb had already shown Doc Grannis a map of the Cripple Creek region showing that the heart of the basin had been formed by volcanic activity, and the professor added to the explanation his own theory that at some time a massive volcanic eruption had swept away the granite crust of the Cripple Creek region leaving deep cracks that were ultimately filled with minerals from deep within the earth. Lamb was more than pleased to announce to Bob and Grannis that the ore from the crack that was the El Paso lode assayed at $250 a ton.

Bob placed the chunks of El Paso ore in the window of a Springs furniture store hoping to attract some investors, knowing full well that it was a long and expensive road between discovery and production. When Ed De La Vergne, a canny, self-taught geologist, spied the ore, he recalled that it looked like some he'd seen up around Boulder that was calaverite, a gold telluride. De La Vergne hitched up his buckboard and rode up to Cripple Creek to see for himself the El Paso layout. When he satisfied himself that Womack's vein was calaverite, he came back to the Springs to organize six square miles around the El Paso into the Cripple Creek mining district.

When the news spread that a mining district had been formed, small as it was, second-hand furniture dealers Horace Bennett and Julius Myers, who a few months earlier had acquired four of the homesteads neighboring that of Will Womack's, platted a town they called Frémont on their property. Downstream a few yards, a competitor group of townmakers laid out another. After a battle of words and high promotion lot sales, Bennett and Myers saw their rival capitulate and their camp was the winner. They quickly consolidated the two camps and named the whole Cripple Creek.

Throughout the spring and summer of 1891 a steady file of prospectors made their way to the foot of Mount Pisgah. Men of all descriptions, tenderfeet and crackajacks, took a chance that all the to-do about the El Paso Lode meant something. Of the hodgepodge assortment of men who beat their way into the granite-rimmed basin, the oddest was Winfield Scott Stratton.

When he first set foot on the trail that led to Poverty Gulch in 1891, Stratton was in his forty-third year. Lean, with sharp, deep-set eyes, a long thin nose, and hair "white and shiny as bleached silk," Stratton carried with him the reputation of a master carpenter and the personality of a sorehead. He was his own man, and if people objected to his short temper, moody withdrawals, or other behavioral eccentricities it was their problem. Stratton was a loner. He did as he pleased, when it pleased him.

After finishing school and his apprenticeship in his father's boat works in Jeffersonville, Indiana, on the Ohio River, it pleased the young artisan to gather up his blanket roll, his carpetbag, and the $300 earned as a shipwright, and head for the booming West. He trudged into Colorado Springs in 1872, one year after General William J. Palmer inaugurated his Rocky Mountain counterpart of the East's fashionable Saratoga. Carpenters were in great demand as the Springs became a town and Stratton did well. So well that he sold his business in two years to follow the silver-seekers who swarmed into the San Juans. But as a prospector he made a fine carpenter. After eighteen months of wandering with not one satisfactory claim to show for it, the dour and dejected Stratton was back in Colorado Springs ready to go to work again with hammer and nails.

After the day's work Stratton chummed with a fellow prospector, D. K. Lee, who also had no luck in countless forays into the hills. One evening his friend Lee, who expanded his prospecting to include staking out pretty girls, came to Stratton and asked him to help him squire a pair of young ladies to a dance at Odle's Hall. The carpenter was unenthusiastic but he agreed and the foursome set out. Stratton's date was a nondescript creature who interested him not at all and a few weeks later when Lee suggested they trade girls the carpenter was delighted. Lee's girl was a sight more attractive despite the fact that she carried the forbidding given name of Zeurah. Before many more weeks passed Lee married his new girl and with the contagion which that sort of thing creates in virile bystanders, Stratton promptly proposed and was accepted by the seventeen-year-old Zeurah. Their marriage, however, was far from idyllic. The twenty-one-year difference in their ages and the casualness of their acquaintance might have been sufficient obstacles to their compatibility over the years— no one will ever know. Within two months, in searing rage, Stratton shipped his bride off to her stepfather in Illinois when he found that she had been pregnant when she married him. He never saw her again. Two years later he got a divorce on the grounds that Zeurah had "willfully deserted and absented herself."

Embittered now more than before, Stratton struck out again, this time for Leadville. Again his luck was poor and the nearest he got to a bonanza was to build a vault in H. A. W. Tabor's Bank of Leadville. Once more Stratton ambled back to the Springs, to settle into a not unpleasant routine—in the winter he carpentered, in the summer he prospected. During one fall and winter when business was slow he took Professor Lamb's advice and journeyed up to Golden to the Colorado School of Mines to take a course in metallurgy, learning the technique of blowpipe analysis.

When he got back to the Springs he soon learned of Bob Womack's El Paso discovery and the organization of the mining district around it promoted by Ed De La Vergne. Gold in the Springs' backyard? Stratton stilled a derisive chuckle. But in the June of the following year, at Bob Womack's insistent urging, he threw his saddlebags on the back of his burro and went up to Poverty Gulch to see for himself. As he clambered through the scrubby underbrush peering into prairie dog holes and the deep clefts in the soil made by torrential runoffs Stratton saw nothing that sparked his intuitive sense of geology. "Crazy Bob" was way off if he thought this was gold ground. The El Paso must be a fluke.

At the end of one long day of gophering, Stratton lay down in his blanket roll a few miles from the tent and log cabin camp that was Cripple Creek and felt welcome sleep pervade his body. And then he dreamed…"[T]he impression came to me," he recalled later, "that on the south slope of Battle Mountain where the altered dike ran up among the briars, was a mine." When he awoke, the first hint of dawn edged the horizon. Throwing off his blanket Stratton jumped up, hopped around a couple of times to take the dullness of the night air out of his legs, and climbed on his burro, turning its head south. As his burro ambled along the trail the carpenter cast his mind over the features of the terrain he had passed on the way up to Poverty Gulch. And then he remembered— there was a purplish ledge of rock jutting out from the side of the hill that Cripple cowmen called Battle Mountain. In a wave of mixed fear and elation—fear that someone had gotten there ahead of him, and elation at his vision—Stratton dug his heels into the flanks of his dutiful carrier. When he was under the ledge he tied his burro to a spruce seedling and broke out his pick, scrambling up the hillside as fast as he could break through the briars and tangled vines and roots of the ground cover. There were no stakes and no signs. He was the first. Quickly he paced off two claims, side by side. After he scratched out his name on the stake tags, he paused. What was the date? And what should he call his claims? Counting back over the days, he suddenly realized that it was the Fourth of July. Stratton smiled. He would call one the Independence and the other the Washington. But when the first hundred pounds of ore from the two claims were put to the furnace in Professor Lamb's basement laboratory, Stratton had reason to grimace. The tally was a paltry $11.80 a ton. Some dream, thought Stratton bitterly. As soon as he heard the disappointing results, he tried to unload the Independence on a newcomer, consumptive young Albert E. Carlton, for $500.

"Sorry, Stratton," answered the twenty-five-year-old Carlton, "I expect to be dead of T.B. in six months."

Then Stratton tried to sell the Washington to the Colorado Midland promoter, James J. Hagerman, for the same amount. Hagerman also turned down the carpenter's offer. So Stratton went back to his two-by-fours and cornices, fully out of patience with the whole idea of riches stored beneath the alluvial overburden of Cripple Creek.

But the signs of gold persisted. In the three gulches south of Poverty that drained into Cripple Creek and at the head of Wilson Creek to the east, prospectors diligently applied their picks and after days of back-breaking effort turned up ore with just enough gold content to whet their curiosity. Unlike Gregory Gulch, where the oxidized high-grade blossom rock lay a few inches below the surface, and South Park, where the streams and their banks were lined with float, the ore veins at Cripple were squeezed between folds of igneous rock. Picks and shovels hardly made a dent in the unyielding granite. Blasting powder, teams of drillers and muckers, deep shafts and long drifts, hundreds of feet of timbering, and mechanical hoists and trams were the key in converting a promising lead to a productive mine. All of these necessities boiled down to dollars required. And where was that capital to come from? From carpenter Stratton, from blacksmith Bill Shemwell, from tailor C. C. Hagerty, from ash-hauler Smith Gee, all of whom had located promising claims and were typical of the economic class of prospectors who roamed the area? Not likely. By the end of summer the camp that Bennett and Myers had so enthusiastically platted in spring wallowed in gloom. Most of its inhabitants were willing to believe that Cripple was going to end up as another Mount Pisgah hoax. Bob Womack was as discouraged as the rest, and he sold his share of the El Paso to Doc Grannis for $300.

Only Ed De La Vergne was undeterred. Long before the reality of Cripple Creek's requirements dawned on the Smith Gees and Bill Shemwells, he had grasped the situation and was busy promoting sources of investment.

Cripple could hardly ask for a better advocate. Articulate, knowledgeable, and reasonable, De La Vergne by virtue of a good head, a fair education, and a white-collar occupation could command the ear of Colorado Springs' most prominent citizens. The one who listened the hardest to his pitch scarcely gave the appearance of being the man to launch the new gold camp on its money-making way. Fun-loving, puckish James M. Pourtales had come to Colorado Springs to woo his cousin Berthe de Pourtales who was recovering from an ill-conceived and recently terminated marriage with a Boston Schlesinger. Pourtales' real work in life, however, was the maintenance of Glumbowitz, the family seat in far off Silesia. To keep the property viable, the thirty-year-old count trotted around the world looking for sound investments with a steady return of at least 8 percent. In the Springs, with German-bred frugality, he did not spend all his time courting but used some of it and his knowledge of selective breeding to bring back to life the Broadmoor Dairy Farm, an enterprise floundering hopelessly toward bankruptcy from a want of proper management. Pourtales' "Gilt Edged Crown Butter" proved a bonanza in its own right, and when he finally got the captivating Berthe to say yes the count moved his bride into a big house in the fashionable north end of town and the pair took their proper place in Springs society. Still, Glumbowitz required bolstering. So genial James decided what he and Colorado Springs could use was a Monte Carlo style casino whose roulette wheels would collect enough coin to keep both Berthe and Glumbowitz in handsome fashion. The Broadmoor Casino, elegant in every respect from its graceful verandas to the blue and gold Hussar style uniformed bandsmen, opened in June, 1891, and closed shortly thereafter owing to a series of mishaps not the least of which was an open letter in the Colorado Springs *Gazette* in which the Reverend L. L. Taylor took the count to task for staging Sunday concerts accompanied by a libation of wine.

Pourtales lost money on his casino but not so much that the idea of a gold mine that returned a high-grade profit failed to capture his ever-ready investment sense. After listening to De La Vergne's heart-felt assertions about Cripple Creek's potential, Pourtales donned his green Bavarian hunting coat and his porkpie hat and took the trail that led to Mount Pisgah. For company he brought along his artist friend Tom Parrish. While Parrish sketched prospectors' cabins and burro trains, Pourtales gathered sackfuls of rock specimens for assay as he and the painter moved through the crackling underbrush and rustling bunch grass on the sun-seared hills and gulches of Cripple. At nightfall the two tyros were greeted warmly at every campfire. Any diversion that made the lonely bands of prospectors forget that they had no money to develop the promising leads they found was a welcome change. Toward the end of October Pourtales and Parrish packed their frypan, bedrolls, ore specimens, and sketchbooks and returned to Cripple. On November 10, to the utter astonishment of the gold-hunters ferreting around the foot of Mount Pisgah, the popular and highly respected dairyman let it be known in the Colorado Springs *Gazette* that he was negotiating to purchase carpenter Steve Blair's Buena Vista claim, and that he had seen with his own eyes a million dollars worth of high-grade ore in plain view in the mine.

The count, as catalyst, succeeded when all other forms of persuasion failed. From all over the state came interested investors to buy claims already located. Others sent their agents to prospect and record claims as fast as they could find promising pay dirt. Springs' Judge

Colburn paid a hefty $10,000 for one-tenth of the El Paso, Stratton found Pourtales' partner in the Broadmoor Dairy was ready to write out a check for $80,000 for the Washington, and when the count, after slightly over-extending himself, discovered that his Buena Vista needed one more monetary infusion to make it pay, the bantam businessman J. J. Hagerman was now more than happy to buy in for a round $225,000.

Along with capital came waves of greenhorns caught up by the effervescent stories in the papers. They all prospected, and some were lucky. Two gulches south of Womack's Poverty, John Bernard, a tenderfoot of the first degree, walked all night up the trail from Colorado Springs, and at dawn he paused when he caught sight of a pair of elk horns on a pile of rocks. After contemplating the scene for a while, something prompted him to hack away at a purplish crumble of rock sandwiched between layers of hard granite. After several days' work Bernard carried a parcel of samples to the assayer. They showed enough to let him settle his $36.40 grocery bill by exchanging a half interest in his Elkton claim for a paid receipt. A day or so later the grocers sold half of their interest to a teacher who laid aside his books, trotted up to the Elkton, spat in his palms, and swung a pick to open up a vein worth $13 million.

"Don't jostle the man in the street," sang the Cripple Creek *Crusher,* the camp's first news sheet, "he may be a millionaire tomorrow and resent the insult."

Druggists A. D. Jones and A. J. Miller put away their mortars and pestles and scampered up to Cripple to join the fun. Their chemistry was first class, but their geology was nonexistent. When it came to deciding where to start they were stymied. They couldn't tell one rock from another. But in the spirit of the game that all Colorado Springs seemed to be playing, Jones and Miller agreed that they would toss a hat into the air and would dig where it fell. They did and the Pharmacist mine was the second mine in the district to ship paying loads of ore.

Two Irishmen stalking hidden pay dirt had worked Bull Hill over as carefully as they could when they suddenly found their stake consisted of one last dollar and not so much as one piece of likely ore. Over their campfire that night they formulated a plan for the next day.

"Tell you what, dogs are born lucky," said Mike, his eyes resting on the stretched-out form of their hound toasting his back near the coals. "The first place he stops tomorrow, we dig."

"Shake," said Pat.

In the morning with first light, the two turned out of their bedrolls and got underway. True to their bargain, the Irishmen sunk in their picks at the first stop their dog made. By noon they had dug away the overburden to find sparkling calaverite. After three weeks work they sold the Last Dollar for $100,000.

The word of Cripple's bonanzas spread fast. In Iowa a group of farmers leased a dud claim ironically called the Jack Pot and sent Billy Davenport, their county sheriff, to oversee its development. Billy dumped $13,000 into finding the illusive vein stashed away in a blanket of hard rock and when he reported no progress his bosses called him home. But Billy asked for one more chance. He got it, and that was all it took. A vein of gold two-and-a-half-inches wide paying $60 to the ton was the reward Davenport, now sporting "diamonds as large as hazelnuts and a watch chain of gold nuggets," telegraphed his backers. Ultimately the Jack Pot paid $1,250,000.

From a one-boardinghouse camp with a makeshift bar consisting of a plank resting on two beer barrels, Cripple swelled to city style proportions. The main streets were appropriately named Bennett and Myers. The Palace and Windsor hotels on Bennett Avenue rented rooms by the day and night and even rented chairs to sleep in at $1 a night. Shacks rented for $15 a month. Lots inflated to $3,000 and $5,000 apiece from their original modest tags of $25 and $50. Water sold for 5 cents a bucket and beer for 20 cents. Wiry, black-eyed Joe Wolfe, after being run out of Ouray for fleecing one too many residents on get-rich-quick schemes, scuttled into town to work his games but suddenly decided legitimacy was the best policy. In Cripple, he hung his thumbs in the fob pockets of his garishly printed vest and oversaw the building of a new hotel, the Continental, to which Pourtales, still after capital for mine development, brought William Lidderdale, ex-governor of the Bank of England, and French milling magnate G. de la Bouglise on an inspection trip of the burgeoning Cripple gold mines. Wolfe's hotel, sporting paintings donated by his cousin, tobacco heiress Catherine Lorillard Wolfe, could sleep two hundred and it was so popular that the beds hardly had time to cool before the next occupancy and there was no time to wash the silverware between settings. New arrivals coming via stage to Cripple Creek were met at the terminal by the town marshal who relieved them of their shooting irons. This was a unique method of combining crime prevention with school taxation. Periodically, the marshal gathered up his appropriated arsenal and tumbled it down to Denver to be sold. The money was then used to help pay Cripple Creek's teachers.

It fell to Myers Avenue to house the gambling dens, taxi dance halls, opium dens, pawnshops, and parlor houses such as henna-haired Pearl de Vere's two-story brick, affectionately labeled the "Old Homestead." Up and down the red-lit street between the showier establishments were shanty saloons, and one-girl cribs where the height of the window blind indicated a girl's availability.

On the boardwalks of the two streets by day stiff-collared millionaires rubbed shoulders with sleazy procurers, red-faced miners reeled out of saloons through swinging doors into the paths of bustled matrons picking their way through the crowd to the new circulating library. Occasionally, a pair of gamblers, flush with a night's winnings, sat against a building in the sun, stacks of gold pieces at their sides, spitting at a knothole in the boardwalk at $20 a spit, while a prospector in tattered denim gazed in the window of a cutrate clothing store advertising "Schradsky's pants are always down." Periodically, delivery boys dashed through the throngs of shoppers spooking horses and upsetting stacks of mining and geological texts displayed outside bookstores. Men in soft felt fedoras hurried to catch the Midland for their daily trip down to the Springs' busy mining exchange while others stood on street corners discussing the latest ore strike on Bull Hill, letting their eyes follow one of Pearl de Vere's girls as she waltzed by en route to the drugstore for her weekly purchase of laudanum. When night fell there was no let up in the hurly-burly of the street. Arc lights shone on crowds of revelers who caromed from one pleasure palace to another in the carnival atmosphere of drumbeats and notes from brassy cornets and tinny pianos blaring from open doors.

By 1892, five thousand people lived in Cripple and another five thousand clustered in the nearby camps of Victor, Elkton, Goldfield, Independence, and Altman. Victor, platted on Wilson Creek just south of the Portland, Independence, and Strong mines by a trio of brothers, threatened to surpass Cripple Creek's boom. Warren, Frank, and Harry Woods knew what they were about. They had grown rich in Leadville, Kokomo, and Robinson at the start of the silver rush. In Victor, the brothers gave free lots to anyone who would open a business and if a man did not have the ready cash to buy a $25 lot, he need only sign his name to the new town's roll to get it. With enough names on the town clerk's roll, Victor got a post office. When the Woods boys set about to build a hotel for their town, the first shovel of dirt from the excavation revealed a rich lead of sylvanite, and the Gold Coin mine, paying $50,000 a month, was born.

Despite competition from Victor, Cripple remained the hub. A look at its business district was evidence enough to prove it. Bennett Avenue was lined with sturdy buildings housing all of the commercial and professional necessaries of a thriving city, including three bathhouses, forty-four law offices, and ten meat markets.

Dave Moffat opened the Bi-Metallic Bank, taking Horace Bennett, now a millionaire, in as vice president. With his share of the profits Moffat went back to railroading, in particular the building of a route west through Colorado's Rockies to the Pacific. The notion obsessed him, and he would squander his millions on the project. On the day of his death in 1911 he was still trying to round up money to go on with the project. Eventually, in 1928, his idea came to fruition when the Moffat Tunnel through the main range of the Rockies was opened to carry the first train directly west from Denver to Salt Lake. But in the nineties Cripple Creek residents were far more interested in a railroad to their town than one to the West Coast. Stage travel and burro back were adequate transport in the district's very earliest days but everyone knew that the town would soon need a railroad to handle the spiraling mine production. As soon as he digested the significance of Pourtales' acquisition of the Buena Vista, Dave Moffat got busy and built the Florence and Cripple Creek, a narrow gauge line from the south, connecting with the Denver and Rio Grande at Florence on the Arkansas River and from there winding its way up to the gold camp.

On the north the Colorado Midland from the Springs to Leadville passed within eighteen miles of Cripple. When the gold camp started its rise, the line quickly established stage service to cover those eighteen miles and advertised for passengers and freight in eastern papers with the inviting slogan, "Cripple Creek is not only a HEALTH but a WEALTH resort." Despite the advertising promotion, the directors of the Santa Fe, which now owned the Colorado Midland, were not convinced that there was enough business in Cripple to convert the eighteen miles from a stage road to a railroad. Harry Collbran, the Midland general manager, was convinced, however, and so was W. K. Gillett, passenger auditor for the Santa Fe. In a succession of deft moves Collbran supplied the labor and Gillett found the capital for the construction of what came to be known as the Midland Terminal Railroad. The year was 1893 and traffic on the Colorado Midland was at a low point owing to the nationwide depression. So Collbran put idle Colorado Midland hands to work on the Midland Terminal. When track was laid a third of the way to Cripple, Collbran borrowed some Colorado Midland rolling stock and ran a daily schedule over the six miles. When the freight shipments arrived at the end of the six miles, Gillett collected the freight charges, and on outgoing passengers and freight he collected charges in advance. All but 5 percent of this revenue belonged to other railroads and would customarily be remitted to the roads within ninety days. But Gillett used the money for laying the last lap of track to Cripple and stalled settling his accounts with other railroads for some five years.

The ingenuity of the builders of the Midland Terminal gave Cripple two railroads, but it was to have three more before the end of its boom days. When freight rates rose too high to suit the major mineowners, they built their own Cripple Creek short line to Colorado Springs. And some time later, L. D. Ross, a Springs real estate man, put in two interurbans, the High Line circling Cripple on the tops of the hills from one camp to another and the Low Line running along the shoulders of the hills. With efficient and comparatively low-cost transport to the smelters available, the development and production of Cripple Creek's gold mines shot upward.

A couple of hillsides north of Stratton's Independence mine a Mutt and Jeff pair of Maine Irishmen laid claim to a mine they called the Portland after their home town. At forty-one, after spending his salad days thrashing around sugar cane plantations in Cuba and South America, cocky little Jimmie Burns joined his three maiden sisters in Colorado Springs in 1886. When the sisters had moved west they brought with them a lanky, irrepressible, good-natured orphan in his teens named Jimmy Doyle, who became the errand boy in their dressmaking establishment. Anyone handy with tools could find a job waiting for him in the Springs' building trades as more and more people took up permanent residence in the resort, and when the time came for Jimmy Doyle to learn a trade he took up carpentering. When he arrived to live with his sisters, Jimmie Burns fell back on his old trade of plumbing and the two Jimmies often worked together on construction jobs. In time, despite the nearly twenty years difference in their ages, the two grew to be fast friends; where one was so was the other. When the Cripple Creek excitement hit the Springs, the Jimmies, together as usual, upped and followed the crowd. Their knowledge of minerals and mining was as abysmal as that of most other tenderfeet. When they stepped off their Portland claim, using a borrowed clothesline, their only reason for selecting the site was that someone else had once claimed it and then abandoned it and that it was not far from where Stratton was doggedly working his Independence claim. After a week or so of fruitless digging the Jimmies found nothing on their ridiculously tiny claim that encompassed only one-sixth of an acre. It bothered the pair somewhat that they were not entirely sure what pay dirt looked like. After a few weeks of digging they decided to ask for help. When they approached their friend, Johnny Harmon, on the subject, he took one look at the pile of rock sitting beside the thirty-foot shaft, picked up a piece of sylvanite, smiled slightly, and said, "What'll you give me if I do?"

"One-third interest," piped Jimmie Burns.

"You're on," said Harmon as he started down the ladder in the shaft. Unlike the Jimmies, Harmon knew what he was looking for. He had helped his brother locate the Hilltop, a high-grade producer in Fairplay some years before, and had kept his eyes on mother Earth ever since. He soon found the vein, and within a few days the three were busily sacking ore that assayed at $640 a ton.

But the trio's exhilaration gave way to the heebie-jeebies when they realized that once the word got out that the little Portland was a bonanza, owners of surrounding claims would slap them with a briefcase full of lawsuits declaring that the Portland vein surfaced or "apexed" on their property and therefore they had a right to ownership by the law of apex. In mining, the right of apex establishes that a vein belongs to the claim on which it surfaces or apexes even though the vein extends underground outside the boundaries of the claim. To forestall word of their bonanza getting out as long as possible, Burns and Doyle and Harmon lugged the ore out at night on their backs to a waiting wagon at the foot of Battle Mountain. One night the axle of the wagon broke and there in the morning for all to see was the load of Portland ore and the secret of the Irishmen's discovery was out.

The "Three J's," as Cripple called Burns, Doyle, and Harmon, quickly patented their claim, and with $90,000 in profits already earned from the Portland they braced themselves for the onslaught of legal writs. They were well aware that lawsuits, as well as machinery and labor, were an integral part of the mining game. When the deluge came there were twenty-seven different suits against the Portland. All that the three of them could hope was that the $90,000 would see them through the voracious appetites of the lawyers who would fight the cases. Burns, the worrywart of the combine, decided to ask Winfield S. Stratton for advice. Stratton after all was one of the fraternity of building tradesmen and knew a hatful about mining. Burns knew he would get a straight answer from him.

Straight to the mark were Stratton's words. It seemed that the canny carpenter had been hoarding a secret of his own these past months. After methodically hacking out the original shaft in the Independence putting in four crosscuts at the thirty-foot level, three of which he had worked with inauspicious results, Stratton optioned his dream mine for thirty days to a representative of a San Francisco mining company. On the day before the option began, Stratton went down into the Independence to remove his gear. Clambering over the debris in the fourth and unworked crosscut, he picked up an old drill and idly hammered away at a rock jutting out from the wall. Suddenly it fell from its place and there in front of his eyes was a sparkling vein of gold. The vein ran about thirty feet along the tunnel wall and he figured it must be nearly ten feet wide and could well extend downward for a hundred feet. He broke off some samples, replaced the debris in the drift as he found it, and scurried to the assayer. When assayer N. B. Guyot announced "$380 to the ton," Stratton began to shiver.

He continued to shiver and shake for the entire thirty days for fear the lessees of the Independence would find his secret vein. But they did not, and when the San Francisco company announced to him that they had decided not to renew the option Stratton ducked into Johnny Nolan's saloon, bought a bottle of aged bourbon, and shut himself up in his cabin to celebrate. His dream-vision had come true. The Independence would make him Cripple's first millionaire.

After Jimmie Burns poured out his story of the Portland, Stratton told him about the Independence and then and there offered to back the Portland's owners.

"Let me buy in for $70,000," said Stratton, "and we'll pick up every mining claim on Battle Mountain and let the shysters earn their money some other way."

Burns and his two partners quickly accepted the carpenter's offer. And then Stratton went to work. His first move was to corner Verner Z. Reed, a young man busily turning one Springs real-estate deal after another into golden pocket pieces for himself. Stratton wanted Reed to see what he could do to turn the $3 million worth of lawsuits into profit for the Portland. Reed, who left the Chicago *Tribune's* reporting staff to help his father operate a livery stable in Colorado Springs, had absolutely no legal experience, but his ability as a strategist of property and money was phenomenal. In double-quick time he proved that Stratton knew how to pick the right man for the job. After a whirlwind series of options let and redeemed on the Portland, purchases of nearby small claims, partnerships and dissolutions, and incorporations and dividends that left bewildered plaintiffs casting vainly around for a corpus to name in their suits, Reed brought the Portland Gold Mining Company out of the storm with a grand total of 182 5/6 acres. Added to the Jimmies' original one-sixth acre, the property became the largest and strongest gold mine in the country. The Portland, called "The Gold Factory" in Cripple Creek, produced $62 million worth of ore in its heyday.

With more money than he could think of things to spend it on, Stratton did not become expansive in the style of H. A. W. Tabor and Tom Walsh. No flamboyant, jewel-studded clothes for him. He stuck to his plain sack suits and cream-colored Stetson. And no magnificent mansion in Washington or even in Colorado Springs did he build as a suitable backdrop to his new status. Instead, Stratton bought a comfortable and unpretentious house only a block or so from the Springs' business district. It was a known quantity. He had worked on it when it was being built and he could vouch for the workmanship. When the papers announced his new residence Springs' society was prepared to lionize their carpenter-turned-Croesus. But Stratton ignored their invitations and would have none of the pomp and circumstance that is the by-product of wealth. Yet he was not a miser. No one, not even the most flighty and effusive society matron who approached Stratton for a donation, was turned down. His charitable works were legion, if peculiarly of his own devising. He bought bicycles for all of Colorado Springs' laundresses. He established a fund that he gave to the Salvation Army to provide destitutes with meal tickets and places to sleep. In recognition of the help Henry Lamb had given him, Stratton gave Colorado College $20,000 and the School of Mines at Golden $25,000. When H. A. W. Tabor came to him and asked for a loan to develop a claim he managed to salvage from the debacle of 1893, Stratton gave him $15,000 and protested when Tabor made him draw up a legal contract; Stratton wanted it to be a gift. And when the child that Zeurah bore showed up one day in the Springs, Stratton arranged to pay for the boy's tuition at the University of Illinois and threw in an allowance of $100 a month if he would stay out of the carpenter's sight forever. In his will, Stratton endowed the Myron Stratton Home for "poor persons who are without means of support, and who are physically unable by reason of old age, youth, sickness or other infirmity to earn a livelihood." But the wealth that poured into his bank account brought no particular pleasure to the solitary Stratton. He lived a spartan existence, almost always dining alone. For companionship he chose bourbon bottled in bond, and now and again for conversation he called upon his old bootmaker friend Bob Schwartz or Verner Reed, and with some consistency heavily veiled women slipped into the 115 North Weber Street house to spend the weekend. He bullied one housekeeper after another until he found Swedish immigrant Anna Hellmark. She devoted her life to making him comfortable, and at the moment of his death at fifty-four of hypertrophic cirrhosis, she was at his bedside to receive his one and only embrace, given in delirium.

But several years before that happened Stratton's concern was for Cripple Creek. Above all he wanted it to experience an orderly expansion that might guarantee its survival. To that end he refused to mine more than $2,000 a month from the Independence. He had seen too many boom and bust ghost camps in his seventeen years of prospecting around Colorado. He would make Cripple last as long as he could.

His worries about Cripple Creek were groundless. Of all of Colorado's bonanza towns it would have the longest continuous life. This was clue in part to the fortuitous combination of native resources and well-bred native business talent in the form of a covey of young men the rest of Cripple dubbed "the Socialites."

The first on the scene was spectacled Charlie Tutt who, unlike the rest, was thrust into the world to make a living long before he finished high school owing to the early death of his father, a prominent Philadelphia physician. Charlie launched himself on a career in Springs real estate and did rather better than he had hoped. Enough better in any case to give him a grubstake and time to locate the C.O.D. mine in the Cripple Creek stampede.

When the C.O.D. threatened to be a moneymaker Charlie wrote his childhood chum Spencer Penrose who ran the Mesilla Valley Fruit and Produce Company down in Las Cruces, New Mexico, to come on up and try his luck. Penrose decided to at least take a look at Cripple Creek. His career so far had been something less than a source of pride to his well-to-do father, Dr. Richard A. F. Penrose, a medical school classmate of Charlie Tutt's father, and Spec was on the lookout for a break.

The Penrose prominence began in the late 1600s when one of the family allied himself in business with William Penn. The Penroses had been a fixture of Philadelphia society ever since. Spec, like his three older brothers, Boies, Charles, and Dick, graduated from Harvard. But unlike his brothers who copped nearly every scholastic honor available, his was an undistinguished record. Not even his transgressions of an alcoholic and pugilistic nature lived up to those of his esteemed brothers. His failure to compete drove him out of the traditional arena of his blue-blooded tribe's endeavors and into the anonymous West where luck played as much a role in a man's success as ability. But until Charlie Tutt's letter came, Spec, at an advanced twenty-seven, was enjoying the fruits of neither luck nor ability. His search for a silver mine was unrewarding and his fruit business was a bust.

When the towering, aristocratically handsome Spec arrived in the Springs on December 10, 1892, he was broke and game to try anything. When Charlie Tutt offered him a partnership in his real estate and mining interests for $500, Spec quickly put out his strong right hand to shake on it. Where he was going to get the $500 neither young man discussed. Before Charlie sent Spec up to Cripple to handle the enterprises of Tutt and Penrose in the mining camp, Charlie introduced his friend to Springs' society. It was a milieu that Penrose with his archly proper rearing perfectly understood.

Colorado Springs was the result of the presence in southern Colorado of a goodly flock of younger, bastard, or profligate sons of a number of genteel English families and the burning desire of General William J. Palmer to provide his bride, Long Island's beauteous and headstrong Queen Mellon, with a suitable home. Regrettably for the general, Queen Palmer never accepted the Springs, but hundreds of others did. From the organization of the first town company in 1871 by Palmer and Dr. William Bell, a young Englishman less interested in homeopathy than in homesteading, the Springs and Manitou, its adjacent watering place, drew a select crowd of accented, finely bred residents.

In the seventies the Rocky Mountain region experienced a surge of British immigration that began some years earlier when English cattle and sheep growers, worried about the diminishment of their own pasture land, sought additional land in the American West. Typical of the newcomers were Wilson Waddingham and William Blackmore who bought one half of the prodigious, one-million-acre Sangre de Cristo grant from Colorado's ex-Governor Gilpin. Blackmore immediately backed Palmer's investment. With the Bells and Waddinghams and Blackmores came lesser luminences to run the ranches and businesses of the new town. These people drew their friends, and when the restorative value of the mineral springs at Manitou were advertised new flocks of American and British visitors—chiefly the well-to-do ill and those to whom taking the waters was a social ritual—cascaded into Palmer's duchy at the foot of Pikes Peak. Among the early arrivals were the Canon of Westminster, Charles Kingsley, and Helen Hunt Jackson who stopped over so long they were considered residents.

The architecture of the town was Tudor mixed with English Gothic, including General Palmer's Glen Eyrie, a copy of the Duke of Marlborough's pile. The pleasures provided were the same as those to be found in European and eastern centers of culture—polo, pigeon shooting, drinking at the Cheyenne Mountain Country Club, and riding to the hounds. In the absence of foxes, coyotes were found to be acceptable substitutes. "Little London," as the Springs came to be called, vibrated with Oxford, Harvard, Mainline, Back Bay, and Duchess County accents. At the same time it quivered with the manipulations of business. William Palmer's vast network of commercial interests brought him an international reputation. Banker Irving Howbert dealt himself silver mines as freely as gamesters in Pourtales' Broadmoor Casino dealt vingt-et-un. While Canadian James J. Hagerman waited in the high, dry altitude of the Springs to die of tuberculosis he entertained himself by building the Colorado Midland railroad and almost unnoticed got well. When Cripple Creek gold burst over the Springs it was already filled with men whose entrepreneurial instincts had been translated to ability.

After hobnobbing with a few of the Springs' well-heeled elders, Spec Penrose decided to follow their lead. He bought a ticket on the Midland to Florissant and caught the stage to Cripple Creek with one goal firmly in mind—make money fast. On the way up to the gold camp he sat beside the stagedriver, Harry Leonard, another scion of eastern wealth who found his equestrian training at New York's fashionable riding academies sufficiently good to allow him to handle the six fractious mules on the daily round-trip run. Leonard offered to share his bachelor cabin with Penrose, and the two soon became the leaders of a fraternity of ambitious and roistering young bucks who made Cripple Creek and Colorado Springs resound with their exploits.

Other members of the fraternity were burly Horace Devereux, Princeton '81, who found time to manage the bits of his polo ponies at the same time he managed the bits of a mine drill, and Charlie MacNeill, an uncommonly nervous young man in his twenties, who could not relax until he bought Cripple Creek's first chlorination mill and put into practice his newly learned knowledge of ore reduction. Occasionally, Spec's good-looking but sober-sided brother, geologist Dick Penrose, happened by, but he was too interested in calcification and thrust faults to join in the high, wide, and handsome life of the group who periodically tried to drink up their assets, fell in and out of love with the regularity of the sunrise and the sunset, and kept cocked fists at the ready to emphasize their robust masculinity. By day the energetic pack of youngbloods labored untiringly at their jobs in the mines, on the stages, and in the heat and stink of the mills, but by night they were transformed to boulevardiers. Often they donned white tie and tails to coast down to the Springs, to down a few at the Cheyenne Mountain Country Club before squiring pretty, well-bred Springs girls to a dinner dance in the worldly atmosphere of the Antlers Hotel. For one of the pack, the tall, aesthetic-looking Albert E. Carlton, one such dance led to complications of the heart that would have bankrupted a poorer man.

When it became obvious that the invigorating climate of the Springs had arrested Bert Carlton's galloping consumption, his grateful and doting father settled $10,000 on him as a grubstake. While he decided how best to increase the principal of his windfall, young Bert clerked in a store and drove one of Stratton's horsecars that served Springs' residents as public transportation. Then he heard about the Midland Terminal railroad building to Cripple Creek, and that was his cue. Like the rest of the Socialites, Carlton could see clearly that the odds of his making a fortune were considerably better if he stuck to endeavors ancillary to mining rather than if he tried to discover a high-grade bonanza. Acting on this conviction he persuaded the mine-owners on Cripple's Bull Hill and Battle Mountain to give him contracts to haul their ore the twelve miles from the mines to the Midland loading yard. Then he bought a fleet of wagons and mules and hired a crew of muleskinners and went to work. As a sideline he negotiated with the Midland Terminal's Collbran to give him an exclusive franchise to carry the railroad's freight from the terminal to Cripple Creek, thereby giving him a profitable cargo to the railhead and back to Cripple.

At thirty-one Bert Carlton was executive officer of the Colorado Trading and Transfer Company, the first of a very few steps that brought him status as a millionaire. Carlton, who thought nothing of signing checks that ran to three and four figures but who agonized over the expenditure of 5 cents for a cigar, went on to corner a monopoly on coal delivery to the high-altitude gold camp. At first his sales were meager because there was plenty of wood around to power the steam hoists used by the larger mines. But the wood supply began to be depleted when some of the mines, including Charlie Tutt's C.O.D., began to fill with water as the shafts reached deeper into the volcanic basin that was Cripple Creek and pumps had to be set to work twenty-four hours a day to keep the mines dry. Soon coal was needed for fuel to operate the hoists and the pumps, and Bert Carlton, charging what the traffic would bear, was ready to supply all the coal the camp needed.

Bert's enterprises expanded in spite of himself. He was a generous creditor. When a coal bill and an ore-hauling bill came due he often took it out in stock. Soon he owned a dozen mines, including the high-paying Jack Pot and the Findlay. In the great game of business Bert played fair and hard. If the vein of one of his mines ran over into another's property he was quick to start suit on the basis of the apex law. The frequency and success of his suits were such that those who met defeat at his hands grimly swore that Carlton's initials, A.E., stood for Apex Everybody. In 1898, Carlton bought the First National Bank of Cripple Creek, and at thirty-three could point to personal assets of $500,000. Not a bad parlay for a man who six years before was supposed to have been a terminal T.B. patient.

Covetous female eyes followed Bert Carlton wherever he went, but the pensive, preoccupied young man was comfortably unaware of them. However, when his mother suggested he invite the appealing Miss Mary Quigley to a dance at the El Paso Club Bert was quick to oblige. As the twosome became better known around the Springs, Bert's mother let it be known that the pair were engaged. Bert didn't mind. He was fond of his mother and he did not dislike girls. In fact he liked them so well that he had, unknown to his parent, married one some months previously on a trip to Illinois to visit relatives. All the while Bert was money-making in Colorado, Eve Stanton, his Illinois bride, patiently waited for him either to make her his widow or make public their marriage. But in all the excitement of creating his business empire it slipped Bert's mind. Nor did he remember when Mary Quigley became his fiancée.

The Carlton enterprises kept Bert in Cripple Creek more often than Mary liked, and she began to simmer in resentment at his frequent and prolonged absences from the Springs social whirl. But she boiled over when she heard that her single-minded affianced had been suddenly taken desperately in love with a pert seventeen-year-old girl Friday named Ethel Frizzell who kept her father's mining office in the gold camp. To add to the ferment Eve Stanton Carlton arrived waving a fistful of newspaper clippings heralding her husband as well and wealthy, and she loudly demanded that he assume his husbandly duties.

Poor bumbling Bert. One wife, one betrothed, and one true love—all fighting mad.

After a series of deliciously unrestrained stories in the papers of Cripple and Colorado Springs and copious withdrawals from Bert's ample bank account, the rectangle was solved. Eve accepted a divorce for a peck of cash. Mary Quigley accepted a large house in the Springs. And Ethel accepted Bert. Bert's mother, to whom the whole affair was an incomprehensible nightmare, went abroad on a grand tour to try to forget. Meanwhile the newlyweds moved into an apartment above Bert's bank and the happy bridegroom got on with the business of making money.

While Bert Carlton was making his millions and setting straight his affairs of the heart, the rest of the Socialites kept busy. When Spec Penrose looked over Cripple he nightly saw throngs of miners crowd into the Topic Dance Hall on Cripple's booming Myers Avenue and he decided that Tutt and Penrose should have a piece of the profit. Once that lease was signed, Spec telegraphed his brother Boies for money to buy into Tutt's C.O.D. mine. Boies wired $150 and instructions to come home to Philadelphia where Spec belonged. Instead, Spec borrowed the additional money from a Springs banker, and Tutt handed over a one-third interest in the mine. Everyone except the two owners thought the C.O.D. was finished because it was filling with water. But Tutt and Penrose were handsomely vindicated when a French outfit one day plunked down $250,000 for the mine. When he got his share of the sale price Spec couldn't resist sending his conservative brother, Boies, a check for $10,000 as a return on his $150 investment.

While Carlton, Penrose, and others of the shrewd young men of their crowd were making hefty profits on services, Cripple's mines continued to pour forth gold. In the six square miles that comprised the mining area, the Vindicator, the Mary McKinney, C.O.D., Anaconda, Portland, and Gold King alone of the mines in the district produced ore at the rate of three-quarters to a million dollars worth a month. 'Round-the-clock shifts were standard in the mines, and there were jobs for all out-of-work miners who crowded into Cripple after the bottom dropped out of silver in 1893. But the rollicking prosperity that swept over the camp was soon to be interrupted. When miners looked around at the huge profits Stratton and other mine-owners were pulling down and then looked at their paychecks—$3 for nine hours of back-breaking work—they began high-grading on a large scale. Miners surreptitiously hacked small pieces of high-grade ore from rich seams and hid it in boot tops, in pockets sewn into their clothes, and in their lunch buckets. Crooked assayers reduced the smuggled ore in their little furnaces and paid the miners who brought it in one-half its value. With some of the ore running as high as $40 of gold to the pound, high-grading was a welcome bonus to the weekly paycheck. In time, mine-owners caught on to the system of "dinner pail shipments," and set up change rooms, ordering miners to strip and be searched before leaving the mine. Resentment flared among the miners who were already disgruntled at the vast abyss that separated them economically from the district's overnight millionaires. Watching the spread of grumbling among the miners was John Calderwood, a scholarly Scot who was an organizer for the newly formed Western Federation of Miners. He worked fast, beginning at Altman, a camp two gulches over from Cripple Creek at the foot of Bull Hill housing the men who worked the heavy-producing Vindicator, Findlay, Independence, and Golden Cycle mines. By January, 1894, every miner and mucker in Altman was a member of the W.F.M.

On February 5, Calderwood pulled five hundred Altman miners out on strike, demanding $3 for eight hours of work. Bull Hill mine-owners shrugged. The consensus of opinion among them was that when the miners got hungry they would pick up their tools again. But the miners did not grow hungry. Calderwood saw to that. He set up a community kitchen for strikers and their families, and with funds collected from sympathetic merchants and nonstriking miners, plus a stake from the union's treasury at headquarters in Butte, he kept the strikers' rents paid and clothing on their backs. Gradually mine-owners Irving Howbert, Bert Carlton, Ed Giddings, and James Hagerman began to realize with some concern that they were dealing with something more than a flash-in-the-pan, impetuous rebellion. Their complacence was further jarred when Winfield S. Stratton and

Jimmie Burns, who remembered that their origins were as humble as those of most of the miners, said in public that miners had the right to bargain collectively with their employers. When their statement was reported in the Cripple Creek *Crusher,* the writer of the story added that when Burns's words were read at the El Paso Club, the exclusive preserve of Cripple's mining millionaires in the Springs, "three members collapsed on the pool table and died of apoplexy." A few more popped off when a few days later Stratton signed a compromise contract with his Independence crew providing $3.25 for nine hours' work. The mine-owners howled for Governor Waite to call out the state militia to put down the strikers. In their rage, they failed to remember that Populist Waite had run and won on a platform calling for an eight-hour working day. Only Irving Howbert remembered and when he tried to remind his fellow mine-owners that Waite was on the strikers' side and that if his militia settled the strike it might bring enough votes to reelect Bloody Bridles and ruin the Republicans' chances in November they shouted him down. The rest of the owners then declared that they would take over the management of their struck mines themselves and operate the properties with scabs.

On March 15, Sheriff Bowers of El Paso County served an injunction on Calderwood calling for him as leader of the strike to cease obstructing the operation of the struck mines. Calderwood's union men responded by arming themselves and turning Bull Hill into a fortified camp. Miners with loaded guns guarded every road and footpath into Altman. On March 16 when Bowers attempted to arrest the strike ringleaders he and his handful of deputies were run off the hill by pistol-waving miners. Forthwith the sheriff telegraphed Waite to send the state militia.

The troopers arrived on the morning of March 18 under the leadership of braggart attorney Thomas T. Tarsney, Waite's adjutant general. Tarsney, with a great show of military pomp and protocol, informed the strike leaders that if they resisted arrest his troops would march on Altman and whip the lot of them. Meanwhile, in private, Waite and Tarsney reached an understanding with Calderwood in which the strikers would appear to acquiesce, submit to arrest, stand trial, and be guaranteed acquittal while Calderwood organized miners in other regions to join in a sympathy strike to achieve the eight-hour day. A mock peace would deflate the pompous Springs mine-owners and give some time for labor men to rally support. After a period of quiet settled over Cripple Creek that saw the militant miners putting away their guns Waite sent the militia home.

But when Calderwood left on his statewide swing in April to round upsupport, all hell broke loose in Cripple. Led by two tough agitators, the miners got out their guns again and staged a series of raids throughout the district. Every night roving bands shot up saloons, surged into the parlor houses on Myers Avenue, reducing the girls and the furnishings to shambles, and broke into the homes of non-union men and beat them senseless in front of their terrified families. With it all they shouted deprecations and threats they hoped would reach down the mountain to Colorado Springs. They did, and Springs residents grew anxious, their sleep troubled by recurrent visions of a crazed mob of miners spilling out of Ute Pass to ravage the genteel spa. After observing a day or two of mayhem in Cripple the mine-owners met in urgent caucus to find a way to protect their holdings and homes. The state milita was obviously in the enemy camp and would not lift a rifle barrel to help a Springs resident. When the heated outbursts of indignation and fury died down someone proposed an idea that met with immediate and universal approval. Why not deputize an army of El Paso County sheriffs? Why not? But who shall they be? Preferably they should be professionals, but where could you find a brigade of out-of-work policemen?

"Denver," piped up Irving Howbert. "Why not recruit the bunch Waite turned out on the streets after his City Hall war?"

The poetic justice of the idea readily appealed to the mine-owners and they sent off telegrams hiring 125 ex-Denver police and firemen who were more than willing to have a chance for revenge and get paid for it at the same time. The "deputies" rode Palmer's D&RG to Florence, where they were armed and loaded onto a fleet of Florence & Cripple Creek flatcars for the trip up to the gold-camp. As the cars carrying the deputies rounded the hill into Victor there were two ear-splitting explosions as the shaft house of the Strong mine rose several feet and disintegrated into a million pieces of wood and iron from shattered windlasses, hoists, frames, and boilers. The engineer in the cab of the engine pulling the deputies' cars was ordered to reverse his direction. As he did, the striking miners set out after the retreating brigade of deputies to finish the job. Scrambling down the grade on foot, the miners caught up with the deputies at midnight and in a lightning-fast exchange of shots two men were killed and five strikers were taken prisoner.

When the word of the battle got to Cripple Creek the camp burst with venom leveled against Colorado Springs and all it stood for—wealth, Protestantism, and Republicanism. The Springs likewise turned in fury on the Populist-loving and Popish (one-third were Roman Catholics) inhabitants of Cripple Creek. When Calderwood got back to Cripple he found more miners ready to join the strike. In the Springs, Irving Howbert, and even Stratton, whose dander was up because the strikers seized his Independence mine despite his compromise contract, called for more deputies. Among the first to volunteer were the Socialites, led by Spec Penrose, who formed a dashing company decked out in a uniform of polo britches and riding boots. The young Philadelphian's Company K had the presence of mind to adhere to the military adage that cautions against cutting oneself off from one's supplies. To assure that his troops had enough whiskey on hand to meet the demand for an anesthetic should one occur during the campaign, Penrose provided his men with hot water bottles filled with aged bourbon to attach to their saddles. Barrels, he decided, were too heavy and reduced his men's mobility. When the new deputies, now some twelve hunched strong, went into camp they refused to acknowledge Sheriff Bowers as their leader and instead elected a firebrand and former shoe clerk, W. S. Boynton, to head them. With his leadership challenged and the prospect of an irresponsible and bloody encounter facing him, Bowers, in a frantic plea, once more sent to Waite for the state militia.

Tarsney and the militia arrived the next day but not before Boynton sent his battalion of deputies on a night sortie against the strikers on Bull Hill. Several rounds of shots sang in the night air and the deputies fled, some getting lost and shooting at each other. Miraculously, no one was hurt, and when the militia arrived the miners happily turned over their guns to Waite's sympathetic troopers.

A formal peace treaty on June 10 concluded the long and bitter strike. The miners agreed to surrender and the deputies were disbanded. Calderwood and three hundred strikers agreed to stand trial for any criminal acts committed during the hostilities. And the mineowners agreed to Calderwood's original demand—$3 for eight hours' work. Calderwood and most of the miners won acquittal at their trials. The man who defended them was none other than Thomas Tarsney. For his efforts Tarsney was treated to a tar and feathering by a night-riding, masked group of Springs mine-owners, none of whose identity to this day is certain.

The union victory gave strength to the Western Federation of Miners but it left wounds of spirit that would heal only on the surface. Cripple Creek would see a bloodier microcosmic class war only ten years later.

With the strike of 1894 over, however, the real business of Cripple Creek resumed with a vengeance, and incomes from gold production, transport, and milling activities soared.

In December 1895, when Charlie MacNeill's chlorination plant burned down, Spec Penrose and Charlie Tutt joined forces with MacNeill to found the Colorado-Philadelphia Reduction Company. Penrose and the two Charlies worked hand in glove with Bert Carlton who by now, with Henry M. Blackmer, controlled the Midland Terminal Railroad. In a few months the Socialites, now all advanced to their early thirties, dominated the smelting and transport business of Cripple Creek and were multimillionaires.

Cripple itself did not fare so well. On an unusually warm Saturday morning in April, 1896, one of Spec's bartenders at the Topic left work after his night shift and strolled up to Myers Avenue past the saloons and parlor houses and climbed the stairs to a room on the second floor of the Central Dance Hall that he shared with a tempestuous taxi dancer. No sooner had he entered the grubby little room than the woman, who was heating a flat iron on the coal oil stove, lit into him about something he had done that goaded her. Angry words bounced back and forth, until the bartender suddenly slapped the girl's face. She whirled, picked up a knife, and lunged at him. As she did, he reached for her arm and they swung against the stove knocking it over. In an instant the fire enveloped the floor and one wall. The bartender and his girl fled.

The entire building was in flames when the volunteer hose companies arrived and so quickly did the fire spread that before they could bring hoses to bear it leaped the alley to ignite buildings on the south side of Bennett Avenue and raced north and east burning everything in its path until the wind died at sundown. Carlton lost the First National Bank and Dave Moffat's Bi-Metallic Bank burned with such intensity that "twin streams of molten gold and silver cascaded from its cash drawers and into the gutter outside." Penrose and Tutt's Topic was a smoking pile of embers and so were three of Cripple's popular parlor houses and six of its one-girl cribs. A business block, the opera house, and Nolan's saloon were also gone.

Fires were nothing new to miners in Colorado's money-metal camps. Almost every one of them sooner or later experienced a ruinous conflagration. But two devastating fires within five days was an unmatchable catastrophe. On the following Wednesday another cry for volunteer firemen carried over the town as flames spurted from the ramshackle Portland Hotel. A gusty wind spread the fire to raze buildings spared by Wednesday's flames. In the midst of the racing, crackling flames the Palace Hotel boiler exploded, scalding a half dozen firemen, while down the street dynamite in hardware stores detonated to send flaming shards of wood into the streets crowded with fleeing people. When the fire finally burned itself out five thousand people clustered on the hillsides, cold, hungry, and homeless.

Confirmation of the second fire to hit Cripple traveled to Colorado Springs by telephone, but Springs residents had already been alerted that something ominous was happening in the gold camp when they saw the funnel of yellow-black smoke that continued to rise all afternoon behind Pikes Peak. As soon as the word came, a relief committee led by Stratton, Howbert, and Verner Reed went to work. Two special trains, crammed with cases of food, blankets, cots, tents, medical supplies, and clothing, chugged up the Midland tracks to the still smouldering Cripple Creek.

Like Denver, Central City, and Creede, Cripple Creek shook itself out of its despair at the loss of people and property and built a town cleaner and sturdier than before. The disaster had a beneficial byproduct—in the face of the aid unhesitatingly contributed by Springs residents, some of the enmity between the two towns diminished as the miners saw that beneath the protective armor of wealth of the mine- and mill-owners lay qualities of compassion and generosity.

But Cripple Creek, no matter how fresh and substantial its new facade, was still a mining camp. Despite exotic attractions such as Joe Wolfe's promotion of an authentic bullfight, the only one ever held in the United States, complete with matadors, picadors, fighting bulls, an imported band to play stirring pasodobles, and cases of champagne and barrels of oysters trundled up to grace the tables of Cripple's well-to-do, the town with its rough-hewn atmosphere had trouble holding on to its newly rich.

One who left Cripple to establish his permanent residence amid the gentility of "Little London" was Spencer Penrose. Once away from the gold camp, however, Penrose did not stem his consuming drive to make money. With his inveterate business and social pals, Charlie Tutt and Charlie MacNeill, he invested in copper mills in Utah, and for the rest of his life the fourth and supposedly lack-a-brain son of the perspicacious Dr. A. F. Penrose pocketed a million dollars a year, after taxes.

With some of his money Spec bought land at the foot of Cheyenne Mountain where Count Pourtales' once-mailglittering casino had stood and developed the world-famous Broadmoor Hotel. Then his restless itch to create something of permanence led him to build an auto road to the top of Pikes Peak. No longer would those who wanted to savor the superlative and awe-inspiring panorama from the summit of the symbol of Colorado's beginning have to make the arduous climb on foot as did bebloomered Julia Holmes in 1858. But perhaps the finest memorial created by the shrewd, impulsive, yet humanitarian milling king was the El Pomar Foundation. Founded to be used for "charitable uses and purposes (including public, educational, scientific and benevolent uses and purposes)...to assist, encourage and promote the general well being of the inhabitants of the State of Colorado," El Pomar has contributed millions of dollars to philanthropic projects since Penrose's death in 1939.

One who helped to channel Spec Penrose's constructive urges was a demure, small-statured widow with bright blue eyes and the forthright manner of those to whom no one says no. Mrs. James H. McMillan, nee Julie Villiers Lewis, was the daughter of the mayor of Detroit. Her marriage to McMillan, one of Detroit's prominent lawyers and the son of Senator McMillan of Michigan had been without a care until he contracted tuberculosis in Cuba during the Spanish-American War. Hoping to find a restorative for her husband's health in the dry air of Colorado, Julie packed up the ailing McMillan and their two children and moved to Colorado Springs. But the climate failed to help Jim McMillan and he died soon after their arrival. Not long afterward Julie's twelve-year-old son succumbed to appendicitis. Instead of fleeing the reminders of her personal tragedy, Julie

McMillan stayed on. She liked Colorado Springs, and after the blinding grief of her bereavement wore off she found she also liked the cut of forty-year-old bachelor Spencer Penrose.

The Penrose boys had assiduously avoided marriage. Their mother died when Spec was fifteen, and their father, in assuming the role generally left to mothers in schooling boys on how to discern nice girls from the other kind, injected his sons with an overdose of caution. Julie McMillan, however, did not know that and went forthrightly ahead to capture Spec's heart. After two years of being subjected to a strategy of domestic infiltration—she had her laundress do his laundry, her cook prepared his meals—Spec Penrose's resistance began to crack. Soon it was crumbling. When he realized his plight, he did what most people did at the turn of the century, if they could afford it, to cure a romantic urge. He went abroad. He cabled brother Dick that he would join him in New York to sail aboard the fast, sleek *Kaiser Wilhelm der Grosse,* holder of the Blue Riband and the approved transatlantic steamship for socialites of the period. Penrose felt more at ease once he was safely on board and the ship was underway. But when he stood at the rail watching the Statue of Liberty fall astern his eye caught sight of a familiar figure clad in furs to the top of her handsome head watching him from the deck above. When Julie in mock surprise smiled a greeting, Spec knew he'd lost the battle. There then began a procedure remarkable for its illumination of a period when filial obligations were taken as seriously as, and sometimes more seriously than, life and death. Spec's first move was not to propose to the lady but to get Dick Penrose to write a letter to their father. From Paris, Dick wrote to Dr. Penrose that "A couple of ladies from Colorado Springs, friends of Spec's, came over from New York on the same boat with us three weeks ago, and they have, at Spec's invitation, gone south with him in his automobile. One of them is a widow named Mrs. McMillan." After describing her physical characteristics, her family background, and her financial status (which was more than comfortable), Dick got down to the heart of the matter:

> Speck seems very much devoted to Mrs. McMillan, and she equally so to him. He has talked to me about proposing marriage to her and has asked me to write to you about it, and pave the way for him to write to you. Mrs. McMillan seems to understand Speck thoroughly, and the impression I have gotten of her is that she is a thoroughly sensible woman, whom a man ought to get along with if he can get along with any woman whatever. The fact that she is a widow has given her an experience with her first husband that lets her know what men are; and her thirty-five years of age has probably removed all the obnoxious ambitions of many modern women, that she might have had. I doubt, however, if she ever had any, as she seems very sensible.

Speck tells me that for two years, he has carefully considered the proposition of marrying her, and feels that he is not now deciding on a snap judgment, but after due consideration. He seems to have been with her a great deal and to know her thoroughly. Speck is peculiarly situated. He can't read much on account of his eye and, as he himself says, he is not interested in any particular subject that would lead him to seek amusement from literary or scientific sources. He is, therefore, peculiarly dependent on social intercourse. As he himself said to me the other day, he "cannot sit down at eight o'clock in the evening and read until bed time," nor can he go on forever drinking rum at clubs. Therefore he seems to think his only refuge is to get married.

It seems very hard for him to get courage to write to you, because he fears you may think him foolish, but I told him you would do no such thing and would only give him such advice as you thought best for his own happiness...

Spencer Penrose, the hard-headed, burly, cool lion of business was beneath it all a lamb. When a letter arrived from his father containing whole-hearted sanction of the match, Spec did not go flying to his love and drop on one well-tailored knee to ask her for her hand. Not he. No bleating lovesick proposals for him. He merely waited until the occasion presented itself and then he dropped his father's open letter over a beach umbrella onto Julie's lap. They were married almost immediately thereafter in London. After one of what would be a number of European tours in their lives, Spec and his bride returned to Colorado Springs to form a long-lived and happy combine of creativity—she attending to the aesthetic, he to the commercial.

By the time Spec and Julie Penrose exchanged their marriage vows in 1906 the vibrant days of Cripple Creek were over. Gold still spewed forth from its mines at an average $25 million a year, exceeded only by the production of the Transvaal of South Africa. But the period of individualism was passing. It was hurried on its way by another violent strike in 1904 climaxed by the death of thirteen miners and the maiming of a score of others in an explosion at the Florence and Cripple Creek depot at Independence. The dynamiter was Western Federation of Miners' terrorist Harry Orchard. In the mass meeting that followed, mine-owners accused the union of murder, a fight broke out, and two more men died. The state militia brought an end to the strike. This time there were no deals, no negotiations. The state government was firmly in the hands of the Republicans whose sympathies lay with the mine-owners and not, as the Populist Waite's were, with labor. A total of 225 union miners were rounded up and marched down Bennett Avenue to the railroad station. They were then taken by boxcar to the Kansas-Colorado border and told to start walking east. The struck mines reopened with miners who carried identification cards stating they were nonunion men.

The advertised crux of the strike, as it had been in 1894, was the eight-hour day. In 1902, however, when the issue went to the people of Colorado the eight-hour day was defeated and the mine-owners happily went back to a nine-hour day. This played directly into the hands of the W.F.M.'s militant socialist, Big Bill Haywood, who set about to use the heavy union membership at Cripple as a lever to break what he considered was the autocratic rule of the mine-owners. But the union overshot its aims and lost for the rank and file membership a chance to achieve better pay and working hours. Between the radical theories of Haywood and the terrorist tactics of Harry Orchard, the Western Federation of Miners in Colorado suffered a setback from which it never recovered. The attempts to organize men working in the metaliferous mines of the state ceased altogether after the strike.

In 1899, gloomy Winfield Stratton, more and more a crotchety recluse, let Verner Reed engineer the sale of the Independence for $11 million, the largest price received in any mining transaction to that time. Then the ex-carpenter became obsessed with the idea that Cripple Creek's veins of gold all met in one enormous core deeply buried within the ancient volcanic basin. He set up a company to find out if his notion was true, but the effort had only scarcely gotten underway when Stratton died in 1902. His death put an end to the exploration and it was never resumed.

In 1914 something close to what Stratton surmised stunned the mining world of Cripple Creek. Richard Roelofs, the ruddy, mustachioed superintendent of the Cresson, a low-grade mine first discovered at the beginning of Cripple's boom, telephoned Ed De La Vergne and in a shaky voice asked him to come up to Cripple the next morning but declined to tell him why he wanted to see him. When the puzzled De La Vergne appeared, the solemn faced Roelofs took him down in the cage to the 1,200-foot level of the Cresson, and then along a new drift tunneled off the main shaft. At the end, behind an iron door fitted to the rock sides and guarded by two armed miners, was an enormous cavern, called a geode, or vug. When De La Vergne stepped into the cavern and shone his carbide lamp ahead of him his eyes bulged. The chamber was forty feet high, about twenty feet long, and nearly fifteen feet wide. The ceiling and sides glittered and sparkled with segments of pure gold between the myriad crystals of quartz. The floor was graveled to a height of one and two feet with the same shining treasure. In four weeks, ore worth $1,200,000 was literally shoveled out of the vug and the Cresson overnight became one of Cripple's largest producers. In its time, it gave up over $45 million. But the Cresson vug was a freak. No other bonanzas of the same proportions turned up and Stratton's notion was still unproved and would remain so.

After Bob Womack sold his El Paso lode interest for a paltry $300 he never struck another lead. He became the town drunk, full of tales of hidden bonanzas, patronized by his friends, and ridiculed by his critics. When his sister moved to the Springs in 1893 to open a boardinghouse Bob shaped up to become her Number 2 cook. He took the Keeley cure for alcoholism, which involved drinking carrot juice and taking injections of bichloride of gold. For Bob the cure worked. He never touched a drop afterward. But his reformation came too late. A year later a stroke left him paralyzed and bedridden. When Charles Wilder of the Colorado Springs *Gazette* pointed out that Cripple Creek gold had made the Springs what it was and called for a subscription of $5,000 for medical treatments for "Crazy" Bob Womack, whose perseverance made it all happen, the citizenry responded with something less than boundless ardor. After four weeks of soliciting, the fund reached $812, and when Bob died in 1909 at the age of sixty-six his passing was hardly noticed in mining circles.

Gold production in Cripple peaked in 1901. From then on it declined. Gold-bearing veins pinched out gradually, and the smaller mines, one after the other, shut down. For some, the loss of revenue during the strike of 1904 was enough to close them forever. Dick Roelofs' magical vug evoked country-wide interest in the camp but it was soon dissipated, and Cripple Creek slid toward decay.

In 1914, the same year in which Roelofs made his discovery, Cripple was the focus of another brief but brilliant shaft of the limelight. *Collier's Weekly* ran a series of articles on the panorama of America by the popular and capable Julian Street. When the gold camp heard that the young writer was going to do an essay on Cripple Creek and Colorado Springs there was universal enthusiasm. Cripple could use a feature story to stimulate its flagging economy. The jaunty Street arrived in the Springs on a cold day in March where Spencer Penrose took him in tow for a round of lionizing that wound up very early the following morning at the festive board of the Cooking Club, the select association of gourmands headed by Penrose himself. During the evening one of the Socialites offered Street the use of his private railway car for the trip up to Cripple the following afternoon. Giddy from the synergistic effect of the high altitude and too much champagne, Street shakily climbed aboard the ponderously ornate car. On the way up to the gold camp the writer had plenty of time to absorb the scenery. The private car was so heavy that the engineer was compelled to halt the train at the foot of each climbing switchback to get up enough steam to make it up the grade. When at last the train pulled into the depot Street had less than sixty minutes before the return trip. To further complicate his task, through some misunderstanding, there was no one to meet him and he had no idea where to begin. So he shoved his hands into his overcoat pockets and started walking up one street and down the other, the brittle icy air numbing his ears and nose. When he came to Myers Avenue he stopped in front of a ramshackle crib where Madame Leo, a hoary old harlot, gave him the eye. When she found that all Street wanted to do was talk she cheerfully obliged him with a half hour's conversation in which she told all she knew about Cripple Creek. Afterward Street caught his train back to the Springs and went on his way. For weeks all Cripple waited anxiously for the appearance of Street's article. When it came there were anguished and indignant outbursts. In the pages of *Collier's* the Springs was eloquently portrayed as the oasis in the desert that stretched between those western poles of sophistication, Chicago and San Francisco. Tacked onto the end of Street's story were eight hundred graphic words describing Madame Leo and her summary of Cripple Creek's greatest need: a dance hall to draw some men. Business was so bad, said Madame Leo, that she made only $2 or $3 in a day. The mighty and unremitting protests from Cripple Creek residents to *Collier's* editors and to the state legislature came to nought. But irate Cripple Creekers finally got the vengeance they wanted. In a meeting of the City Council Mayor Hanley came up with a proposal that met with everyone's approval. The next morning newspapers across the country carried the word that the name of Cripple's notorious Myers Avenue had been officially changed to Julian Street.

As the years went by the character of Cripple Creek changed. Individual ownership of the mines and mills gave way to corporate agglomeration. The once close camaraderie between owners and employees disappeared in the limbo separating the rarified echelons of management and the everyday, grimy-coveralled hard-rock scrabblers and smelter workers. The bitter strikes of '94 and '04 did nothing to bridge the gap. Among the first to sense the change and leave Cripple for palmier boom towns were the gamblers, the saloon owners, prostitutes, and taxi dancers. When these people left the color went out of life at Cripple and whatever vigor was left revolved around the hard, mechanical, and unglamorous business of ore production.

Over the years in the larger mines that still disgorged high-grade ore, shafts went deeper and deeper until flooding became a major problem, threatening to cut off production altogether. Pumps were both inadequate and too costly. In a meeting of mine-owners, it was Bert Carlton who came up with an ambitious project to save the mines. He outlined his plan for an extensive drainage tunnel to encircle the larger mines well below the eight-hundred-foot level basin to release 8,500 gallons a minute from entrapment in the granite-rimmed basin. The plan was adopted, and when it was completed in 1910 the Roosevelt Tunnel successfully unwatered the mines for twenty years. However, shortly after Carlton died in 1931, of a kidney ailment at the age of sixty-five, once again the gold mines of the high, wind-brushed volcanic basin were jeopardized by the ever-rising water. Now Ethel Carlton, who had absorbed enough mining talk over her thirty years with Bert to know what should be done, led the fight for another, deeper drainage tunnel. In 1941, the seventeen-mile,

3,500-foot-deep Carlton Tunnel was opened to successfully drain all the Cripple Creek mines then in operation.

After a hiatus through World War II, Cripple Creek once again came alive when a $1,500,000 mill to reduce gold ore using the cyanide process was built near the Cresson. Named for Bert Carlton, the mill made it possible to treat low-grade ore at a profit and Cripple Creek, whose golden riches brought Colorado out of its silver slump, modestly but proudly took on a semblance of its old booming self. But eventually rising production costs forced the closure of all of Cripple's mines. Today, the once "world's greatest gold camp" has a year-round population perhaps numbering four hundred. But in summer Bennett Avenue and Myers Avenue with their brick and mortar skeletons of a livelier day are again as crowded as they were in the late nineties as hundreds of tourists come to sample vicariously Cripple's rousing past.

16

The End of the Beginning

DISCOVERY OF GOLD AT CRIPPLE CREEK, FOLLOWING AS IT DID ON THE HEELS OF THE panic of 1893, saved the day for Colorado. The unemployed began to find jobs. Businesses began to show a profit. And by 1895 Denver, the gauge of the state's welfare, had perked up considerably, enough in fact to risk staging a weeklong carnival to celebrate the easing of the depression.

The Festival of Mountain and Plain featured a street full of decorated floats and a masquerade ball over which reigned the mystic body of the Silver Serpent. The presence of the symbolic, snakelike creature served to remind the populace of the issue that was still close to its heart—free silver.

In January, 1894, two months after the repeal of the Sherman Silver Purchase Act, L. M. Keasby declared in *The Forum,* a national periodical of current commentary, that "The silver campaign is not nearly over. The Bull's Run campaign only has been fought. The Gettysburg victory comes next year, and in 1896 the gold bug's Appomattox."

Keasby was half right. The silver campaign was by no means over and the year 1896 would be climactic. Contrary to the assurances of the gold standardites in advocating its repeal, the scrapping of the Sherman Act, as it turned out, did not bring instant prosperity. If anything, conditions in the nation worsened. During 1894 commodity prices continued to fall and farm foreclosures mounted. The cost of living and unemployment continued to rise. The nation's poor clamored for relief, but Grover Cleveland refused to espouse a policy of more and cheaper money to bail out the debtor class.

Nor did repeal of the Sherman Act do anything to solve the problem of the country's dwindling gold reserves. Between European nations demanding payment of US debts in gold and hoarding in this country, the gold in the Treasury fell to a precarious level of $52 million in 1894. To stop the flow, Cleveland further alienated the public by making a private deal with J. P. Morgan and his European associates by which the bankers bought $62 million in

government bonds at 4 percent interest, thereby saddling the country
with an additional and painfully high public debt.

To Henry Teller, who had been returned to the Senate in 1885
after his tour in Chester Arthur's cabinet, the plain answer to the
dilemma was a return to bimetallism. A majority of westerners agreed
with him. Free silver coinage at the old ratio of 16 to 1, claimed the
silverites, would put a stop to the spiraling appreciation of gold; it
would get it out of the hands of Wall Street hoarders and would provide the masses with the inexpensive money they needed to transact
the business of their daily lives. Although Colorado was not as badly
off as it might have been owing to the gold discoveries at Cripple
Creek, a return to bimetallism would provide a welcome additional
surge of growth to the state's economy that had suffered such griev-
ous reverses when the price of silver, formerly its major product,
sagged to unprofitable levels.

Throughout 1894 and 1895, therefore, the question uppermost in
the minds of Colorado's public and private citizens was what could be
done to restore free coinage of silver. The local branches of the American Bimetallic League and the National Bimetallic Union kept up
the cry of free silver with a barrage of literature and public lectures.
Among the lecturers who toured Colorado and the West, captivating
his audiences with ringing turns of speech on the essential need of
free coinage was a young and romanesquely handsome former congressman by the name of William Jennings Bryan. Sounding the same
trumpet as Bryan were Colorado's Populist Governor Waite and disenchanted Democrat Tom Patterson.

As the presidential campaign of 1896 drew closer, the issue of
free silver grew in prominence. On a national basis, large numbers
of Republicans and Democrats held firmly to the principle of a single gold standard, claiming that bimetallism could only be tenable
if major European gold standard nations also adopted it. Over this
principle, however, party lines in Colorado and other western mining
states dissolved. In Colorado, Patterson's *Rocky Mountain News* daily
made it clear that local Democrats were heel and toe in step with
the Populists and the purveyors of unilateral bimetallism. Colorado
Republicans, unfortunately for the party, could not boast of the same
unity. Debonair, eastern establishment-oriented Senator E. O. Wolcott followed the fuzzy national Republican line of a gold standard
until "international bimetallism" could be achieved. The venerable
Henry Teller, on the other hand, vehemently shook the shock of his
black hair now shot through with gray and vociferously declared that
the only hope for the return to prosperity was for the United States
to embrace the principle of the free coinage of silver immediately and
independently of all other nations. He worked tirelessly throughout
the state and in Congress advocating his views. Feeling fully the
weight of his sixty-four years and subject to racking attacks of asthma

to boot, Teller made a long and exhausting train trip to Mexico to observe first hand the operation of a monetary system based on a silver standard alone. He came home to argue effectively that even silver as a monometallic standard would provide a more stable price level than a gold standard.

For months seesaw rumors about what the silver bloc in the Republican party would do made their way into the columns of the daily press. The Chicago *Daily News* quoted a "Western Republican of national prominence" as saying that western silverites would walk out of their convention if no free-coinage man were nominated. What's more, said the *Daily News*, the silverites would then ask the Democrats to put up a silver candidate, and failing that they would form a third party and nominate one of their own men in the hope of throwing the election into the House of Representatives in order to bargain among the candidates. Others regarded the silver bloc threat as poppycock. "They will threaten and bluster and then they will surrender," ridiculed the *Nation*. "Such senators as Teller and Wolcott have no idea of disrupting their party."

As the pressure mounted between free silver and gold standard factions of the Republican party, battle lines and strategies of the upcoming campaign took rudimentary form. Wall Streeter Mark Hanna was busy glooming his friend, former congressman and present governor of Ohio William McKinley, to run for the presidency. The issue was to be advocacy of a high tariff to protect American products. On this issue Hanna and his associates knew that they could hold together the Republican party. On the issue of silver they knew the party would disintegrate with the possible result a Democratic victory. Republican strategy called for pushing a new protective tariff bill through Congress, forcing the archly anti-tariff incumbent Cleveland to veto it, and in the accompanying outcry carry the issue to the people. The tariff bill, as it came out of committee, not incidentally happened to contain provisions for protective tariffs on hides, wool, and lead, all aimed to woo westerners away from their fixation about silver. The strategy began to work when the Dingley tariff bill passed the House in a breeze, but when it got to the Senate in January, 1896, where the Republicans had to rely on the good will of the Populists for a majority, Hanna's scheme faltered. When the finance committee attempted to extract pledges from the members of the Senate that they would not amend the bill from the floor, Henry Teller sounded the knell for Hanna's strategy. He announced flatly that he would never vote for another tariff bill unless it contained a proviso for the free coinage of silver. To his side rallied a group of western senators, and the party faithful knew that silver, not the tariff, would be the issue of the campaign.

Back in Colorado, Teller's stand and that of the rest of the "silver senators" was greeted with enthusiasm. Editorials of the *News*, the

Denver *Republican,* and Moffat's Denver *Times* loudly called for unity of the silverites against the eastern faction. However, Senator Ed Wolcott on April 28 declared in an open letter to Colorado Springs' Irving Howbert, the chairman of the state Republican Committee, that he would remain faithful to the party no matter what stand it took on silver. The hue and cry that met his words were typically summed up in the snide words of the Red Cliff *Times:* "Colorado Silver is generally all right, and it is rare indeed that such rank counterfeit specimens as Silver Ed are traced home to this state."

When Teller, in Washington, read Wolcott's letter challenging his leadership Teller got up on the Senate floor and declared that he could not support the Republican party if it did not have a free-coinage platform. The airing of Colorado's family wash on the national clothesline brought quick reaction among the eastern McKinley and Hanna crowd. "Senator Teller," scornfully reminded the Philadelphia *Inquirer,* "is a member of the band of highwaymen who refused to permit the tariff revenue bill to pass without a free silver proviso."

Nevertheless Colorado stood firm behind Teller, sending him as head of the Republican delegation to the national convention in St. Louis. He left for Missouri on June 13. So muggy was the summer weather that he had to stop over for a day en route to allow his asthmatic bronchial tubes to relax. Once in St. Louis, however, his vitality returned and he immediately began a series of conferences with other delegations pledged to silver to force a candidate and a free-coinage plank on the convention.

The battle for a candidate was lost before it began. When Teller arrived it was obvious that Hanna's eastern gold would elevate McKinley to the role of standard-bearer. He was nominated on the first ballot by a plurality of over five hundred votes. McKinley had once espoused the cause of silver but now under the grooming of Hanna he talked facilely of "international bimetallism." To Colorado silverites he sounded as phony as their own Ed Wolcott of whom Mr. Dooley smirked, "I wouldn't call him two-faced. That would be an injustice. Like his own beloved Pike's Peak, he catches the sun on iv'ry side and on iv'ry side he's beautiful."

But Teller and his silver friends still hoped to get a free-silver plank in the platform. That, too, was a futile hope. Teller was the only free-coinage member on the subcommittee that was charged with wording the money plank. In the main platform committee only ten of the represented states and territories were for free coinage. Nevertheless the scrappy silverites wrote a minority report and stood ready to try to stampede the convention, despite Teller's objection to any histrionics, if the plank as written in committee were not amended.

On the morning of June 18, Chairman Foraker of the Resolutions Committee rose to read the platform. The convention hummed with tension. For the first time, in what William Allen White described as a "hollow" and "dreary" affair, the delegates sat upright on the edge of their chairs. When Foraker reached the passage all delegates were waiting to hear he paused briefly and then plunged ahead.

"The Republican party is unreservedly for sound money...We are therefore opposed to the free coinage of silver..." Here the chairman was interrupted by the shouts and whistles of pro-gold demonstrators who made it next to impossible for the delegates to hear the rest of his words as he said, "except by international agreement with the leading commercial nations of the earth...and until such agreement can be obtained the existing gold standard must be maintained."

At the end of the reading after the tumult subsided, the clerk read the silverites' minority report and Teller was recognized to speak on its behalf. At his name the westerners were on their feet "waving hats, flags, umbrellas, fans, and handkerchiefs and shrieking like mad. The fire spread to the galleries and swept across them until they seemed to be almost unanimously carrying on the cheer."

The aging Colorado senator steadied himself by clutching the edge of the speakers' table and began a brief and clear résumé of the arguments for free coinage, citing that international bimetallism was an impossibility as long as Great Britain remained the creditor nation of the world. "This is the first great gathering of the Republicans since the party was organized that has declared the inability of the American people to control their own affairs," admonished Teller. And when at the end he stood on tiptoe and raised his right hand and shouted, "As a bimetallist, I must renounce my allegiance to my party," there were tears in his eyes. Teller then slowly made his way back to his seat to hear the roll call amid hisses from the floor and cheers from the gallery. The voting went swiftly and when it was over the clerk announced that the minority report was tabled by 818½ to 105½ votes. Solemnly then in the middle of the uproar that saw the normally aloof Henry Cabot Lodge and Mark Hanna screaming catcalls with the rest, Teller led twenty-two delegates from the silver states down the aisle and out of the convention.

Colorado blared its approval. When Teller arrived home in Denver on July 1 he was feted with an enormous parade with a brilliant array of marchers. Scores of organizations—from the Beer Drivers' Union to the Societe Italiana Unione e Fratellone—participated. The billowing bunting, the signs, the posters, the speeches, and the hysterical shouts of thepeople caused one observer to write that had the Coloradoan been an easterner "nothing could save Mr. Teller from the White House."

Not a few others, it soon became clear, were also thinking in terms of Teller for president. In the minds of free-silver advocates in the two major parties and among the Populists and National Silverites the senator from Colorado was the logical man behind whom to rally. He was the best-known free-silver advocate in the nation, and his ability and character, if not all of his ideas, were esteemed by all who knew him, friend and foe. To further this end the bolters quickly founded a new party, calling themselves Silver Republicans, and set about to organize a coalition of free-silverites behind Teller. They were immeasurably aided in their campaign by what was going on in the Democratic party, whose convention was to meet on July 7 in Chicago.

For months, there had been simmering among many Democrats, especially Silver Dick Bland and his friends, a movement to repudiate the current party dogma of a single gold standard which had, to the eyes of many Democrats, brought only a disastrous depression. When the delegates began to gather in the stifling summer heat on the shore of Lake Michigan it was more than evident that a steamroller for silver was primed and ready to roll. Unlikely as it seemed that a man could be seriously considered as a possible candidate of the Democrats who until eighteen days earlier had been a loyal Republican, Teller nevertheless was besieged with private and public pledges of support by Democratic leaders. Teller himself, however, was lukewarm at the prospect. He could see that those among the Democrats who supported him were doing so only to rally the Silver Republicans and the Populists to their side and swamp McKinley and the regular Republicans at the polls. He openly declared that he thought his nomination by the Democrats would be unwise.

Nevertheless, when the convention opened oddsmakers were taking bets on the two foremost figures for the nomination—Bland and Teller. The first piece of business was the election of a permanent chairman for the convention. After a clamorous exchange of arguments, a silver man won the gavel. Shortly after that furor died down, the Credentials Committee gave official and, as it turned out, fateful recognition to Nebraskan William Jennings Bryan as a delegate. On Thursday morning, July 9, the majority and minority reports of the Committee on Resolutions were read. In the exact reverse of the Republicans, the Democrats demanded in their majority report "the free and unlimited coinage of both silver and gold at the present legal ratio of 16 to 1 without waiting for the aid or consent of any other nation." Their minority report called for maintenance of the gold standard until international bimetallism could be achieved. Each side was allotted one hour and twenty minutes to argue for their plank. Senator Tillman of South Carolina was the first to speak for silver. It was a fiery, garbled speech, full of sectional prejudice. The lead-off man for the gold bugs was David B. Hill of New York whose sensible approach tarnished further the cause of silver after Tillman's tawdry effort. Next came Bryan to speak for the majority report. Lithe and handsome in a plain black suit, the Nebraskan, whose name had been mentioned as a dark horse candidate, strode down the aisle toward the rostrum. Crumpled in his fist was a note from Clark Howell, editor of the Atlanta *Constitution*. "Make a big, broad, patriotic speech..." Howell had scribbled. "You can make the hit of your life."

In a carefully worded and clearly delivered speech, Bryan persuasively and patiently answered the foes of silver, outlined the virtues of unilateral bimetallism, and with his audience sitting in rapt stillness he conjured up the ogres of the gold bugs, climaxing his oration with the stunning words, "...we are fighting in the defense of our homes, our families, and posterity...We beg no longer; we entreat no more; we petition no more. We defy them! Having behind us the producing masses of this nation and the world, supported by the commercial interests, the laboring interests, and the toilers everywhere, we will answer their demand for a gold standard by saying to them: 'You shall not press down upon the brow of labor this crown of thorns, you shall not crucify mankind on a cross of gold.'"

As Bryan finished he stood as if crucified, his eyes cast down, his arms outstretched at his sides. The stillness in the hot, steamy auditorium turned to bedlam as the convention threw itself at the orator's feet.

For the Silver Republicans and Populists and National Silverites sitting in the gallery the effect of Bryan's peroration was the abrupt and final curtain on their hopes for Teller. It was in fact the end of the hopes for Bland and any others for the nomination in the Democratic party. The names of Bryan, Bland, and a few others were placed in nomination, and after the delegations performed their perfunctory acknowledgment of the favorite sons, Bryan won handily on the fifth ballot. Teller's name, to his relief, was never presented to the convention. The Populists and the splinter silver parties followed the Democratic lead and also nominated Bryan when Teller declined to oppose the Nebraskan.

Colorado voters, junking their party alliances for the time being, rallied by the hundreds behind Bryan. Even the gold Midas, Winfield S. Stratton, was for Bryan and for free silver. McKinley Republicans in the state were aghast when Stratton announced that not only was he for Bryan but that he would post a $100,000 bet on the Democrat against anyone who would bet $300,000 Bryan would not win. Said Stratton in an interview in the Colorado Springs *Gazette*: "I realize that the maintenance of the gold standard would be best for me individually, but I believe that free silver is the best thing for the working masses of this country." For his part Henry Teller wasted no time pitching into the campaign for Bryan. Despite his infirmities, Teller set out upon a grueling tour stumping the crucial Midwest for his candidate. But by October it was apparent that Bryan would lose. McKinley forces, with unlimited campaign funds, flooded the voters with counteractive arguments promising to initiate action toward achievement of international bimetallism and at the same time offering the lure of protective tariffs to strengthen America's internal economy.

When the ballots were counted on November 3, McKinley captured 271 electoral votes to Bryan's 176. Bryan's strength came from the South and West. Colorado went down the line for the Nebraska spellbinder and elected to its state offices Silver Republicans and Democrats. Teller, whose election to the Senate was the responsibility of the state legislature, was reelected to office by 92 out of the 98 votes cast, dangerously close, quipped one man, to a ratio of 16 to 1. Those who voted against him were all Republicans.

After the election of 1896 the mud pie of political fusion in Colorado began to dissolve as the causes once espoused by the "radical" Populists such as an eight-hour day, an income tax, and free silver were adopted by the Democrats. With the decline of the Populists as a party there was a gradual realignment of avid free-coinage supporters, among them Tom Patterson who rejoined the Democrat's fold. The Silver Republicans did not rejoin the regular Republican party but clung to their alliance with the Democrats.

During McKinley's first term he gratified his electors by keeping one campaign promise. Through a series of conclaves abroad, with Ed O. Wolcott as his chief negotiator, the president tried to work out a plan for bimetallism among the American and European nations. But the attempt, as the silverites predicted it would, failed. In 1900, the silver bloc made one last campaign for their cause but when it was stillborn even the most dedicated of free-coinage backers admitted the issue was dead forever.

The real reason for the death of the free-silver movement, however, lay in a more fundamental cause than in fluctuating political fortunes. While silverites in Colorado clamored for relief of the depression through free coinage of the hordes of the white money metal laced through the Rocky Mountains, the world was treated to a rash of new gold discoveries in Alaska, in Australia, in the Transvaal, and ironically, in Colorado itself, in Cripple Creek. World gold production tripled from 1895 to 1905, owing both to new discoveries and to the introduction of the cyanide process of ore reduction which decreased the cost of processing. The effect of these developments was to increase the supply of money. The economy took a turn upward. After 1895, the beginnings of an inflationary period settled in to relieve the burden of the debtor class, and the nation embarked on a binge of prosperity.

Denver sparkled in the new era of good times. British visitors recorded with smug satisfaction that "streetcars now moved over much-improved, well-lighted streets whose sides were lined with excellent private and public buildings." A new generation of elite resided in the stone and brick mansions on Capitol Hill. Most of Denver's affluent still drove in soft-sprung carriages, but a few were seen in new steam Locomobiles ($750 each) lurching down the broad streets of town. Among them was the self-appointed but widely accepted leader of the new society, a petite brunette of southern birth who had won the son of Colorado's smelter king, Nathaniel P. Hill,

When Mary Louise Sneed of Memphis, Tennessee, came to Denver in the doldrum summer of 1893 to visit relatives and friends, she was aristocratically good-looking, wittily assertive, and above all thirty-two years old and ready to marry. Of the eligible and well-financed bachelors she met it was young Crawford Hill who claimed her attention. It was not because of his ardently romantic overtures. On the contrary, Hill was inordinately shy and placid. But to Miss Mary Louise he presented the ideal prospect—rich, adoring, and uncomplaining. At the end of the summer their betrothal was announced. The engagement followed the accepted code of not too long and not too short, and in January, 1895, Crawford Hill claimed Mary Louise as his wife in a fashionably *fin-de-siècle* ceremony in Memphis. The only one at the wedding who was aggrieved at the union was the benedict's mother, telling friends that she was "sick over Crawford's marriage to Louise Sneed." She went on to say that she had nothing against the young lady's rearing or social status. What worried her was the discrepancy in the young couple's ages. Mary Louise was a number of years Crawford's senior and it boded ill that a girl had not found a husband by the time she was thirty-two. But Hill *mère* need not have worried. The marriage was a good and solid one and it brought young Crawford to a prominence that his retiring nature might not have otherwise allowed.

Soon after their occupation of a stately white home on Denver's Sherman Street, Mary Louise embarked on the kind of campaign that suited a southern belle. She set out to become the queen of Denver society. She wore the most tasteful ensembles, generally jet black or all white to set off her dark hair and fair complexion; she was seen at the most select of social gatherings under the most complimentary of circumstances; and she gave the most exquisite dinners for the most felicitous people. She had not many to choose from—sixty-eight, she confided to a gossip columnist, was the number of people whose backgrounds would qualify them for invitations to her table.

All Denver followed her appearances in the daily social columns. Since Crawford Hill and his father were publishers of the Denver *Republican,* Mary Louise had no trouble keeping her name prominently before her public. When asked how one reached the pinnacle of the social world, the attractive mistress of Denver's mores answered in cultivated tones that "first you have to have money. Then you must have the knowledge to give people a wonderful time."

She had both in spades. When a freshet of wacky dance steps tripped their way west from Newport and Saratoga, Mrs. Hill quickly met the craze with morning, afternoon, and evening sessions of the Angle Worm Wiggle, the Turkey Trot, Grizzly Bear, and Bunny Hug in the ballroom of the Sherman Street house known among Denverites as the "great white square." So epidemic was the dance fad that the men begged their hostess to begin the morning parties at 7 AM so that they could get in a good hour or two of dancing before having to go to their offices. Fully aware that even a monarch by divine right must now and again acquiesce to her subjects' wishes in order to keep her throne, Mrs. Hill's invitations thereafter reflected the earlier hour.

In 1906, Mary Louise Hill brought out a Denver bluebook in which she listed the names of those she considered the city's social leaders. The magic number was thirty-six, and among them were such rags-to-riches families as the Thomas Walshes.

Conspicuously absent from the "Sacred 36," as Mrs. Hill liked to refer to them, was Mrs. J. J. (Leadville Johnny) Brown. For thirty years, from 1879, the time when the Little Jonny mine made her a rich woman, Molly Brown was ostracized from Denver society. Too gauche and garrulous was the verdict. Molly provided endless copy for *Polly Pry,* the weekly gossip digest of Denver's exclusives. One of the stories was often cited by the "Sacred 36" as prima facie evidence of her total unsuitability as one of the inner coterie. It was incriminating not because of its content but because it got into print. Molly, it seemed, let her elaborately appointed Denver home to one of the state's ex-governors on one of her periodic sorties abroad. When she returned she was annoyed to find the governor was among "The Expectorates," and when she took his wife to task for the distasteful residual signs of his habit, the wife, said Molly, merely answered, "You didn't provide any cuspidors, so what did you expect?"

But at last on May 1, 1912, three weeks after Molly earned the sobriquet of "Unsinkable" by loudly and efficiently assuming command over the fear- and shock-stunned occupants of lifeboat No. 6 of the ill-fated *Titanic,* she was entertained at luncheon at the Denver Country Club by Mrs. Crawford Hill along with a small, select group of mesdames of the "Sacred 36." The event drew half-inch type on a column heading in the news section of Tom Patterson's *Rocky Mountain News,* archrival of Hills's *Republican,* and the wry comment that perhaps "to one who has looked death calmly in the face and forgotten her own peril in order to comfort the weak and stricken...it does not seem so great a thing to Mrs. Brown to be 'taken up' by prominent society folk as it does to those who are doing the gracious act."

Watching with rueful envy the carefree cavortings of the "Sacred 36" were the once silver-rich H. A. W. Tabor and his Baby Doe. In the collapse of silver, the Tabors lost everything except the worthless Matchless and a few other undeveloped claims dotted around the state. Creditors foreclosed on the Tabor Block, the grandiose Tabor Opera House, and the house on Sherman Street not far from the Crawford Hills'. Baby Doe's jewelry was pawned to pay the rent on a small cottage in west Denver to which she and Tabor and the children slunk in mortification at the cataclysmic decline of their fortunes.

Tabor's boundless good spirits however did not decline. Every day, in his one black business suit so carefully brushed by his devoted Baby, Tabor made his way downtown to join his former business associates in long confabs in the lobby of the Windsor. People noted that the ex-silver king smoked fewer cigars and wore plain gold links instead of blindingly large diamonds at his cuffs, but it was the same old Tabor. Even at sixty-five, at an age when most men think of peaceful retirement, he was constantly busy conjuring deals and gossamer bonanzas.

If the world wanted gold, very well then, he would find gold. Surely one of his undeveloped claims held a pocket of the glittering yellow ore that would once more make the name of Tabor famous. All he needed was one good grubstake and he'd be on his way to wealth again. And who, pondered Tabor, would give it to him? Not his silver friends. They were all as down on their luck as he was. There was only one man in Colorado who might help him out. Tabor got a little money from Baby Doe and got on the D&RG for the seventy-mile trip to Colorado Springs where he presented one of the last of his business cards to Bill Ramsay, Winfield S. Stratton's secretary. When Tabor was shown into Stratton's office, the gold king and the deposed silver king greeted each other with heartfelt fervor. Unspoken was the knowledge that were it not for fate, their places could be exchanged.

Tabor asked Stratton for the loan of $5,000 so that he could work one of his remaining claims near Ward, Colorado, in the mountains west of Boulder. Stratton without a question wrote out a check for $15,000 and gave it to Tabor, telling him it was not a loan but a gift. Tabor flushed and stoutly refused to accept it unless he signed a note and Stratton took the deed to the claim as collateral. The gold millionaire smiled quickly. "All right, Senator, have it your way," he said.

Tabor beamed, signed the note, left the deed in Stratton's hands, and walked out of the office, his step jauntier than it had been in many months.

But the mine near Ward turned out to be a fluke and before long Tabor and Baby Doe were once more reduced to living off the meager funds they received from selling their few remaining personal possessions of value. In desperation, Tabor, whose income had once been $100,000 a month, took a job hauling slag in a smelter for $3 a day. During this darkest of times Baby Doe, with a courage and devotion that surprised many of her critics, stuck firmly by her husband, a bolster to his failing resolve. She and the little girls made the small rooms ring with cheerfulness and laughter whenever Tabor was at home. Never did she complain and never did she give up hope that Fortune would once more smile on her beloved Tabor.

Help for the unfortunate Tabor, when it arrived, came from an unexpected quarter. Henry Teller and Ed Wolcott buried their personal feud long enough to recognize the plight of the one-time silver king who had provided so handsomely for the Republican party in his halcyon days. They interceded with President McKinley in 1898 to award Tabor the postmastership of Denver. Thus it was that the now tired old man who had spread so many of his millions over Colorado in cultural and commercial monuments came to the last of his days. At the news of his appointment to the 3,500-a-year postmaster's job. Tabor's eyes filled with tears. To Baby Doe it was a promising new start. She fussed over her family as they moved from the cheap rooms on the west side into a modest two-room suite at the Windsor Hotel. The girls, Elizabeth Lillie and Silver Dollar, went to school in fresh new dresses. Tabor's frayed cuff shirts were thrown out for starched white new ones. And Baby Doe allowed herself an occasional purchase of new clothes to suit her new status as the wife of Denver's postmaster.

For a year and three months the Tabors lived a routine and modestly comfortable life. Then one afternoon on his way home to the Windsor from the post office Tabor collapsed. He was rushed to his room where an attending physician diagnosed his illness as acute appendicitis. Surgery, it was decided, would be too risky for the aged bonanza baron. So for six days he lay in feverish pain alternately conscious and comatose. Beside him in a sleepless vigil sat his Baby Doe, bathing his hot, dry face with cotton dipped in soothing rose-water, holding his hand, whispering love and encouragement in his ear. Again and again in his lucid moments Tabor clutched at Baby Doe's smoothly soft fingers. "Hang on to the Matchless! Whatever you do," he cried, "keep it! When silver comes back it will make you rich again."

On the afternoon of April 10, 1899, Horace Tabor died. In death Tabor received the public adulation denied him in life. Federal, state, and city buildings flew their flags at half-mast. Telegrams and tributes from all over the country filled the small suite at the Windsor. Tabor's body lay in state in the Capitol building, where the simple services were held. And as the funeral cortege, accompanied by four bands, slowly made its way through the streets of Denver past the Tabor Opera House and the Tabor Block, ten thousand people lined the sidewalk to pay their last respects to the carbonate king.

After the funeral, the disconsolate Baby Doe settled her affairs in Denver and clinging to Tabor's last words moved to Leadville to be near the Matchless. She was still a young woman, barely thirty-four years old, but the idea of marrying again repulsed her. Instead she became obsessed with the mystique of the mine that had once brought her to the summit of riches, and she tried every way she knew to bring it back into profitable production. But times were hard. Once Tabor was gone few of his friends sought to help his Baby Doe and the girls. Once, when she was on the verge of losing the Matchless, W. S. Stratton again came to the rescue, paying off the claims against it and seeing to it that Baby Doe held title to the mine. To own the Matchless free and clear was comforting but there was no money coming in and comfort could not buy food. And so once more Baby Doe pawned her valuables to feed and shelter herself and her two daughters. In time, Elizabeth Lillie grew tired of the gray existence in Leadville and her mother's preoccupation with the Matchless. She ran away to live with Baby Doe's brother, Peter McCourt, in Chicago. Baby Doe did not try to bring her back. Her youngest child, Silver Dollar, was her solace—warm, witty, and beautiful, she was the antithesis of the cold and sullen elder girl. Silver Dollar loved Leadville. The rough and ready life of a mining town, even one in a decline, suited her open nature. But Baby Doe had visions of wider horizons for Silver. She saw her as a figure the whole country would admire, and when the girl showed some literary talent she encouraged her to follow it. Silver went to Denver and then to Chicago where she worked for a time as a reporter and wrote *Star of Blood,* an unsuccessful novel of sensation and little artistic merit. She grew discouraged and in letters home to her mother she spoke of going into a convent.

Never in her letters to Silver did Baby Doe reveal how reduced were her own circumstances. To conserve what little money she had, the widow of H. A. W. Tabor moved out to the drafty shack adjacent the hoist house of the moribund Matchless. Here, with her feet clad in rags and old miners' boots, a shapeless rough wool dress covering her once fulsome body, and her hair, still pathetically lustrous, hidden in an oversized laborer's cap, Baby Doe lived out her lonely days cadging food from wherever she could find it, seeing no one except Sue Bonnie, a neighbor, and endlessly reading her favorite book, *The Lives of the Saints.*

In 1925, a Chicago newspaper gave frontpage space to a sordid story of degeneracy and death. Under the name of Ruth Norman, said the newspaper, Silver Dollar Tabor, the youngest daughter of Colorado's famous mining king, was found scalded to death in a rundown boardinghouse on the south side. The investigation of her death revealed her as a chronic drunk, a drug addict, and a prostitute. When she was shown the article, Baby Doe refused to acknowledge that Ruth Norman was her child. When asked she would say only in steely tones that her daughter was in a convent in the Midwest.

The saga of Colorado's legendary Tabors came to an end one wintry hour in February, 1937, when Baby Doe, alone, freezing, and weak from the lack of food, slumped unconscious to the floor of her fireless cabin. Some days later Sue Bonnie found her body, stiff with cold, lying in the shape of a cross with the arms outstretched. No fanfare attended her burial. To save her from a pauper's grave Leadvillites donated funds for Baby Doe to be laid beside her beloved Tabor in Denver's Mt. Olivet Cemetery.

With the return of good times to Denver in the late nineties there also returned to town its favorite bunko artist, Jefferson Randolph Smith. During his self-imposed exile from Colorado's capital Soapy Smith had not been idle. He ran the rollicking short-lived silver camp of Creede for two years and when its inhabitants turned respectable in the midst of the town's decline, Soapy moved on. St. Louis and Houston were his next stops. In each he was run out of town for brawling. He then struck south into Old Mexico, where with his facile art of persuasion he convinced President Diaz he had been a colonel in the US Army and would be the ideal man to organize a mercenary army to bolster Diaz's Guardia Rurales who had their hands full combating the guerrilla tactics of the oppressed peasantry. Diaz gave Soapy a 4,000-peso advance to recruit his mercenary army and Soapy left for the States to gather together some men. However, when Diaz heard it rumored that Soapy might be intending to use the force against and not for him, the Mexican dictator canceled the con man's commission and threatened him with arrest if he came back to Mexico. Soapy shrugged off the failure of his latest con game as one of the fickle fortunes of war and turned again to Denver, his favorite town.

Once more on the corner of 17th Street the soap game worked like a charm. Soon Soapy's coffers bulged with suckers' gold and he expanded his operations to open the dazzling Tivoli Club. A long and richly finished cherrywood bar graced the downstairs. On the second floor were two dozen green baize tables featuring every popular game of chance known to Soapy's avid customers. Over the heavy etched plate-glass front door was carried the stern admonition, "CAVEAT EMPTOR." The Tivoli flourished and, in the spirit of Robin Hood, Soapy saw to it that much of the club's profit found its way into the hands of Denver's needy. The agent for its dispensation was Fighting Tom Uzzell who came down from Leadville to open the People's Tabernacle, a nondenominational spiritual haven for the troubled souls of town. "The Lord loveth a cheerful giver," said Parson Tom, "and if ever there was a cheerful giver, Jefferson is that man." Few knew of Soapy's good works among the poor and even a smaller number knew that several prominent men of commerce in Denver entrusted to Soapy substantial funds to be distributed to the destitute. In the code of that nobler era, donors to charity, be they crooks or bankers, preferred to remain anonymous.

But if Soapy could giveth he could also taketh away. Two real-estate brokers, their pockets crammed with the proceeds of a recently closed deal in California, sauntered into the Tivoli one evening and some six hours later stormed out again, relieved of every penny. The two immediately went to the police and swore out a warrant for Soapy's arrest on a charge of operating rigged games. Hauled before a panel of magistrates, the glib Soapy presented his defense in masterful style if in specious logic. At the Tivoli, he declared, no one was compelled to play his games. Nor was anyone discouraged from playing. "We let experience be the teacher," concluded Soapy. "No player can win at my games, and when a man goes broke at one of my tables he's learned a lesson he'll never forget. I am therefore providing a valuable and moral community service by operating the Tivoli."

Case dismissed.

This episode clearly indicated that the attempt at reform of the city staged by Bloody Bridles Waite in the mid-nineties had been spectacularly unsuccessful. After the Populist was deposed by a regime of Democrats, Denver reverted to its "business as usual" habits of pre-panic days. Graft, bribery, and general corruption were rampant. A new and highly successful defense plea for murder was a "brainstorm"—temporary insanity created by an overdose of alcohol. Fake mining companies, lotteries, and robber-hack drivers did a thriving business. Court cases were quickly and efficiently fixed. In one judge's chambers there was a continuing crap game nightly for the initiate. At elections the town's politicians liberally crossed the public palm to get votes from the quick and the dead.

Not all of these enterprises were controlled by the suave Soapy Smith. During his absence in Creede and in Old Mexico a portly, oily man with half-mast lids from beneath which peered reptilian, gray eyes slid into the gangland throne Soapy had vacated. Lou Blonger's bulbous nose sensed the right moment for a takeover, and in a short while, from the plush office of his gaming and drinking establishment, he ruled Denver's underworld. With the help of his brother, Sam, the wily Lou built an empire of confidence schemes that threatened to gobble up Soapy's profits. Smith stood the infiltration of Lou and Sam Blonger just so long and then one day he decided to assert his rights on the basis of seniority. Warned that the Blongers were no tinhorn pushovers Soapy nevertheless tucked his derringer in his pocket and headed for his rivals' club. As he was about to enter, a squad of police, tipped off by Soapy's men, arrived to persuade the thimblerigger to give up his notion of having it out with the Blongers. Soapy protested loudly and mightily but allowed himself to be conducted away from the Blonger stronghold. As Soapy left, Lou Blonger, who had watched the proceedings from behind the cigar counter inside the front door of his club, carefully put away the loaded double-barreled shotgun held ready in his hands.

Rankling under his deposition as Denver's bunko king, Soapy felt his old restlessness return. And when the news of the Alaska gold strike filtered into Colorado it was the diversion he needed. Gathering up his faithful attendants he headed north to Skagway. Here, with the technique he used in Denver and Creede, Soapy began his takeover—the con games, the kickbacks, the bribes, the rigged elections, the muggings, all in fast order became features of daily life in the boisterous gateway to Klondike gold. But Soapy's luck at last ran out.

Skagway citizens rebelled. They formed a "Committee of 101" vigilantes to clean up their town. On a cold clear night in 1898, among the scurrying rats of Skagway's main wharf, Soapy Smith and one of the vigilantes stalked each other between islands of silent bales and crates. Suddenly flashes of orange gunfire lit the night. Soapy Smith, the thirty-eight-year-old king of the thimbleriggers, fell dead with a bullet in the chest. A few yards away his pursuer lay sprawled on the rough planks with a fatal wound from Soapy's gun.

Soapy's abdication of Denver however did not make any difference in the Blongers' operations. Through the first fifteen years of the twentieth century they were the undisputed rulers of Denver from the shadowy firmament of the underworld. Eventually, however, the Blongers went too far. In 1915 at the hands of a fire-eating city attorney their hold on the town was broken and the ringleader, Lou Blonger, went off to serve a seven-year sentence in Canon City prison where he died before his term was up.

❧⚬❧

In the 1890s, free silver coinage was to some the panacea that would relieve the stressed economic condition of the little man and create true prosperity. Many others, however, argued that a plethora of foreign markets would do the same thing for the country. America was ready, claimed the supporters of overseas expansion, to take her place as a world power. Trade to and from secure foreign ports carried in sturdy American bottoms protected by a first-class navy, declared geopolitician Navy Captain Alfred Thayer Mahan, would make the United States healthy, wealthy, and strong. Implicit in his argument was the need for colonies. On February 15, 1898, Spain conveniently provided the United States with a step toward acquisition of such colonies and an opportunity to put into practice the Mahanite doctrine. On a peaceful visit to Havana harbor the US battleship *Maine* exploded and sank, claiming the lives of 258 enlisted men and two officers. The details of how it happened were sufficiently obscured so that it was not difficult for the sensation-loving press of the day to label it a dastardly deed of the Spanish. This was the climax of a number of years of concern exhibited by the United States over the high-handed constraints Spain exercised on her Cuban subjects and coincidentally on the more than $50 million in American investments in the island. War fever swept the nation, led by lurid editorials and the rabble-rousing slogan "Remember the *Maine!*" McKinley, who did not want a war, tried, but not very hard, to ward off the obviously impending conflict. However, when Spain's queen regent during Easter week ordered Spanish troops to cease action against Cuban insurgents and offered to consider some degree of liberty for Cubans, the president failed to open the door to negotiations but went to Congress on April 11 asking for authority to use military power to intervene in Cuba to restore stability. Congress was more than willing to grant his request and went further—to form a resolution declaring the avowed purpose of the United States in any intervention was to get Spain out of Cuba. In the Senate, Colorado's Henry Teller was appalled by the imperialistic overtones of McKinley's message and the resulting draft resolution, and he hastened to organize a movement for an amendment calling for Cuban independence. The best he could do among the hawks who held the majority was a statement declaring that the United States had no interest or intention of forcing its own rule on its island neighbor. The Teller Amendment became a part of the Cuban resolution. On April 21 Spain severed diplomatic relations with the United States and on April 25 the two countries went to war.

Colorado embraced the call to arms with the same spontaneous fervor with which it had responded to the Civil War. Soon after the muster of volunteers began Colorado outran its quota of 1,600 men—"all the best youths of the State are volunteering," crowed Governor Alva Adams. After the hardiest were winnowed from the bulk in a remarkably short time one regiment of infantry, two troops of cavalry, and a battery of artillery carrying the bloodied and gloried banner of the First Colorado Regiment swung aboard the Denver Pacific on May 17 headed for San Francisco. They sailed for the Philippines on June 15 aboard the fast packet *China*. On August 13, 1898, the Colorado First led the way in the capture of Manila. When word reached Denver that the grand old outfit raised the Stars and Stripes over the Philippine capital to the tune of "There'll Be a Hot Time in the Old Town Tonight," a ragtime melody born in one of Colorado's own silver camps and already an American classic, people streamed into the streets to embrace their neighbors, their eyes brimming with tears of unbounded pride.

The "splendid little war" was short and sweet with the fruits of victory for the United States. After two disastrous naval engagements with the US fleet, one in Manila Bay and another off Santiago, and a half-dozen encounters with the US Army in Cuba and the Philippines, the Spanish reeled to the peace table ceding outright to the United States Puerto Rico and Guam. For $20 million in US gold dollars Spain also agreed to the sale of the Philippines. In the House, Republican Thomas B. Reed retired as Speaker in disgust at his party's position. In the Senate, Henry Teller, sorely troubled at this last of a series of stands by his party that he found untenable, prepared once and for all to change his political allegiance.

The decision was not lightly taken. As a law student, Teller at twenty-six had been one of the ardent group of anti-slavery men—Democrats, Whigs, and Free Soilers—who met in Angelica, New York, in 1856 to form the Republican party in that state. Through the intervening years as America moved westward, and Teller with it, he found no contradictions between his beliefs and his party's. But now he found himself more and more supporting causes unpopular among Colorado's affluent Republicans. He was pro-Indian, pro-women's suffrage, and pro-income tax, and he was anti-gold, anti-big business, anti-imperialism, and anti-war.

In 1900, when Henry Teller announced that he would take his seat on the Democratic side of the Fifty-seventh Congress, it was symptomatic of what had happened in the new Eldorado.

In the heady, unfettered early days of the mineral rush to the Rockies a man's opportunities and the choice of his pursuits were limited only by his capabilities and will to work. Society was essentially rural, with abundant room for individualism. But as swarms of people gravitated to the Rockies the simplistic frontier life underwent a profound and not very subtle change.

From the mid-sixties to the late nineties those who had struck it rich in Colorado were ranged on the side of the Republicans, and they controlled the political destiny of the region. Gradually, the millions taken out of the granite ramparts of the Rockies polarized the haves and the have-nots. Mining camps and trading posts became cities, further separating the haves and the have-nots in the blight of urbanization. As the bonanza years passed, the haves less and less identified themselves with the have-nots and the fissure between the two grew wider and deeper. The result was rebellion. Behind the free-silver movement was the cry of the have-nots for relief from unemployment and debt. Behind the expanding labor movement was the anguish of the have-nots at the loss in the pyramid of corporate control of the once close tie between miner and mine-owner. In the face of growing urbanization and industrialization, Colorado's common man leavened his politics with the principles of populism and turned to those who advocated social reform. In 1901, Coloradoans turned dapper, pro-eastern capitalist, Republican Edward O. Wolcott out of his Senate seat and replaced him with Democrat Tom Patterson, arch foe of oppression and loyal friend of labor. Now he and Teller, both from the same side of the Senate, could wage their fight together for the rights of the common man in an industrial society. Henry Teller's political transformation and Patterson's election to the Senate were straws in the wind. Never again after 1901 would the new Eldorado be the exclusive dominion of a party of privilege.

The whoopee spirit that pervaded Denver at the turn of the century had its fullest and most memorable expression in the unlikeliest pair of business partners the city would ever see. Roly-poly Henry Haye

Tammen was a well-known figure in the capital city. In his blue polka dot bow ties and crumpled fedora, surrounded by a ring of smoke from his expensive cigars, Tammen had been a bartender at the Windsor Hotel and with his accumulations of tips he went on to open a highly successful tourist trap featuring Moon Eye, an embalmed Indian maiden of disputed age, "Game Heads, Fire Minerals," and a "Large Assortment of Opals, and Agates, Tigereye, Topaz." He was a friendly, generous Dutchman whose mind turned naturally to the showy, the shocking, and the profitable. In 1895, when he heard that the *Denver Post,* a paper begun by old-guard Democrats to air their grievances against Bryanism, was about to go under, Tammen was struck by an uncontrollable urge to run a newspaper. The fact that his experience in this field was limited to publishing a poorly paying magazine of western stories was no deterrent. What was an obstacle to the thirty-year-old curio storeowner was the $12,500 price tag on the *Post.* It was however only a momentary setback to Tammen's journalistic urge.

Through a friend, Tammen was put in touch with the devilish and darkly handsome thirty-four-year-old Frederick Gilmer Bonfils, who had managed with remarkable ingenuity through the depression of the early nineties to gather unto himself a considerable fortune. Son of a prominent member of the Missouri bench, Bonfils, so hinted the Kansas City *Star,* was one and the same L. E. Winn who was wanted for questioning in the matter of a lottery staged by Mr. Winn for the express benefit of Mr. Winn. The lottery, it seems, was not the least of his innovations. Under another of several aliases Bonfils supposedly added nicely to his financial status through a land deal involving the sale of lots in Oklahoma that were actually in Texas.

However, the means of a question never worried Henry Tammen. It was the end that counted. Bonfils obviously felt the same way. Once the sparring was over there grew out of the meeting between the jolly, stubby little Dutchman and the ramrod straight, reserved Bonfils a friendship and partnership that lasted twenty-nine hellbent years, and launched on Denver a style of journalism that was in Henry Tammen's own words "yellow...read, and...true blue."

No subject was sacred, no subject was taboo. Tammen, whom Gene Fowler characterized as having been born with his tongue in his cheek, very quickly made it clear what kind of a paper he and Bonfils wanted the *Post* to be.

"Son," said Tammen to his city editor, "you've seen a vaudeville show, haven't you? It's got every sort of act—laughs, tears, wonder, thrills, melodrama, tragedy, comedy, love and hate. That's what I want you to give our readers."

Headlines were big, black, and bold. "Does It Hurt To Be Born?" headed a feature avidly read by all Denver in an age when it was risqué to speak openly of physiological phenomena. When a notorious

murderer of a group of young girls in a Baptist church was hung in
California's San Quentin prison, the *Post* announced, "Demon of the
Belfry Sent Through the Trap." "Bon" and "Tarn," as the town came to
call the irreverent rogues of news publishing, had a hand themselves
in composing the eye-catching headlines that set their paper apart.
When it came over the wire that explorer Henry Stanley had died,
Bonfils suggested that the obituary be capped with "Stanley Goes
to Find Livingston Again." When a copy desk man suggested tim-
idly that Tammen's "Jealous Gun-Gal Plugs Her Lover Low" was not
exactly good grammar, the rotund little Dutchman shifted his cigar
to the other side of his mouth and waved an instructional forefinger
in the newsman's face.

"That's the trouble with this paper," said Tammen, "too God-
damned much grammar. Let's can the grammar and get out a live sheet."

Every page reeked of the sideshow. Bonfils and Tammen ran
notices of Bible meetings alongside ads for astrological clairvoyants.
They delightedly promoted the carryings-on of Dr. Alexander J. Mcl-
vor-Tyndall, a self-styled metaphysician whose pronouncements such
as "Tyndall Says Loving Mothers Make Drunkards of Sons" drew
two-column headlines. Withal, the hoopla style of the irreverent publishers had one natural complement and it was not long in coming.

Henry Tammen had always yearned to own a circus, and as soon
as the paper's treasury showed a surplus he broached the subject to
Fred Bonfils.

"You can do anything you like, Henry," was Bonfils' response, "as
long as it shows a profit."

Tammen did not wait for his partner to have second thoughts.
He started modestly, gathering together a dog and pony act, and then
with typical Tammen flair he hired on a relative of the Sells Brothers, a bespangled family of the high wire who flew through the air
under the Ringling banner. To add a touch of carnival class to the
name Sells, Tammen charged his staff to rack their brains for a name,
scanned the classics and pulp publications, and was still lacking the
right combination when one day he happened to stroll past the desk
of his sports editor, Otto C. Floto.

The little Dutchman's eyes brightened as he stared lovingly at the
green eye-shaded figure pouring over a list of batting averages. Feeling
the lingering eyes of his chief on his back, the sports scribe looked up,
and as he gazed at Tammen realization spread across his face.

"Sells-Floto," Tammen enunciated slowly, a wide smile puffing
out his round cheeks.

"Oh, God," winced the sports editor.

The Sells-Floto Circus survived both a lawsuit brought by the
miffed Ringlings who thought they had a monopoly on Sells and a
rash of bad publicity surrounding a few cases of escaped tigers and
elephants to become one of the West's favorite shows under the big

top. It also provided Tammen with a fringe benefit of sorts. When he decided to live among Denver's exclusives, Tammen found a house that he liked and set about to buy it. There was only one drawback. The owner did not want to sell. Henry pondered a while and then he bought the vacant lot next door and announced to the neighborhood that he intended to tether Princess Alice and Snyder, two of the Sells-Floto Circus elephants, on the property. Henry got the house he wanted in jig time.

The *Post* had no editorial page. Every story was an editorial in itself. Reporters and editors were encouraged to write themselves and the views of the paper into their columns. The closest thing to a masthead editorial was a column written by Bonfils himself, unabashedly headed "So The People May Know." Here the Corsican toasted such dignitaries as Dave Moffat and Mayor Robert Speer, whose administration was marked by an ambitious plan to build parks, boulevards, and public buildings to make Denver the "City Beautiful," and a spree of town-hall corruption that made Soapy Smith's regime look tame by comparison.

Unlike his partner, Bonfils disdained the hoi polloi. His was a superior bearing, aloof, regal, and superconfident. He was more than willing that his public understood that he was a descendent, however tenuous the line, of Napoleon and that he had been at West Point. It was a saving grace that Tammen had a sense of humor. The patrician Bonfils had none.

He saw no reason for the guffaws that greeted the slogan he created to grace the facade of the *Post* building: "*O, Justice, When Expelled from Other Habitations, Make This Thy Dwelling Place.*" In his eyes the *Post* was the guardian of the people. To prove his point he dove into one "cause" after another. Child labor at Denver stores drew his righteous fire in "So The People May Know." "Little girls, poorly dressed, and with pale faces are employed in these department-store sweat shops," wrote Bonfils. "Let every lover of childhood, let every person who has children of his own see to it that such conditions cannot find favor...in the Great West." When that campaign brought the legislation of some humane child labor laws in the state, the dashing corsair of Champa Street flailed his cutlass in the cause of prison reform. With the *Post's* independent-spirited and statuesque female reporter Mrs. Leonel O'Bryan as legman, Bonfils waded into a fight for the pardon of none other than Colorado's favorite cannibal, Alfred Packer. Mrs. O'Bryan and Bonfils turned out a series of columns citing cases of those convicted of murder and rape who were given lesser sentences than life, each story ending with "Why not Packer?" That did the trick. Packer was paroled in 1901.

The aftermath, however, was not quite what the *Post* publishers bargained for. When Bonfils heard that Denver attorney "Plug Hat" Anderson had misrepresented himself to Packer as a member of the *Post* staff and had gotten power of attorney and a $25 fee from the bewildered convict, he called Anderson to account in the Red Room, the plum-hued offices of Bonfils and Tammen known among reporters as the "Bucket of Blood." Never was it a truer nickname than on January 13, 1900, when Bonfils, Tammen, and Mrs. O'Bryan confronted the wily Anderson within its precincts. The upshot was a flurry of fisticuffs and curses of elaborate structure punctuated by several well-aimed shots from a pistol in Anderson's hand. Bonfils fell to the floor with a bullet in his throat and one near his heart. Tammen dove behind a desk, but two shots caught him, one in the wrist and another in the shoulder, as Mrs. O'Bryan grappled with Anderson to keep him from finishing off his adversaries.

"I'll kill you!" raged the crazed attorney at the female columnist clutching his arms.

"Go ahead, shoot!" cried the indomitable Mrs. O'Bryan, "and then hang!" Anderson blinked and suddenly dropped his gun arm. Then he turned without another word and walked calmly out of the office and to the police station to give himself in.

Regrettably, both Bonfils and Tammen, to whom the pure melodrama of the scene would have been the ultimate satisfaction, were both unconscious.

The publishers recovered from their wounds within a few months, with the only lasting injury being to Bonfil's throat. He could no longer smoke. He overcame his handicap however by keeping at hand an aide and a humidor full of expensive cigars. On command the aide lit up and Bonfils leaned back in his chair to savor the aroma of the smoke blown in his face.

Another lingering discomfort in Bonfils' mind was that his life had been saved by a woman. After the shooting and his recovery, he made life so difficult for Mrs. O'Bryan that she left the *Post* to start her own weekly newssheet, *Polly Pry*. Not that either Tammen or Bonfils was a woman-hater. They both had wives and Bonfils had children, but neither family desired to move in the keep or even in the outer breastworks of the "Sacred 36," and were but rarely seen in public. Yet it was Bonfils who at forty-three found in one of Denver's most admired society women a constant and fond acquaintance. It began when Mrs. Madge Reynolds, the thirtyish, good-looking wife of a well-set-up oil man of Denver, took up the crusade to free from prison a youth convicted of murder at the tender age of ten. Mrs. Reynolds came to the *Post* one day to ask for help in her battle. After an hour with the gracious, composed lady, Bonfils and the *Post* were solidly behind the charming supplicant. Under the aegis of the Reynolds-Bonfils campaign, the young prisoner, apparently fully rehabilitated, was duly pardoned. For five years afterward the relationship between the publisher and Mrs. Reynolds, remarkably untainted by any public or private suggestion of the carnal, grew closer and closer. The couple were often seen together, and Bonfils was a frequent visitor in the

Reynolds' modest home. His regard for the lady was unconcealed and around his associates he referred to her as "Dearest." One February day the dark and brooding-eyed Bonfils took stock of his feelings and sought the counsel of his most trusted friend. What, asked Bon of Tam, would he think of a wealthy, powerful man who cast aside all that he had to sojourn with a "beautiful and understanding woman."

"Come down out of the clouds and think it over first," was the verdict of the solid, salt-of-the-earth Henry Tammen who knew immediately what Bonfils was driving at.

A few days later on a brisk afternoon Fred Bonfils and Madge Reynolds went horseback-riding as was their weekly wont. Sometime during the ride, so it is said, Bonfils spoke of his fondness for her but suggested that owing to each of their positions it might be better that their association cease for a while to see how matters resolved themselves. Apparently the lady agreed, however she may have felt within, without any outward show of anguish, and the two parted cordially at the stable. The next morning when Bonfils arrived at his office he picked up Patterson's *Rocky Mountain News* to see two-inch headlines dancing in a ribbon of black before his eyes. "MRS. REYN-OLDS DIES!" When the message finally made its way through his retina to his brain Bonfils collapsed. In popular fiction the death of the lady would have been a suicide. In this the bereaved survivor could exorcise some of his grief in guilt by taking the blame for the act. But not so in life. Bonfils was allowed no such outlet for his feelings. The cause of death, according to the attending physician, was a purely capricious attack of angina pectoris.

For months the specter of his lost idol haunted Bonfils. The fact that he first learned the word of her passing in the *News* served to strengthen his deep enmity toward Tom Patterson and his powerful newssheet. In the midst of his bitterness, however, Bonfils took some grim satisfaction in remembering a day almost a year earlier when he had shown the militant Patterson that it did not pay to joust with the *Post*.

There were three other papers in Denver when Bonfils and Tammen let loose their style of journalism on the unsuspecting popu-lace—the Hills's *Republican*, Dave Moffat's *Times*, and Byers' old paper, the *Rocky Mountain News*, under the leadership of Tom Pat-terson. Very quickly through their flamboyant coverage Tam and Bon made their paper a household fixture. People laughed, stormed, swore, and spit at the *Post*, but everybody read it. Two years after its first edition hit the streets, it had 25,000 subscribers. By 1907, 83,000 people read the *Post*, more than the combined total of its three com-petitors. Only one of the three, the *News*, seriously took up the fight against the brash interloper.

The battle for the affections of advertisers grew to white-hot pitch. Tom Patterson, seething and sputtering at the *News's* diminish-ing subscriptions and the high-handed tricks of Tammen and Bonfils to corral and keep advertisers, leveled his big guns on his rivals and banged away in a series of no-punches-pulled editorials. In the last of the series, Patterson laid it on the line—the publishers of the *Post* were blackmailing their advertisers. To go with the editorial Patterson singled out the debonair Bonfils and commissioned a cartoon showing him in the garb of Captain Kidd.

On the day after Christmas, 1907, the sixty-eight-year-old but still spry Patterson set out from his home for the mile walk to his office at the *News*. His route never varied—north on Logan to the corner of Thirteenth, cattycorner across the intersection to the well-worn path through a vacant lot to Fourteenth and Grant, and then on downtown. So habitual was his walk that he scarcely watched where he was going through the thick-lensed glasses occasioned by his congenital nearsightedness. On this chill morning he strode purposefully along, crossed the intersection, and started down the path through the vacant lot, absently noting a passing bicyclist. Almost at once afterward he heard light footsteps behind him and a cheery "Good morning." As he turned to see who greeted him, first one fist and then another slammed into his head. Patterson staggered and fell, his mouth cut by his broken upper plate. As he lay on the ground more blows pummeled his already bruised and bleeding head. At last they stopped and Patterson groggily felt around for his glasses that had gone flying at the first cuff. When he found them he shakily hooked the bows over his ears and peered up from his couch of defeat to see Frederick G. Bonfils standing over him, white with fury and his usual reserve dissolving in a stream of barroom invective that was as shocking to the dignified Patterson as were the blows to his head.

The trial was short and the outcome in the corruption-ridden state of city politics was predictable. The *Post* publisher was fined $50. In the course of the testimony it came out that there had been tactics on the part of the *Post* that could qualify as blackmail, but Bonfils, smugly certain of the moral character of his large readership, bothered neither to disprove the allegations nor to apologize to Patterson and the upstart *Post* continued on its flagrant and eminently popular way.

When he wasn't sniping at Tarn and Bon, Patterson used the *News* to lambaste the business trusts, wealthy mine-owners, and the well-entrenched Mayor Robert Speer, whose political activities in 1906 had split the Colorado Democrats, causing Patterson to lose his Senate seat. The ramrod of his campaigns was a dynamic writer named George Creel. Creel came to Denver a "militant liberal...filled with outrages against life's injustices and red hot for anything whatsoever that looked like a holy war." He first landed behind the news desk of the *Post*. Before long, however, Bonfils, who would cheerfully pay a boy $3 a day to look for his strayed pet cat but who begrudged his reporters a 12-cent-a-day raise, drove him to the *News*. Here, in a crusade that brought a last and satisfying triumph to the aging Patterson,

Creel's skillful arousal of the public managed to unseat, if only for one term, veteran Mayor Speer, whose City Beautiful machine benignly masked the cesspit that was city government. In the wake of their victory, Patterson, Creel, and their loyal followers made substantial progress behind a reform mayor. Named to be commissioner of police in 1913, Creel, with the help of Sheriff Glen Duffield, proceeded to transform the capital. If Speer had prettied it up, Creel and company cleaned it up. Forced out of business were the palaces of pleasure on Market Street. Shuttered and boarded until new and legitimate tenants moved in were Mattie Silks's memorable address and Jennie Rogers' infamous House of Mirrors. Run out of town were the bunko steerers, the numbers racketeers, boodlers, pimps, and opium sellers. Summarily faced with notices of closure, the gamblers, big time and tinhorn, fled the steamrolling forces of right. With the same fervor with which she erred, Denver laid the ghost of her unsavory reputation and never again was the sin town of her youth.

ꙅ꙰

Though the *Rocky Mountain News* closed up shop in 2009, the *Denver Post* survived as the spokesmen of the region, though considerably tamed by age. And so is the rest of Colorado. Mining is still a major industry. Lead and zinc and a host of other minerals, under syndicated ownership, are worked with steady and placid prosperity. Industrialization in mining has totally replaced individualism. The glamorous days of the money-metal bonanzas have long since passed into history. The once roistering gold and silver camps are all quiet now. Some are ghost towns with their rotting shaft houses and sprawling tailing dumps silent testimony to the throes of the state's birth. Others, overlaid with a new prosperity stemming from the countrywide interest in summer and winter recreation, bask in the comparative peace of tourism, bearing out Bayard Taylor's 1866 prophecy that the region was "destined to become for us what Switzerland is to Europe."

Like the *Post,* the new Eldorado was "weaned on tiger's milk." To the gold- and silver-hungry hordes, to the starry-eyed converts to Manifest Destiny, to the dedicated farmers of the gospel, to the sometimes savage Arapaho and Ute, to the wily mountain men, the slick townmakers, slippery-fingered card men, hurdy-gurdy girls, and ferret-eyed madams, to the farsighted railroaders, to the politicians, honorable and corrupt, to the perceptive entrepreneurs and crusading newsmen, to all of these and more, Colorado owes her stormy adolescence. Out of forty years of boom and bust that began with the bonanzas of Gregory and Jackson on Clear Creek in 1859 there emerged a rich, progressive, and scenically splendored state.

Notes

Note numbers indicate page.

1. Raising Color

3. De Vaca: "Many signs of gold": Bolton, 14.
4. Friar Marcos: "More fear than food": Clissold, 116.
5. Friar Marcos: "Larger than the city of Mexico": Hafen and Rister, 8.
9. William Bent: "Last and best home": Willison, 6.
9. "Good water, grass, and timber": Hafen, *Pikes Peak Gold Rush Guidebooks of 1859*, 35.
12. Green Russell: "You can all go": *ibid.*, 110.
13. "Young, handsome, and intelligent": *ibid.*, 59.
15. "It would have looked like humanity": *ibid.*, 75.
15. "The New Eldorado!!!": Hafen, *Colorado Gold Rush: Contemporary Letters*, 30.
16. "Not over a hundred miles...from here": *ibid.*, 54.
16. "Glorification speeches": *ibid.*, 58.
16. A. D. Richardson: "The excitement which I predicted": *ibid.*, 52.
17. "An unmitigated humbug": *ibid.*, 50.
17. "Taking a slight dose": *ibid.*, 46.
17. "Free as the air": *ibid.*, 83.
18. "Merry times": *ibid.*, 83.

2. Rush to the Rockies

23. "Humbug as to the plenitude of gold": Hafen, *Colorado Gold Rush: Contemporary Letters*, 214.
25. "Broken hopes and blasted fortunes": *ibid.*, 310.
25. "No acts of violence": *ibid.*, 324.
25. "The humbug of humbugs": *ibid.*, 325.
26. George Jackson: "Will quit and try to get back": *Colorado Magazine*, XII, Nov., 1935.
27. George Jackson: "As tight as a No. 4 Beaver trap": Willison, 51.
30. "An entirely new mode of traveling": Hafen, *Colorado Gold Rush: Contemporary Letters*, 259.
31. William N. Byers: "Leave not your character at home": Hafen, *Pikes Peak Gold Rush Guidebooks of 1859*, 293.
31. "With ordinary tools": *ibid.*, 162.
33. "Little rats of mules": Hafen, *Colorado Gold Rush: Contemporary Letters*, 279.
37. "Leavenworth hears the echo": Hafen, *Colorado and Its People*, I, 189.
38. Horace Greeley: "Exploded bubble": Greeley, 70.
39. Horace Greeley: "I had not dreamed": *ibid.*, 70.
39. Horace Greeley: "Decidedly limited": *ibid.*, 137.

3. Gateway to Gold

42. William Larimer: "In high glee": Larimer, 10.
45. "Rope and noose would be used on him": Hafen, *Colorado and Its People*, I, 156.
45. "Half-Mexican, half-Indian": Smiley, 283.
46. "Kill at forty-five rods": Hafen, *Colorado Gold Rush: Contemporary Letters*, 182.
46. Horace Greeley: "The Astor House of the gold region": Greeley, 136.

47. Horace Greeley: "Twenty-five cents for a drink": *ibid.*, 136.
47. William Larimer: "You have no idea of the gambling": Larimer, 168.
50. Matthew Dale: "The political demogogues in Denver": *Colorado Magazine,* XXXII, April, 1955.
54. "Was immediately adjourned": Hafen, *Colorado Gold Rush: Contemporary Letters,* 349.
55. "With his shirttail full of type": Willison, 33.
56. "Where he may be at all times found": *Rocky Mountain News,* April 23, 1859.
58. "Ten cents a glass—froth included": Hafen, *Colorado and Its People,* I, 233.
61. "General, I plumped him": Larimer, 179.
61. Richens Lacey Wootton: "It was as neat and orderly an execution": Conard, 381.
64. "Adonis of the Rocky Mountains": Perkin, 256.
65. Matthew Dale: "Bold shrewd, go-a-head style": *Colorado Magazine,* XXXII, April, 1955.

4. Bonanza on Clear Creek
66. "Foster, Slaughter, and Shanley are in the mountains": Hafen, *Colorado Gold Rush: Contemporary Letters,* 368.
68. "When we reached Clear Creek": Greeley, 107.
69. "Commodious and comfortable": *ibid.*, 104.
70. "Within ten years": *ibid.*, 106.
70. "Unparalleled Richness. Gold! Gold!! Gold!!!": Hafen, *Colorado Gold Rush: Contemporary Letters,* 329.
71. "Give each man a Bible": Gandy, 30.
76. "Nary a church": Hafen, *Reports from Colorado: The Wildman Letters, 1859–1865,* 173.
78. "Size, shape, and mobility": Young, 33.
80. "Yes, the Guard will be there": *ibid.*, 56.
85. "The gamblers could not live": Willison, 68.

5. Fanning Out
86. "Most in the least time": Villard, 180.
89. "Nearly as large as watermelon seeds": *Rocky Mountain News,* Sept. 10, 1859.
89. "A continual stream of miners": *ibid.*, Aug. 13, 1959.
90. "The miners are lying about loose": Hafen, *Colorado and Its People,* I, 240.
95. "Dear Father, I don't know": Dyer, 313.
98. "In the face of every kind of opposition": *ibid.*, 137.
98. "Bachelors hall": *ibid.*, 142.
101. "They were too drunk": *ibid.*, 146.
101. "Leading silly women": *ibid.*, 160.
101. "Convictions are slight": *ibid.*, 328.
107. "WESTERN COLORADO TO CALIFORNIA—GREETINGS!": Eberhart, 141.
110. "The fourth day in the park": Polk, 225.
110. "The men had not arrived": *ibid.*, 227.
112. "Fine gold in an abundance": *ibid.*, 226.
112. "Walled up with log Palaces": Hafen, *Colorado Gold Rush:: Contemporary Letters,* 195.
112. "Spree all night and return": Willison, 135.

6. Bucksin, Broadcloth, and Bullets
114. "Gone to bury my wife": Hafen, *Colorado and Its People,* I, 272.

115. "There is not a single hat in the town": Hafen, *Reports from Colorado: The Wildman Letters, 1859-1865*, 42.

115. "A good dentist's office": *ibid.*, 224.

115. "It keeps me awake": *ibid.*, 54.

116. "An absolute necessity": Zamonski and Keller, 99.

116. "Whereas, the towns at and near": Polk, 63.

116. William Larimer: "They and I are now cheek-by-jowl": Zamonski and Keller, 100.

116. "Restrain, suppress, and prohibit": Perkin, 225.

119. "Instantly decided": *ibid.*, 171.

119. "Drunken devils and bummers": Baker and Hafen, I, 380.

119. "Tenacious of revenge": *ibid.*, 380.

121. "Rivaled Delmonico's": Zamonski and Keller, 119.

122. "We have only to say": *ibid.*, 118.

122. "The act was wanton": *ibid.*, 128.

122. "The rowdies, the ruffians, and bullies": *ibid.*, 129.

125. "Regular and constitutional tribunals": *ibid.*, 146.

125. "Is the Inquisition revived": *ibid.*, 154.

126. Richens Lacey Wootton: "The killing of persons": Conard, 265.

126. Richens Lacey Wootton: "And we allowed them": *ibid.*, 266.

126. "Of all the low cunning, small dealing": Hafen, *Reports from Colorado: The Wildman Letters, 1859-1865;* 140.

129. "6 dollars a week": Hafen, *Colorado and Its People,* I, 220.

130. "The conspiracy of silence": Lamar, 218.

130. "It may be concluded that": Zamonski and Keller, 186.

131. "The fears of the South": *ibid.*, 186.

132. "The Lincoln administration": *ibid.*, 40.

133. "Leaven of treason": Perkin, 234.

134. "253 scalps of warriors": Hafen, *Colorado and Its People,* I, 132.

134. William Gilpin: "The facts collected by me": Hafen, *Pikes Peak Gold Rush Guidebooks of 1859,* 241.

135. "Were much spoken in business": Hafen, *Reports from Colorado: The Wildman Letters, 1859-1865,* 226.

135. "Where two or three are gathered": Polk, 65.

135. "Everyone that thirsteth": *ibid.*, 64.

136. "Clairvoyant physician": Athearn, 39, 137

136. "Diseases of the heart": *ibid.*, 39.

136. "Brevet rank of lady and wife": *ibid.*, 39.

136. "In St. Louis or Baltimore": Hafen, *Reports from Colorado: The Wildman Letters, 1859-1865,* 223.

136. "Theatrical ladies": *ibid.*, 224.

136. William Gilpin: "The stern and delicate duty": Hafen, *Colorado and Its People,* I, 283.

137. "A large number of Denver's fairest ladies": *ibid.*, 281.

137. "Three loud cheers": *ibid.*, 281.

139. "There never was a viler set of men": Hafen, *Relations with the Plains Indians, 1851–1861,* 271.

142. "Balm of a thousand bayonets": Hafen, *Colorado and Its People,* I, 262.

144. "Way out upon the Platte": *ibid.*, 263.

146. "All women were belles": *ibid.*, 562.

146. "When quadrilles were danced": *ibid.*, 562.

146. "I derive no pleasure whatever": Hafen, *Reports from Colorado: The Wildman*

Letters, 1850-1865, 276.

147. "Wanted, a girl to do housework": Polk, 93.

147. "Rich, racy, and refreshing": Schoberlin, 52.

151. "The roaring of Niagara": Wharton, 121.

151. "Broken buildings, tables": Perkin, 213.

151. "Lost, on the night of the 19th": Polk, 69.

153. John M Chivington: "Little and big": Perkin, 269.

153. John M. Chivington: "Nits make lice": *ibid.,* 269.

156. John Evans: "Organize for the defense": *ibid.,* 265.

156. John Evans: "Unless the authority is given": Hoig, 67. (From US War Department. *The War of the Rebellion. A Compilation of the Official Records of the Union and Confederate Armies.* Series 1, vol. XLI, 753–54.)

157. John M. Chivington: "If any of them are caught": Hoig, 83. (From US War Department. *The War of the Rebellion. A Compilation of the Official Records of the Union and Confederate Armies.* Series 1, vol. XXXIV, 151.)

159. "Until troops can be sent out": Hoig, 127. (From US War Department. *The War of the Rebellion. A Compilation of the Official Records of the Union and Confederate Armies.* Series 1, vol. XLI, 915.)

161. John M. Chivington: "At daylight this morning": Hoig, 154. (From US War Department. *The War of the Rebellion. A Compilation of the Official Records of the Union and Confederate Armies.* Series 1, vol. XLI, 948.)

161. "Among the brilliant feats of arms": Hoig, 162. (From the *Rocky Mountain News,* December 17 and 29, 1864.)

161. "Great Indian drama": Schoberlin, 31.

161. "Splendid war bonnet": *ibid.,* 31.

164. "The old mining excitement has ceased": Quiett, 156.

7. The Little Kingdom of Gilpin

166. "The deserted mills": Taylor, 61.

168. "Miserable and expensive failure": Hafen, *Colorado and Its People,* II, 499. (From the *Rocky Mountain Directory and Colorado Gazeteer,* 1871, 231.)

172. "Melo Dramatic Pantomime": Schoberlin, 41.

173. James Thomson: "The hills surrounding us": Salt, 85.

173. James Thomson: "Would be very graceful": *ibid.,* 79.

173. "Hotels, families, and bars"; Hollenback, 88.

176. "The driver never spoke:" Bird, 190.

177. "Such churlishness": *ibid.,* 188.

177. "One is never in doubt"; Cushman, 130.

178. "The first for the gentlemen": Beebe, *Narrow Gauge in the Rockies,* 156.

178. "Might easily be taken for": Hollenback, 20.

179. "One of these days": Bancroft, *Gulch of Gold,* 231.

182. "The young lady": *ibid.,* 298. (From Rhoads, H. M., *Town Talk,* Central City, 1877.)

186. Baby Doe Tabor: "Jake gave me these": *Empire Magazine* in the Denver *Post,* Oct. 21, 1955. (Taken from Baby Doe Tabor's scrapbooks. The dates of entries in the scrapbooks are uncertain. Baby Doe's reference to herself as BT in the quotation suggests that the gentians and the message may have been entered some time after her marriage to H. A. W. Tabor in 1882 when her initials would have been BT, for Baby Tabor.)

187. Jacob Sandelowsky: "Meet me, my darling, at ten": Bancroft, *Gulch of Gold,* 309.

187. Baby Doe Tabor: "Dark, dark hair, very curly": *ibid.,* 314.

189. "No more delays": Hollenback, 35.

8. White Metal

212. "The road pursued the canyon": Bird, 90.
213. "How to Be Healthy, Happy, and Rich": Schoberlin, 200.
216. Louis Dupuy: "Man is a machine": *Colorado Magazine,* XIII, 1936, 214.
217. Louis Dupuy: "Whose beauty of face": *Colorado Magazine,* XXXII, 1955, 11. (From an interview on K.FEL-TV with Joseph E. Smith, Feb. 28, 1954.)
218. Louis Dupuy: "In our crowded societies": quotation posted on the kitchen wall of the Hotel De Paris, Georgetown, Colorado, dated June 5, 1887.

9. The Silvery San Juans

232. Nathan C. Meeker: "Cut every Indian down": Dunn, 592.
233. "Indians off their reservation": Vickers, "History of Colorado" in *History of Clear Creek and Boulder Valleys,* 139.
234. "Slowly and sullenly they filed": Hafen, *Colorado and Its People.* I, 393. (From Wyman, "A Preface to the Settlement of Grand Junction," *Colorado Magazine,* X, 1933, 27.)
235. "How joyful it sounds": Hafen, *Colorado and Its People,* 392. (From the Ouray *Times,* Sept. 11, 1881.)
235. "The whites...were so eager": *ibid.,* 392. (From Report of Brigadier General Pope, Sept. 22, 1881, in Secretary of War Reports, 1881.)
235. "Working to develop the resources": Hafen, *Pikes Peak Gold Rush Guidebooks of 1859,* 75.
236. "San Juan Polka": Wolle, *Stampede to Timberline,* 421. (From the *Silver World,* April, 1877.)
238. "Undesirable element dispersed": *ibid.,* 422.
238. "Neat and tasty uniforms": *ibid.,* 422. (From the *Silver World,* 1878.)
243. "Farmed it out like potatoes": Rockwell, 209.
245. "I'm dead. I'm dead!": Gibbons, 45.
245. "My Dear Mother": Coquoz, *Tales of Early Leadville,* 1966, 20-22.
246. "On the 30th": Ouray *Times,* April 5, 1878.
246. "Rocky Mountain canaries": *Colorado: A Guide to the Highest State,* 72.
248. "Ouray has had the noble red man": Ouray *Times,* Aug. 30, 1879.
248. "There is not enough business": *ibid.,* Sept. 13, 1879.
249. "For opposition or politician": Wolle, *Stampede to Timberline,* 372.
249. "A conglomeration of Rural": *Pioneers of the San Juan Country,* I.
252. "Daughter, we've struck it rich!": Beebe, *The Big Spenders,* 349.
253. "They knew money existed": *ibid.,* 350.
259. "Everything is there": Ouray *Times,* Nov. 29, 1879.
259. "Paw hell out of the ivories": Hafen, *Colorado and Its People,* II, 560.
259. "Thirty pound rails": Beebe, *Narrow Gauge in the Rockies,* 190.
260. "Front seats on the way up": *ibid.,* 97.

10. Leadville— The Biggest Bonanza of Them All

270. "Why, it is plain": Polk, 229.
272. "Intense with woe and agony": Griswold, 50.
273. "The stuff had come off the hill": Davis, 203.
275. "'In The Sweet Bye and Bye'": Griswold, 121.
277. Oscar Wilde: "Miners, men working in metals": Mason, 131.
277. Oscar Wilde: "I read them passages": *ibid.,* 31.
278. "There is no city": Griswold, 151.
278. "Is there no law": *ibid.,* 151.
279. "'Murderous Attack'": *ibid.,* 156.

279. "Notice to all": *ibid.*, 170.
281. "In harmony": Birmingham, 292.
283. General William Palmer: "For 120 miles": Feller, 129.
288. "Cancans, female bathers": *Colorado: A Guide to the Highest State,* 173.
288. "Eligible eight": Griswold, 25.
290. "Over-run with deposits": *Colorado: A Guide to the Highest State,* 175.
290. "Please don't swear": *ibid.*, 178.
291. "Exceedingly vulgar": Parkhill, 152.
292. "Lake County is declared": Griswold, 194.

11. The Carbonate Trail

300. "Queen Anne and modern": Wolle, *Stampede to Timberline,* 43.
300. "A peacock among a lot of mudhens": *ibid.*, 46.
300. "Might well be called": Wallace, 50.
301. "Sacred music to dance to": *ibid.*, 50.
305. "Grim furrows of resolute determination": Woole, *Stampede to Timberline,* 235. (From the Aspen *Times,* Sept. 19, 1888.)
308. "Struck it rich": *ibid.*, 71.

12. Statehood and Status

309. "Denver Crowd": Lamar, 254.
309. "Golden Crowd": *ibid.*, 254.
310. "The politics of business": *ibid.*, 283.
312. "Broadway dandies": Bird, 95.
313. "English sporting tourists": *ibid.*, 95.
314. "Elephantine hands," Sprague, *A Gallery of Dudes,* 100.
314. "Pearl colored gloves": *ibid.*, 100.
315. William Byers: "We have had a ducal ball": *ibid.*, 109.
317. "Meretricious display": Parkhill, 25.
317. "Naughty capers": *ibid.*, 25.
318. "Taking the food": Parkhill, 112.
321. William Byers: "Denver during the past week": Schoberlin, 211.
321. "The Denver banks": Bird, 119.
323. Ulysses S. Grant: "It possesses all the elements": Stone, 359.
325. "Denver, conscious of a shady record": Athearn, 45.
325. "A newer and fashionable den,": Parkhill, 24.
325. Eugene Field: "Colonel K. G. Cooper": Perkin, 364.
326. Eugene Field: "The Reverend George W. Miln": *ibid.*, 375.
326. William Byers: "Leave not your character": Hafen, *Pikes Peak Gold Rush Guidebooks of 1859,* 253.
326. "Eyes of coal gray tint": Perkin, 318.
326. "Oh, infernal villain": *ibid.*, 316.
327. Owen J. Goldrick: "Mr. Byers was a big fool": *ibid.*, 317.
330. Augusta Tabor: "I wish the record to show": Parkhill, 163.
330. "Collation chamber": Karsner, 215. (From the Washington *Post,* March 2, 1883.)

13. "Holy Moses!"

338. Carlyle C. Davis: "That the streets are paved": Mumey, 65.
338. "Faced you everywhere": *Colorado: A Guide to the Highest State,* 79.
339. "Not a brick": *ibid.*, 79.
340. "It's day all day": *ibid.*, 89.

340. "Assessments: saloons, $5.00": Lavender, *The Big Divide,* 133.
341. "I don't like this quiet": Mumey, 98.
343. "J. J. Dore Finds the Petrified Body": *ibid.,* 130. (From the *Rocky Mountain News,* April 11, 1892.)
345. "Free from the law": Collier and Westrate, 106.
350. "Lulu Slain, a frail daughter": Wolle, *Stampede to Timberline,* 325. (From the Creede *Candle,* Sept. 15, 1893.)

14. Populism and Panic

354. Grover Cleveland: "Dangerous and reckless": Perkin, 385.
354. "A cowardly makeshift": Porter and Johnson, 88.
355. Richard Harding Davis: "Thoroughly eastern": Perkin, 393.
356. "Western empire": *Colorado Magazine,* X, 1933, 44.
356. Davis H. Waite: "Monarchy and monopoly": *ibid.,* 45.
356. Davis H. Waite:" And if the money power": *ibid.,* 45.
357. Henry M. Teller: "Rabid frothings": *ibid.,* 45. (From the Denver *Republican,* July 18, 1893.)
357. "Criminal folly": *ibid.,* 45.
358. "If a single shot is fired": Collier and Westrate, 170.
359. "Serfs of the men": *Colorado Magazine,* X, 1933, 45.
360. "Economists were beginning to say": Stone, 371.

15. Cripple Creek— "The World's Greatest Gold Camp"

369. "White and shiny": Waters, 112.
369. "Willfully deserted and absented herself": *ibid.,* 42.
370. Winfield S. Stratton: "The impression came to me": *ibid.,* 123.
373. "Don't jostle the man": *Colorado: A Guide to the Highest State,* 106.
375. "Schradsky's pants are always down,": Porter, *Pay Dirt,* 13.
376. "Cripple Creek is not only": Sprague, *Money Mountain,* 98.
379. "Poor persons who are without means": Waters, 250.
386. "Three members collapsed": Sprague, *Money Mountain,* 138.
389. "Twin streams of molten gold": Beebe, *Narrow Gauge in the Rockies,* 170.
390. "Charitable uses and purposes": Sprague, *Newport in the Rockies,* 273.
391. "A couple of ladies from Colorado Springs": *ibid.,* 245.
391. "Speck seems very much devoted": *ibid.,* 246–47. (This and the quotation immediately above are excerpts from a letter written by R. A. F. Penrose, Jr., to his father and printed in *Life and Letters of R. A. F. Penrose, Jr.,* by Helen R. Fairbanks and Charles P. Berkey, The Geological Society of America, New York, 1952. The excerpts are reproduced here with the permission of The Geological Society of America.)

16. The End of the Beginning

397. "The silver campaign": Ellis, 228 (From *The Forum,* January, 1894.)
398. "International bimetallism": *Bulletin of the University of Wichita,* No. 46, 1960, 8.
399. "Western Republican of national prominence": Ellis, 239. (From the St. Louis *Republican,* July 7, 1895.)
399. "They will threaten": *The Nation,* May 14, 1895.
400. "Colorado Silver": *Bulletin of the University of Wichita,* No. 46, 1960, 7. (From the Denver *Republican,* May 5, 1896.)
400. "Senator Teller": Ellis, 254. (From *Public Opinion,* May 21, 1896.)
400. "I wouldn't call him two-faced": Ellis, 250.

401. "Hollow" and "dreary": Glad, 111.
401. "The Republican party is unreservedly for sound money": Porter and Johnson, 104.
401. "Except by international agreement": *ibid.*, 104.
401. "Waving hats, flags": Ellis, 259.
401. Henry M. Teller: "This is the first great gathering": *ibid.*, 260.
401. Henry M. Teller: "As a bimetallist": *ibid.*, 260.
401. "Nothing could save Mr. Teller from the White House": Ellis 228. (From *The Nation*, Nov. 23, 1893.)
402. "The free and unlimited coinage": Porter and Johnson, 98.
403. "Make a big, broad, patriotic speech": Coletta, 136.
403. "We are fighting": Waters, 166.
403. Winfield S. Stratton: "I realize that the maintenance": *ibid.*, 169.
404. "Streetcars now moved": Athearn, 46.
405. "Sick over Crawford's marriage": *Rocky Mountain News*, April 19, 1959.
405. "First you have to have money": *ibid.*
406. "The Expectorates": Denver *Post*, Feb. 18, 1968.
406. "To one who has looked death": *Rocky Mountain News*, May 2, 1912.
410. "The Lord loveth a cheerful giver": Collier and Westrate, 148.
412. "Committee of 101": *ibid.*, 209.
413. "All the best youths": Ginger, 204.
413. "Splendid little war": *ibid.*, 203.
415. "Game Heads, Fire Minerals": Fowler, 64.
415. "Yellow...read, and...true blue": *ibid.*, 99.
415. "Son, you've seen a vaudeville show": Martin, 13.
415. "Does It Hurt To be Born?": Perkin, 406.
416. "Demon of the Belfry": Fowler, 139.
416. "Stanley Goes To Find Livingston": *ibid.*, 276.
416. "Jealous Gun-Gal": *ibid.*, 100.
416. "That's the trouble with this paper": *ibid.*, 100.
416. "Tyndall Says": Fowler, 222.
417. "O, Justice, When Expelled": Perkin, 407.
417. "Little girls, poorly dressed": Fowler, 138.
417. "Why Not Packer?": Denver *Post*, Feb. 18, 1968.
419. "Dearest": Fowler, 225.
419. "Beautiful and understanding woman": *ibid.*, 267.
419. "Mrs. Reynolds Dies": *ibid.*, 267.
420. "Militant liberal": Perkin, 414.
421. "Destined to become for us": Taylor, 11.
421. "Weaned on tiger's milk": Fowler, 88.

Bibliography

Books

Athearn, Robert G., *Westward the Briton*. Scribner's Sons, New York, 1953.

Baker, J. H., and Hafen, LeRoy, *History of Colorado*. State Historical Society of Colorado. Linderman Publishers, Denver, 1927.

Baggs, Mae L., *Colorado, The Queen Jewel of the Rockies*. The Page Company, Boston, 1918.

Bancroft, Caroline, *Gulch of Gold: A History of Central City*, Sage Books, Denver, 1958.

Bancroft, Hubert H., *History of the Life of William Gilpin*, The History Company Publishers, San Francisco, 1889.

Barker, W. J., *Boldly They Rode*. Lakewood, Colorado, 1949.

Barney, Libeus, *Letters of the Pike's Peak Gold Rush: Early Day Letters from Auraria, 1859–1860*. The Talisman Press, San Jose, California, 1959.

Beebe, Lucius, *The Big Spenders*. Doubleday & Company, Inc., New York, 1966.

———, and Clegg, Charles, *Narrow Gauge in the Rockies*, Howell-North Books, Berkeley, California, 1958.

———, and Clegg, Charles, *Rio Grande, Mainline of the Rockies*. Howell-North Books, Berkeley, California, 1962.

Bird, Isabella L., *A Lady's Life in the Rocky Mountains*. G. P. Putnam's Sons, New York, 1875.

Birmingham, Stephen, *Our Crowd, The Great Jewish Families of New York*. Harper and Row, New York, 1967.

Bliss, Edward, *Brief History of the New Gold Regions of Colorado*. John W. Amerman, New York, 1864.

Bolton, Herbert E., *Coronado, Knight of Pueblos and Plains*. Whittlesey House, New York, 1949.

Boner, Harold A., *The Giant's Ladder*. Kalmbach Publishing Company, Milwaukee, 1962.

Brown, Robert L., *An Empire of Silver*. Caxton Printers, Ltd., Caldwell, Idaho, 1965.

Byers, William N., and Kellom, John H., *Hand Book to the Gold Fields of Nebraska and Kansas*. D. B. Cooke and Company, Chicago, 1859.

Carver, Jack, Vondergeest, Jerry, Boyd, Dallas, and Pade, Tom, *Land of Legend*. Caravon Press, Denver, 1959.

Clissold, Stephen, *The Seven Cities of Cibola*. Clarkson N. Potter, Inc., New York, 1962.

Coletta, Paolo E., *William Jennings Bryan: I, Political Evangelist, 1860–1908*. University of Nebraska Press, Lincoln, Nebraska, 1964.

Collins, J. N., *Principles of Metal Mining*. G. P. Putnam's Sons, New York, 1874.

Collier, William Ross, and Westrate, Edwin Victor, *The Reign of Soapy Smith*. Doubleday, Doran, Inc., New York, 1935.

Colorado: A Guide to the Highest State. American Guide Series, WPA Writers' Program, Hastings House, New York, 1941.

Conard, Howard L., *Uncle Dick Wootton: The Pioneer Frontiersman of the Rocky Mountain Region*. The Lakeside Press, Chicago, 1957, reprint.

Cushman, Samuel, and Waterman, J. P., *The Gold Mines—Gilpin County, Colorado*. Register Steam Printing House, Central City, Colorado, 1876.

Dallas, Sandra, *Gaslights and Gingerbread*. Sage Books, Denver, 1965.

Davis, Carlyle Channing, *Olden Times in Colorado.* Phillips Publishing Company, Los Angeles, 1916.

Davis, E. O., *The First Five Years of the Railroad Era in Colorado.* Sage Books, Denver, 1948.

Davis, Herman S. (editor), *Reminiscences of General William H. Larimer and of his Son Wm. H. H. Larimer.* Privately printed, Lancaster, Pennsylvania, 1918.

Dawson, Thomas F., *Life and Character of Edward Oliver Wolcott.* Knickerbocker Press, New York, 1911.

Dunn, J. P., *Massacres of the Mountains: A History of the Indian Wars of the Far West, 1850–1875.* Archer House, Inc., New York, 1886; reissue 1958.

Dyer, John L., *The Snow-Shoe Itinerant.* Cranston and Stowe, Cincinnati, 1890.

Eberiiart, Perry, *Guide to the Colorado Ghost Towns and Mining Camps.* Sage Books, Denver, 1959.

Ellis, Elmer, *Henry Moore Teller: Defender of the West.* Caxton Printers, Caldwell, Idaho, 1941.

Emmitt, Robert, *The Last War Trail: The Utes and the Settlement of Colorado.* University of Oklahoma Press, Norman, Oklahoma, 1954.

Fetler, John, *Pikes Peak People: The Story of America's Most Popular Mountain.* Caxton Printers, Ltd., Caldwell, Idaho, 1966.

Fossett, Frank, *Colorado: Its Gold and Silver Mines.* C. G. Crawford Company, New York, 1879.

Fowler, Gene, *Timberline.* Covici, New York, 1933.

Fritz, Percy S., *Colorado, the Centennial State.* Prentice-Hall, Inc., New York, 1941.

Gandy, Lewis C., *The Tabors.* Press of the Pioneers, New York, 1934.

Gard, Wayne, *Frontier Justice.* University of Oklahoma Press, Norman, Oklahoma, 1949.

Ghost Towns of Colorado. WPA Writers' Program, American Guide Series, Hastings House, New York, 1947.

Ginger, Ray, *Age of Excess: The United States from 1877 to 1914.* The Macmillan Company, New York, 1965.

Glad, Paul W., *McKinley, Bryan, and the People.* J. B. Lippincott Company, Philadelphia and New York, 1964.

Greeley, Horace, *An Overland Journey from New York to San Francisco in the Summer of 1859.* S. M. Saxton, Barker and Company, New York, 1860.

Greever, William, *The Bonanza West.* University of Oklahoma Press, Norman, Oklahoma, 1963.

Grinnell, George B., *The Fighting Cheyennes.* University of Oklahoma Press, Norman, Oklahoma, 1956.

Griswold, Don L. and Jean H., *The Carbonate Camp Called Leadville.* Smith Brooks, Denver, 1951.

Hafen, LeRoy R., *Colorado and Its People.* Lewis Historical Publishing Company, New York, 1948.

————, *Colorado Gold Rush: Contemporary Letters and Reports.* The Arthur Clark Company, Glendale, California, 1941.

————, *Colorado, the Story of a Western Commonwealth.* Peerless Publishing Company, Denver, 1933.

————, *Overland Routes to the Gold Fields, 1859.* The Arthur Clark Company, Glendale, California, 1942.

————, *Pikes Peak Gold Rush Guidebooks of 1859.* The Arthur Clark Company, Glendale, California, 1941.

————, and Ann W., *Colorado, A Story of the State and Its People.* The Old West Publishing Company, Denver, 1943.

————, and Ann W., *Relations with the Plains Indians. 1857–1861.* The Arthur Clark Company, Glendale, California, 1959.

————, and Ann W., *Reports from Colorado: The Wildman Letters, 1859–1865, with Other Related Letters and Newspaper Reports, 1859.* The Arthur Clark Company, Glendale, California, 1961.

————, and Rister, Carl C, *Western America.* Prentice-Hall, Inc., New York, 1941.

Hall, Frank, *History of the State of Colorado.* Blakely Printing Company, Chicago, 1890.

Hall, Gordon L., *The Two Lives of Baby Doe.* MacRae, Smith, and Company, Philadelphia, 1962.

Henderson, Charles W., *Mining in Colorado; A History of Discovery, Development and Production.* US Geological Survey Professional Paper 138, Washington, DC, 1926.

Henderson, Junius, et al., *Colorado: Short Studies of its Past and Present.* University of Colorado, Boulder, Colorado, 1927.

Hill, Alice Polk, *Tales of the Colorado Pioneers.* Pierce-Gardner, Denver, 1884.

————, *History of Clear Creek and Boulder Valleys.* O. L. Baskin & Company, Chicago, 1880.

Hoig, Stan, *The Sand Creek Affair.* University of Oklahoma Press, Norman, Oklahoma, 1961.

Hollenback, Frank R., *Central City and Blackhawk Colorado: Then and Now.* Sage Books, Denver, 1961.

Hollister, Ovando J., *The Mines of Colorado.* Samuel Bowles and Company, Springfield, Massachusetts, 1867.

Horner, John W., *Silver Town.* Caxton Printers, Ltd., Caldwell, Idaho, 1950.

Howbert, Irving, *Memoirs of a Lifetime in the Pike's Peak Region.* G. P. Putnam's Sons, New York, 1925.

Howlett, Rev. W. J., *Life of the Right Reverend Joseph P. Machebeuf.* Franklin Press, Pueblo, Colorado, 1908.

Jocknock, David, *Early Days on the Western Slope.* Carson-Harper Company, Denver, 1913.

Karsner, David, *Silver Dollar; The Story of the Tabors.* Crown Publishers, New York, 1958.

Kemp, Donald C., *Colorado's Little Kingdom.* Sage Books, Denver, 1949.

Lamar, Howard R., *The Far Southwest, 1846–1912: A Territorial History.* Yale University Press, New Haven, 1966.

Larimer, William Henry Harrison, *Reminiscences.* Privately printed, Lancaster, Pennsylvania, 1918.

Lavender, David, *Bent's Fort.* Doubleday & Company, Inc., New York, 1954.

————, *The Big Divide.* Doubleday & Company, Inc., New York, 1948.

————, *One Man's West.* Doubleday &: Company, Inc., New York, 1956.

Lee, Mable Barbee, *Cripple Creek Days.* Doubleday & Company, Inc., New York, 1958.

McLean, Evalyn W., *Father Struck It Rich.* Little, Brown, Inc., New York, and Boston, 1936.

Marshall, Thomas M., *Early Records of Gilpin County, Colorado.* University of Colorado, Boulder, Colorado, 1920.

Martin, Lawrence, *So the People May Know.* Denver *Post,* Denver, 1950.

Mason, Stuart (editor), *Oscar Wilde, Impressions of America.* London: Sunderland, 1906.

Miller, Max, *Holladay Street.* Signet Books, Inc., New York, 1962.

Mitchell, Wesley C., *A History of the Greenbacks.* University of Chicago Press, Chicago, 1903.

Monroe, Arthur W., *San Juan Silver*. Grand Junction *Sentinel*, Grand Junction, Colorado, 1940.

Moody, Ralph, *The Old Trails West*. Thomas Y. Crowell Company, New York, 1963.

Morgan, H. Wayne (editor), *The Gilded Age, 1865–1890, A Reappraisal*. Syracuse University Press, Syracuse, New York, 1963.

Mumey, Nome, *Creede: The History of a Colorado Silver Mining Town*. Art-craft Company, Denver, 1949.

———, *Professor Oscar J. Goldrick and His Denver*. Sage Books, Denver, 1959.

Newton, Harry J., *Yellow Gold of Cripple Creek*. Nelson Publishing Company, Denver, 1928.

Oakes, D. C., *Traveler's Guide to the New Gold Mines in Kansas and Nebraska*. Pacific City, Iowa, 1859; Denver, 1949, facsimile.

Old, R. O., *Colorado: United States; America; Its History, Geography, and Mining*. British and Colorado Mining Bureau, London, 1869.

Parkhill, Forbes, *The Wildest of the West*. Henry Holt and Company, New York, 1951.

Paul, Rodman W., *Mining Frontiers of the Far West, 1848–1880*. Holt, Rinehart, and Winston, Inc., New York, 1963.

Perkin, Robert L., *The First Hundred Years: An Informal History of Denver and the Rocky Mountain News*. Doubleday & Company, Inc., New York, 1959.

Pickel, Hugh E., Jr., *John H. Picket, Hugh C. McCammon and The Silver Strike at Caribou Hill*. Privately printed, Seattle, Washington, 1963.

Porter, Kirk H., and Johnson, Bruce J., *National Party Platforms, 1840–1964*. University of Illinois Press, Urbana, Illinois, 1966.

Quiett, Glenn C., *They Built The West: An Epic of Rails and Cities*. D. Appleton-Century Company, Inc., New York, 1934.

Richardson, Albert D., *Beyond The Mississippi*. Bliss and Company, Hartford, Connecticut, 1867.

Rockwell, Wilson, *Uncompahgre Country*. Sage Books, Denver, 1965.

Salt, Henry S., *Life of James Thomson*. Reeves and Turner, London, 1889.

Schoberlin, Melvin, *From Candles to Footlights: A Biography of the Pike's Peak Theater, 1859–1876*. Old West Publishing Company, Denver, 1941.

Smiley, Jerome C, *History of Denver*. Sun Publishing Company, Denver, 1901.

Sprague, Marshall, *A Gallery of Dudes*. Little, Brown and Company, Boston, 1966.

———, *Money Mountain: The Story of Cripple Creek Gold*. Little, Brown and Company, Boston, 1953.

Steinel, Alvin T., *History of Agriculture in Colorado*. Colorado State University, Fort Collins, Colorado, 1926.

Stone, Irving, *Men To Match My Mountains: The Opening of the Far West, 1840–1900*. Doubleday & Company, Inc., New York, 1956.

Taylor, Bayard, *Colorado: A Summer Trip*. G. P. Putnam's Sons, New York, 1867.

Villard, Henry, *Past and Present of the Pike's Peak Region*. Princeton University Press, Princeton, New Jersey, 1932.

Waters, Frank, *Midas of the Rockies*. Covici, New York, 1937.

Wharton, J. E., *History of the City of Denver from the Earliest Settlement to the Present Time*. Byers & Bailey, Denver, 1866.

Willard, James F., "The Gold Rush and Afterward," in *Colorado: Short Studies of its Past and Present*. University of Colorado, Boulder, Colorado, 1927.

Williams, Albert N., *Rocky Mountain Country*. Duell, Sloan &. Pearce, Inc., New York, 1950.

Willison, George F., *Here They Dug the Gold*. Reynal and Hitchcock, New York, 1946.

Wolle, Muriel S., *Stampede to Timberline*. University of Colorado, Boulder, Colorado, 1949.

————, *The Bonanza Trail*. Indiana University Press, Bloomington, Indiana, 1958.

Young, Frank C., *Echoes from Arcadia*. Privately printed, Denver, 1903.

Booklets

Bancroft, Caroline, "Augusta Tabor, Her Side of the Scandal." Johnson Publishing Company, Boulder, Colorado, 1965.

————, "Denver's Lively Past." Johnson Publishing Company, Boulder, Colorado, 1964.

————, "Famous Aspen." Johnson Publishing Company, Boulder, Colorado, 1964.

————, "Glenwood's Early Glamor." Johnson Publishing Company, Boulder, Colorado, 1958.

————, "Historic Central City." Johnson Publishing Company, Boulder, Colorado, 1964.

————, "Silver Queen, the Fabulous Story of Baby Doe Tabor." Johnson Publishing Company, Boulder, Colorado, 1965.

————, "Six Racy Madams." Johnson Publishing Company, Boulder, Colorado, 1965.

————, "Tabor's Matchless Mine and Lusty Leadville." Johnson Publishing Company, Boulder, Colorado, 1964.

Burton, Sarah J., and Smith, Doris B., "The Tabors." The Victorian Shop, Leadville, Colorado, 1949.

Coquoz, Rene, "New Tales of Early Leadville." Leadville, Colorado, 1964.

————, "Tales of Early Leadville." Leadville, Colorado, 1961.

————, "The Leadville Story." Leadville, Colorado, 1959.

Dallas, Sandra, "Gold and Gothic: Story of Larimer Square." Lick Skillet Press, Denver, Colorado, 1967.

Draper, Benjamin P., "Georgetown Pictorial." Old West Publishing Company, Denver, Colorado, 1964.

Feitz, Leland, "Cripple Creek! A Quick History of the World's Greatest Gold Camp." Dentan-Berkeland Printing Company, Inc., Colorado Springs, Colorado, 1967.

————, "Myers Avenue, A Quick History of Cripple Creek's Red-Light District." Dentan-Berkeland Printing Company, Inc., Colorado Springs, Colorado, 1967.

Gibbons, Rev. J. J., "In The San Juan." Calumet Book & Engraving Company, Chicago, 1898.

Gilfillan, George and Ruth, "Among the Tailings, A Guide to Leadville Mines." *The Herald Democrat*, Leadville, Colorado, 1964.

"Hotel De Paris and Louis Dupuy in Georgetown, Colorado." State Historical Society of Colorado, Denver, Colorado, 1954.

Howe, Hazel M., "The Story of Silver Plume." Dentan Printing Company, Colorado Springs, Colorado, 1960.

Mazzula, Fred and Jo, "Brass Checks and Red Lights." Denver, Colorado, 1966.

————, "The First 100 Years: Cripple Creek and the Pikes Peak Region." A. B. Hirschfield Press, Denver, Colorado, 1956.

LaBaw, Wallace L., "Nah-Oon-Kara, The Gold of Breckenridge." Big Mountain Press, Denver, Colorado, 1965.

Olsen, Mary Ann, "The Silverton Story." Beaber Printing Company, Cortez, Colorado, 1962.

Pinckert, Leta, "True Stories of Early Days in the San Juan Basin." Hustler Press, Inc., Farmington, New Mexico, 1964.

"Pioneers of the San Juan Country," compiled by the Sarah Piatt Decker Chapter DAR, Durango, Colorado, 1942.

"Pioneer Potluck, Stories and Recipes of Early Colorado," collected by the Volunteers of the State Historical Society of Colorado, Johnson Publishing Company, Boulder, Colorado, 1963.

Porter, Rufus L., "Gold Fever." Cascade, Colorado, 1954.

———, "Pay Dirt." Cascade, Colorado, 1961.

Wallace, Betty, "Gunnison, A Short, Illustrated History." Sage Books, Denver, 1964.

Zamonski, Stanley W., and Keller, Teddy, "The Fifty Niners: A Denver Diary." Sage Books, Denver, 1961.

Periodicals and Newspapers

Arps, Louisa W., "Letters from Isabella Bird." *Colorado Quarterly,* University of Colorado, Boulder, 1955–56.

Athearn, Robert G., "Life in the Pike's Peak Region: Letters of Matthew H. Dale." *Colorado Magazine,* XXXII, April, 1955.

Bahmer, Robert H., "The Colorado Gold Rush and California." *Colorado Magazine,* VII, 1930, 222–29.

Baillie, William, "Silver in Colorado." Colorado School of Mines *Mineral Industries Bulletin,* Golden, Colorado, 1962.

Cooper, Ray H., "Early History of San Juan County." *Colorado Magazine,* XXII, 1945. 205–12.

Cummins, D. H., "Toll Roads in Southwestern Colorado." *Colorado Magazine,* XXIX, 1952, 98–104.

Degitz, Dorothy M., "History of the Tabor Opera House at Leadville." *Colorado Magazine,* XIII, 1936, 81–89.

Denison, L. G., "Tales of an Early Pioneer." *Telluride Journal,* December 10, 1937.

Dinkel, William M., "A Pioneer of the Roaring Fork." *Colorado Magazine,* XXI, 1944, 133–40.

Fellows, Milo, "The First Congressional Election in Colorado, 1858." *Colorado Magazine,* VI, 1929, 46–47.

Fuller, Leon W., "Governor Waite and His Silver Panacea." *Colorado Magazine,* X, 1933, 41–47.

Gerhard, Paul F., "The Silver Issue and Political Fusion in Colorado, 1896." *University of Wichita Bulletin,* No. 46, Wichita, Kansas, November, 1960.

Gressley, Gene M., "Hotel de Paris and Its Creator." *Colorado Magazine,* XXXII, 1955, 28–42.

Hafen, LeRoy R., "George A. Jackson's Diary, 1858–1859." *Colorado Magazine,* XII, 1935, 201–14.

———, "Currency, Coinage and Banking in Pioneer Colorado." *Colorado Magazine,* X, 1933, 81–90.

———, "Otto Mears, Pathfinder of the San Juan," *Colorado Magazine,* IX, 1932, 71–74.

Hagie, C. E., "Gunnison in Early Days." *Colorado Magazine,* VIII, 1931, 121–29.

Hale, Jesse D., "The First Sucessful Smelter in Colorado." *Colorado Magazine,* XIII, 1936, 161–67.

Hill, Emma S., "Empire City in the Sixties," *Colorado Magazine,* V, 1928, 23–32.

Kingsbury, Joseph L., "The Pike's Peak Rush, 1859," *Colorado Magazine,* IV, 1927, 1–6.

Kinkin, L. C, "Early Days in Telluride." *Colorado Magazine,* XXVI, 1949, 14–26.

Lonsdale, David L., "The Fight for An Eight Hour Day." *Colorado Magazine,* XLIII, 1966, 33–53.

Major, Mrs. A. H., "Pioneer Days in Crestone and Creede." *Colorado Magazine,* XXI, 1944, 212–17.

Morris, John R., "The Women and Governor Waite." *Colorado Magazine,* XLIV, 1967, 11–19.

Murray, Robert B., "The Supreme Court of Colorado Territory." *Colorado Magazine,* XLIV, 1967, 20–34.

Ourada, Patricia K., "The Chinese in Colorado." *Colorado Magazine,* XXIX, 1952. 273–84.

Perrigo, Lynn I., "The Cornish Miners of Early Gilpin County." *Colorado Magazine,* XIV, 1937, 92–101.

———, "The First Two Decades of Central City Theatricals." *Colorado Magazine,* XI, 1934, 141–52.

Poet, S. E., "The Story of Tin Cup, Colorado." *Colorado Magazine,* IX, 1932, 30–38.

Russell, James E., "Louis Dupuy and the Hotel de Paris of Georgetown." *Colorado Magazine,* XIII, 1936, 210–15.

Sanford, Albert B., "Life at Camp Weld and Fort Lyon, 1861–62: An Extract from the Diary of Mrs. Byron N. Sanford." *Colorado Magazine,* VII, 1930, 132–39.

Sayre, Hal, "Early Central City Theatricals and Other Reminiscences." *Colorado Magazine,* VI, 1929, 47–53.

Spence, Clark C, "The British and Colorado Mining Bureau." *Colorado Magazine,* XXXIII, 1956, 81–92.

Storey, Brit, A., "William Jackson Palmer, Promoter." *Colorado Magazine,* XLIII, 1966, 44–55.

Tabor, Mrs. H. A. W., "Cabin Life in Colorado." *Colorado Magazine,* IV, 1927, 71–75.

Thomas, Chauncy, "Ouray, the Opal of America." *Colorado Magazine,* XI, 1934, 17–22.

Tischendorf, Alfred P., "British Investment in Colorado Mines." *Colorado Magazine,* XXX, 1953, 241–46.

Willard, James F., "Spreading the News of the Early Discoveries of Gold in Colorado." *Colorado Magazine,* VI, 1929, 98–104.

Williams, Francis S., "Trials and Judgments of the People's Courts of Denver." *Colorado Magazine,* XXVII, 1950, 294–302.

Newspapers consulted: Central City *Register-Call, Harper's Weekly, The Nation,* Denver *Post,* Ouray *Times, Park County Republican Fairplay, Flume, Rocky Mountain News, Silverton Standard, Solid Muldoon, Summit County Journal, Telluride Journal.*

Manuscripts

Bancroft, Hubert H., *Life of H. A. W. Tabor, 1889–91.* Prepared for The Chronicles of the Builders of the Commonwealth. Bancroft Library, University of California, Berkeley, California.

Chivington, John M., *The First Colorado Regiment.* Denver, October 18, 1884. Bancroft Library, University of California, Berkeley, California.

Gilpin, William, *A Pioneer of 1842.* 1884. Bancroft Library, University of California, Berkeley, California.

Perrigo, Lynn I., *A Social History of Central City, Colorado: 1859–1900.* Ph.D. dissertation. Boulder, 1936. Denver Public Library, Denver, Colorado.

The Scrapbooks of Baby Doe Tabor, State Historical Society of Colorado, Denver, Colorado.

Index